ENDORSEMENTS

Applying contract management tools directly from the agreed contract is a necessary but also demanding exercise and requires detailed knowledge of both the contract and the effectiveness of the tools being considered. This book is well written for contract management practitioners of all levels. Great value for the international construction world.

Robert Aloys Werth, *certified adjudicator and publicly appointed FIDIC Expert*

Innovative Project Management with FIDIC Contracts is a must-have book for those acting as project managers, contract managers, contract administrators or FIDIC Engineers on projects using one of the FIDIC suite of contracts. It is a very practical guide with flowcharts explaining processes under the FIDIC forms such as disputes progressing to the Dispute Board for their decision. It breaks the project lifecycle down into manageable chunks that the reader can read and progress through whilst being assisted with project delivery and issues that arise on all projects such as dealing with procurement at the start and claims during the delivery phase. It's a really practical book that I would encourage everyone to own.

Sean Gibbs, *BSc LLB (Hons) PG Dip Arb LLM FICE FCIOB FRICS FCIARB, CEO Hanscomb Intercontinental Ltd*

Adriana's wealth of practical project management experience and in-depth knowledge of the FIDIC forms make this book a valuable addition to the FIDIC library. She is absolutely correct to remind us that the FIDIC contract contains many tools for good project management which can complement the innovation needed today to guide sustainable development.

Jeremy Glover, *Partner, Fenwick Elliott LLP*

INNOVATIVE PROJECT MANAGEMENT WITH FIDIC CONTRACTS

This book provides comprehensive guidance for the implementation of the legal requirements of FIDIC contracts with a detailed application of modern project management principles. It demonstrates that the FIDIC suites for works and services contain many tools for good project management, which can complement the innovation needed today to guide sustainable development.

The author's experience in construction management, construction law, and the application of FIDIC contracts as a FIDIC engineer, a procurement/claims consultant, and a FIDIC accredited/certified adjudicator and trainer underpins this unique and much-needed book linking FIDIC contract management with project management practices and concepts. It is structured around project management knowledge areas, and details not only the FIDIC works contracts (Red/Yellow/Silver 2017 Reprint 2022, Emerald 2019, Gold 2008, and Green Book 2021) but also the project management requirements of the contract between the Employer and Engineer (White Book 2017) and the FIDIC supply-chain contracts, and explains how these all interface in project governance. Linking the project management knowledge areas from the PMBOK® and its Construction Extension with the FIDIC Body of Knowledge (FBOK), the book provides recommendations for integration of best practice and innovations. It illustrates these processes with 54 flowcharts and 16 case studies, as well as a railway scenario demonstrating the application of the book's recommended procedures.

This will be an invaluable source of practical guidance for employers, contractors, and engineers responsible for procuring and managing construction projects. It provides useful insight for technical, legal, and financial experts on the management of processes, contracts, and digital tools, with recommendations on how to tailor and use the FIDIC contracts for project success.

Adriana Spassova is a civil engineer with an MSc in Construction Law and Dispute Resolution at King's College London and a PhD in Construction Project Management, a FIDIC Certified/Accredited Adjudicator and Trainer, and presently Vice Chair of the FIDIC Contracts Committee, acting FIDIC Engineer, and Dispute Avoidance/Adjudication Board.

PRACTICAL LEGAL GUIDES FOR CONSTRUCTION AND TECHNOLOGY PROJECTS

Series Editor: Donald Charrett

The books in this series provide lawyers and construction professionals with the relevant knowledge, resources and practical guidance to understand the contemporary construction industry and the law necessary to achieve successful outcomes for construction projects. The topics covered include front-end issues such as project procurement, the production, negotiation and administration of contracts, and the avoidance of disputes, as well as backend issues such as taking part in, managing or resolving disputes. The series aims to keep abreast of the technologies that have increasing impact in changing traditional ways of working, contracting and resolving disputes.

FIDIC Contracts in Asia Pacific
A Practical Guide to Application
Donald Charrett

Contracts for Construction and Engineering Projects
Second Edition
Donald Charrett

Contracts for Infrastructure Projects
An International Guide to Application
Donald Charrett

FIDIC Contracts in Europe
A Practical Guide to Application
Donald Charrett

FIDIC Contracts in the Americas
A Practical Guide to Application
Donald Charrett

FIDIC Contracts in Africa and the Middle East
A Practical Guide to Application
Donald Charrett

Decision-making in International Construction Arbitration
Haytham Besaiso

Innovative Project Management with FIDIC Contracts
Adriana Spassova

For more information, please visit: www.routledge.com/Practical-Legal-Guides-for-Construction-and-Technology-Projects/book-series/PLGCTP

INNOVATIVE PROJECT MANAGEMENT WITH FIDIC CONTRACTS

ADRIANA SPASSOVA

informa law
from Routledge

Cover image: ITER Cryostat - photograph taken by the author during visit to the ITER Site in Saint-Paul-lez-Durance, France 23 May 2023

First published 2024
by Informa Law from Routledge
4 Park Square, Milton Park, Abingdon, Oxon OX14 4RN

and by Informa Law from Routledge
605 Third Avenue, New York, NY 10158

Informa Law from Routledge is an imprint of the Taylor & Francis Group, an informa business

British Library Cataloguing-in-Publication Data
A catalogue record for this book is available from the British Library

Library of Congress Cataloging-in-Publication Data
Names: Spassova, Adriana, author.
Title: Innovative project management with FIDIC contracts /
Adriana Spassova.
Other titles: Innovative project management with Fédération
internationale des ingénieurs-conseils contracts
Description: Abingdon, Oxon [UK]; New York, NY: Routledge, 2023. |
Series: Practical legal guides for construction and technology projects |
Includes bibliographical references and index. |
Identifiers: LCCN 2023030292 (print) | LCCN 2023030293 (ebook) |
ISBN 9781032225227 (hardback) | ISBN 9781032225234 (paperback) |
ISBN 9781003272892 (ebook)
Subjects: LCSH: Construction contracts. | Engineering contracts. |
Architectural contracts. | Project management.
Classification: LCC K891.B8 S69 2023 (print) | LCC K891.B8 (ebook) |
DDC 343.07/8624–dc23/eng/20230926
LC record available at https://lccn.loc.gov/2023030292
LC ebook record available at https://lccn.loc.gov/2023030293

ISBN: 978-1-032-22522-7 (hbk)
ISBN: 978-1-032-22523-4 (pbk)
ISBN: 978-1-003-27289-2 (ebk)

DOI: 10.4324/9781003272892

Typeset in Times New Roman
by Deanta Global Publishing Services, Chennai, India

This book is dedicated to my granddaughters Alice and Eva, wishing them to be healthy and happy and to pursue their dreams!

Adriana Spassova, Sofia, June 2023

CONTENTS

ABOUT THE AUTHOR

Dr Adriana Spassova has 36 years' professional experience as a designer, construction manager, and consultant, including 26 years' FIDIC experience in energy, railway, industrial facilities, waste management, and water infrastructure projects. In the last 6 years, she has delivered FIDIC accredited trainings in Europe, Asia, and America for more than 1,600 participants from 50 countries. Dr Spassova is presently serving on standing Dispute Boards in Europe and Asia and is FIDIC Engineer on water projects. From 2021, she has been a member of the FIDIC Contracts Committee, and was involved in the review of the Green Book 2021 and the 2022 Reprint with amendments of the Red, Yellow, and Silver Books 2017. She is the liaison for TG9, developing six new subcontract forms, and TG23, drafting a net zero guidance for all FIDIC works contracts and agreements for services. In 2022, she became a Vice Chair of the FIDIC Contracts Committee.

REMARKS BY THE SERIES EDITOR

The Practical Legal Guides for Construction and Technology Projects series aims to provide lawyers and construction professionals with relevant knowledge, resources and practical guidance to understand the contemporary construction industry and the law necessary to achieve successful outcomes for construction projects. Adriana Spassova's book is a welcome addition to the series, as it ticks all of these boxes. She has provided a comprehensive amalgam of the legal requirements of FIDIC contracts with a detailed application of modern project management principles. The book lives up to its name with its discussion of various aspects of modern innovations in project management.

Other books on FIDIC focus on the contract and explain the legal implications of the clauses in detail, and how the contract should be managed. Uniquely, this book focuses on the project and how the detailed FIDIC provisions mesh with the requirements of management of construction projects from their inception. The book is not confined to the FIDIC works contracts (Red, Yellow, Silver, Emerald, Gold, and Green Books) but also details the project management requirements of the contract between the Employer and Engineer (White Book) and how these interface with project management of the works contracts.

Adriana has successfully undertaken the challenge of converting the research she carried out for her PhD thesis into a practical and comprehensive guide for practitioners. The result is a book structured around project management knowledge areas, not the arrangement of the works contracts. Each chapter discusses the FIDIC provisions relevant to a specific project management knowledge area, illustrated with diagrams and tables.

This book will be an invaluable practical source of guidance for Employers, Contractors, and Engineers responsible for procuring and managing construction projects. The case studies and the Notices and logs applied to a scenario in the Appendix provide context and illustrate the application of the book's recommended procedures.

Donald Charrett, Melbourne, May 2023

FOREWORD

One of the key messages I have always tried to instil in students of construction law is that contracts are not only about enforcement of rights and obligations; still less should they be regarded primarily as weapons to be deployed in trying to win a victory against an opponent. Rather, they should be seen as vital tools for the successful delivery of major projects. There can hardly be another book that demonstrates more clearly the validity of this belief than Adriana Spassova's *Innovative Project Management with FIDIC Contracts.* 'Management' is in the title of every substantive chapter. and almost every page emphasises and examines the linkage between contract provisions and management activities. But, it is also a 'contracts' book, and specifically, a FIDIC contracts book. A notable feature of the FIDIC material, not just in Chapter 4, is how up to date it is, referencing not merely recent developments such as the 2019 Conditions of Contract for Underground Works (the Emerald Book) and the Short Form of Contract second edition 2021 (the Green Book) but the 2022 reprints of the 2017 second editions of the Red, Yellow, and Silver Books. A glance at the Acknowledgements page of the 2022 Guide to these last contracts, which records Dr Spassova's assistance, helps to explain this: as Vice Chair of the Contracts Committee, she is well placed to provide the in-depth knowledge that is displayed throughout her own book. Indeed, there is even a projection forward into the future in the work of TG 14 on an Off-Shore Wind-Farm contract (historically, of course, the Yellow Book has often been used in the absence of a specialist form) and TG 20 on Joint Venture Agreements for Contractors, as well as the six new sub-contracts being prepared.

It was not a coincidence that I received last summer my first briefing on the work of TG 18 on the Public Private Partnership form of contract in Sofia, at the Annual Conference of the European Society of Construction Law (ESCL). As ESCL President with strong FIDIC connections, Dr Spassova possessed the knowledge and contacts to arrange a valuable update for the many FIDIC users present. Producing this book has required very significant understanding of law and legal analysis. It was gratifying that such a senior professional as Adriana Spassova chose to study at King's College, London, obtaining her Master's degree, with Merit, from the Centre of Construction Law and Dispute Resolution in 2014 and winning a major prize in the 2015 ESCL Competition for her thesis. Nevertheless, *Innovative Project Management with FIDIC Contracts* is an engineer's book, though one that will be used with great benefit by many construction lawyers. The structure is clear and logical, as one would expect. After laying the foundations in Chapters 1–4 by introducing features of modern project management and the FIDIC contracts, the chapters that follow consider the relationship between those contracts and all the management disciplines that have to be utilised in a complex project.

The treatment of management is structured broadly chronologically, from Project Procurement Management (Chapter 5) and Project Integration management (Chapter 6) at the inception stage, via Project Scope Management (Chapter 7), Schedule Management

(Chapter 8), Financial and Cost Management (Chapter 9), Quality Management (Chapter 10), Resource Management (Chapter 11), Communications Management (Chapter 12), Risk Management (Chapter 13), Stakeholder Management (Chapter 14), and Health, Safety, Security, and Environmental Management (Chapter 15), all the way through to Claims Management (Chapter 16) and Dispute Management (Chapter 17). It is not a criticism – a compliment, if anything – to observe that an engineer devotes 52 pages to the Procurement Management chapter as contrasted with a Dispute Management Chapter of 12 pages. Dr Spassova is able to draw on a long career in engineering consultancy and her experience of very large-scale international projects, principally funded by development banks, in deciding what is really important. This is apparent from the Case Studies, which are a welcome feature of the text. Of particular interest to me was the Chapter 7 Case Study on ITER, the remarkable nuclear fusion energy project in south east France, which I had the privilege to visit. I am proud to have taught Adriana Spassova at King's and to have seen at first hand in Bulgaria the distinction she has achieved since, with her PhD, her presidency of ESCL, and her role with FIDIC. I am equally happy that I am now able to learn from her in turn through this book, which I predict will be influential in project management and FIDIC circles for many years to come.

Anthony Lavers LL.B M.Phil Ph.D D.Litt FRICS
Visiting Professor of Law, King's College
London Research Consultant,
Crown Office Chambers, Inner Temple

ACKNOWLEDGEMENTS

I am grateful to my father, Kolio Spassov, who always encouraged me to write. I am thankful to Rumen Mladjov (chief executive officer of a steel structures engineering, procurement, construction contractor), who appointed me after university and encouraged me to design modern methods of construction with preassembly of building structures with services and bridges in huge blocks, to be mounted with two heavy cranes or by hydraulic jacks. After 3 years, his deputy Grigor Grigorov invited me to become head of logistics, a challenging position with responsibilities for the supply-chain management. Later, I became head of contracts and tenders, and was fascinated by FIDIC contracts. After 10 years working for a contractor, I decided to become a consultant, and had the chance to be appointed in EQE to use the FIDIC Red Book 1987 reprint 1992 in the Kozloduy nuclear power plant. I am grateful to our team leader from Carl Bro, Friedrich Hengstenberg, who taught me how to apply FIDIC contracts in complex projects with many stakeholders, sharing 40 years' experience in 40 countries.

I am grateful to EQE Bulgaria's management, entitling me to found and manage the daughter company EQE Control to enhance the project management and FIDIC consultancy services when new legislation made design and construction supervision mandatory in Bulgaria. Thanks to the new managers of EQE Control, Evgeni Tsvetanov, Valeri Matneshliev, and Lubomir Petrov, now I can allocate my time mostly to FIDIC consultancy, trainings, adjudication, and presently volunteering work for the FIDIC Contracts Committee and its Task Groups 9 and 23.

I would like to share my gratitude to all the fantastic tutors at King's College London, with special thanks to Prof. Anthony Lavers for his foreword and encouragement after checking my first assignment, and also to Prof. David Mosey, with whom I continue to work for collaborative procurement.

I am grateful to the University of Architecture and Civil Engineering in Sofia, where I became a civil engineer, because in 2015 they agreed to host the Bulgarian Society of Construction Law, and gave me the opportunity to develop a discipline 'Project Management with FIDIC Contracts' for the Bachelor and MSc in Construction Management programmes. This led to my PhD research, linking the PMBOK Knowledge Areas with FIDIC Contracts, and I would like to thank my PhD supervisors, Prof. Dr. Julieta Mancheva and Associate Professor Dr. Rosiza Petrova, for their support.

I am thankful to the previous FIDIC director, Enriko Vink, for making me part of the FIDIC family and allowing me to become a FIDIC accredited adjudicator and trainer. A big thank you to the present FIDIC chief executive officer, Dr. Nelson Ogunshakin, for taking FIDIC on a new journey with the launching of the certification programmes, the FIDIC Academy, the Global Leadership Forum, the Climate Charter, signing licence agreements with the Multilateral Development Banks for the use of the new FIDIC forms, and developing a bigger and more diverse FIDIC network, where I can share all my ideas for sustainability and innovations. I am grateful to the chair of the FIDIC Contracts Committee, Vincent Leloup, for inviting me to be part of this group of distinguished professionals, and for his exceptional leadership and guidance. I am thankful to the FIDIC Secretariat for their continuous support, especially to Daduna Kokhreidze, Ieva Liaugaude, Silvia Fossati, Chanel Holder, Oro Kohen, and Lise Kroeyer. FBOK material has been reproduced with the kind permission of FIDIC.

I am extremely thankful to the more than 1,600 participants from 50 countries in the FIDIC accredited trainings I delivered in the last years. Their numerous questions related to the use of FIDIC in different jurisdictions urged me to give answers in this book on how to apply the FIDIC provisions in the most flexible and efficient manner for project success.

Thanks to Karoly Tamas Olajos, JD/MBA/LLM/PMP, International Thermonuclear Experimental Reactor Site & Buildings Legal Affairs Group Leader from Fusion for Energy, who worked with me on the text for the International Thermonuclear Experimental Reactor case study, for his insights on the application of FIDIC to this unique international innovative project.

PMBOK material from the following publications has been reproduced with the permission of the Project Management Institute:

- Project Management Institute Inc., *A Guide to the Project Management Body of Knowledge (PMBOK® Guide)* (sixth edition 2017). Copyright and all rights reserved.
- Project Management Institute Inc., *Construction Extension to the PMBOK® Guide* (2016). Copyright and all rights reserved.

I am grateful to Dr. Donald Charrett and Guy Loft from Routledge, who embraced at once my idea for the book. I am honoured to share my experience in this series. Donald's profound knowledge of FIDIC and project management led to many useful improvements.

Adriana Spassova, Sofia, June 2023

LIST OF FIGURES

LIST OF TABLES

TABLE OF CASES

TABLE OF LEGISLATION

TABLE OF STANDARDS

Standard	*Section*
Eurocodes: a series of ten European Standards	10.3.2
ISO 14001:2015 Environmental Management Systems	15.2.3
ISO 14064-1:2018 Greenhouse gases – Part 1: Specification with Guidance at the Organization Level for Quantification and Reporting of Greenhouse Gas Emissions and Removals	15.4.3
ISO 14067:2018 Greenhouse Gases – Carbon Footprint of Products – Requirements and Guidelines for Quantification	15.4.3
ISO 19650-2:2018(en) Organization and Digitization of Information about Buildings and Civil Engineering Works, Including Building Information Modelling (BIM) – Information Management Using Building Information Modelling – Part 2: Delivery Phase of the Assets	5.12.4.1, 6.5
ISO 19650-3:2020 Organization and Digitization of Information about Buildings and Civil Engineering Works, Including Building Information Modelling (BIM) – Information Management Using Building Information Modelling — Part 3: Operational Phase of the Assets	5.12.4.6
ISO 20887:2020 (en) Sustainability in Buildings and Civil Engineering Works – Design for Disassembly and Adaptability – Principles, Requirements and Guidance	3.4.3.2
ISO 21500:2021 Project, Programme and Portfolio Management – Context and Concepts	5.12.4.1
ISO 21508:2018 Earned Value Management in Project and Programme Management	9.2.15.1
ISO 26000:2010 Guidance on Social Responsibility	15.2.3
ISO 31000:2018 Risk Management – Guidelines	13.1
ISO 44001:2017: Collaborative Business Relationship Management Systems – Requirements and Framework	5.12.9
ISO 45001:2018 Occupational Health and Safety Management Systems – Requirements with Guidance for Use	15.2.3
ISO 9000:2015 Quality Management Systems – Fundamentals and Vocabulary	10.1
ISO 9001:2015 Quality Management Systems – Requirements	15.2.3

LIST OF EXAMPLE NOTICES INCLUDED IN APPENDIX 1

Knowledge Area	*Sub-Clause*	*Description*	
Project Procurement Management	3.1	Notice of the natural person authorised to act on behalf of the Engineer	Engineer
Project Integration Management	8.1	Notice of the Commencement Date	Engineer
Project Integration Management	3.4	Notice of delegation of authority to assistants	Engineer
Project Scope Management	13.3.1	Notice of Variation	Engineer
Project Scope Management	**13.1** (c)	Notice of objection to the Engineer's Notice of Variation	Contractor
Project Scope Management	13.1	Notice of confirmation of the Engineer's Notice of Variation	Engineer
Project Scope Management	13.2	Value engineering proposal	Contractor
Project Scope Management	13.2	Notice of consent to the Contractor's VE proposal	Engineer
Project Scope Management	13.6	Notice of adjustment to the execution of the Works as a result of a change in the Laws	Engineer
Project Scope Management	13.3.2 (b)	Notice of objection to the Engineer's request for proposal	Contractor
Project Scope Management	13.1, 13.4 (a)	Notice of confirmation of the Engineer's request for proposal	Engineer
Project Schedule Management	8.3	Notice of non-compliance of the initial programme	Engineer
Project Schedule Management	8.7	Notice for slow progress	Engineer
Project Financial Management	2.4	Request for reasonable evidence for financial arrangements	Contractor
Project Cost Management	12.1	Notice of the date on which and place where measurement will be made	Engineer
Project Cost Management	12.1	Notice of the date on which and place where measurement records may be agreed	Engineer
Project Quality Management	7.3	Notice that reinforcement is ready for inspection	Contractor
Project Quality Management	7.3	Notice that inspection is not required	Engineer
Project Resource Management	4.17	Notice of the delivery of Contractor's Equipment	Contractor

Project Communications Management	1.9	Notice of errors identified in the Employer's Requirements	Contractor
Project Communications Management	1.9	Notice of error/fault/defect identified in the Employer's Requirements	Contractor
Project Risk Management	18.2	Notice of an Exceptional Event	Contractor
Project Risk Management	17.2	Notice of damage to the Works, Goods, or Contractor's Documents	Contractor
Project Stakeholder Management	6.5	Notice that work is required on Site during locally recognised days of rest	Contractor
Project HSSE Management	**6.7** and 6.9	Notice for removal and replacement of Contractor's Personnel	Engineer
Project Claims Management	20.2.1	Notice of Claim	Contractor

CHAPTER 1

Introduction

1.1 Project management

Innovation creates value. Innovation brings sustainable solutions and efficiency in the construction sector.

Project management (**PM**) is the application of knowledge, skills, tools, and techniques to project activities to meet the project requirements.[1] PM enables effective and efficient project execution. Project managers use a set of key skills and applied knowledge to satisfy the stakeholders involved in and affected by the project.

The USA Project Management Institute (**PMI**)[2] has more than 300 chapters in more than 200 countries/territories with more than 680,000 members.[3] PMI is developing PM guidance used globally. PMI defines the PM body of knowledge (**PMBOK**) as a term that describes the knowledge within the profession of PM. The body of knowledge (**BOK**) includes both published and unpublished materials. The PMBOK includes traditional as well as emerging innovative practices.

PM and many of its practices originate from construction projects and formed the foundation of the original 1987 PMBOK® Guide. The awareness, growth, and recognised value of PM to all industries led to broadening of concepts that do not fully cover the current PM practices in construction. While the PMBOK® Guide provides a general foundation for managing projects, its Construction Extension addresses the specific practices in construction.[4] Both documents used concurrently form a subset of the PMBOK that is generally recognised as good practice in construction.

The PMBOK® Guide and its Construction Extension identify **PM Knowledge Areas** by knowledge requirements, described in terms of component processes, practices, inputs, outputs, tools, and techniques. Project managers may apply different PM methodologies to their work.

Construction projects occur in a changing, complex environment with a high degree of risk. Understanding and applying the recognised good practice is not sufficient for effective construction PM. The PM tailoring is necessary because each project is unique.

1 Project Management Institute Inc., *A Guide to the Project Management Body of Knowledge (***PMBOK**

2 www.pmi.org

3 Kimberly Whitby, 'December 2021 PMI Fact File Stats' (*The Critical Path Blog*) <https://www.project-management.com/blog-post/71122/december-2021-pmi-fact-file-stats> accessed 10 January 2022

4 Project Management Institute Inc., *Construction Extension to the PMBOK® Guide* (2016) Copyright and all rights reserved. 1

DOI: 10.4324/9781003272892-1

Besides the knowledge and the soft skills needed, contract documents drafted with proper risk allocation provide important tools for good PM. Those tools may include incentives for innovations and improved value. In separate chapters dedicated to each PM Knowledge Area, this book will explore how FIDIC users may integrate provisions for innovative PM in the agreements for services and works contracts.

1.2 FIDIC standard conditions of contract

It is widely acknowledged that standard conditions of contract offer:

- carefully drafted provisions, integrating the industry best practice;
- certainty: the Parties can be reasonably confident that they will be predictable and will deal sensibly with most of the circumstances that one might experience;
- balanced and fair risk allocation, resulting in greater contractor confidence and lower risk contingencies, and hence, lower costs;
- savings in time and cost for negotiation and re-drafting; tenderers do not need to make financial provision for unfamiliar contract conditions;
- dispute avoidance as a result of accumulated experience and a common understanding;
- facilitated training and PM.

The International Federation of Consulting Engineers, FIDIC (*Fédération Internationale des Ingénieurs-Conseils*),[5] was founded in 1913. The founding principles 'quality, integrity, and sustainability' remain important today, when FIDIC has member associations in more than 100 countries. FIDIC issued its first contract in 1957, further developing the most widely used full suite of contracts for construction and related services.

FIDIC contracts do not stand alone in the industry. They are backed up by many other documents and approaches, and are part of a whole eco-system integrated to support the users.[6] The FIDIC Body of Knowledge (**FBOK**) includes all FIDIC contracts, management systems, manuals, guidance documents, and training materials. Working with the Secretariat, FIDIC Committees appointed by the FIDIC Board establish Task Groups to update the FBOK with the latest trends in risk management, quality management,[7] and sustainable development in construction.[8] The Contracts Committee develops and updates the FIDIC suite of contracts and related documents.[9]

The 1999 FIDIC rainbow suite became very popular, especially the three forms for major works, differing in relation to the risk allocation, not the type of project:

- Conditions of Contract for Construction (**Red Book**);

5 www.fidic.org

6 FIDIC, 'Annual Report 2020/2021' https://issuu.com/fidic/docs/2021_annual_report_final_ok accessed 13 January 2022

7 FIDIC, 'FIDIC Risk, Liability & Quality Committee Terms of Reference 2020-2022' <https://fidic.org/sites/default/files/RLQC_Terms%20of%20Reference_V1.2%202020.pdf> accessed 13 January 2022

8 FIDIC, 'FIDIC Diversity & Inclusion Committee Terms of Reference 2020-2022' <https://fidic.org/sites/default/files/D%26IC_Terms%20of%20Reference_V1.2%202020.pd> accessed 13 January 2022

9 FIDIC, 'FIDIC Contracts Committee Terms of Reference 2020-2022' <https://fidic.org/sites/default/files /CC_Terms%20of%20Reference_V1.2%202020_1.pdf> accessed 13 January 2022

- Conditions of Contract for Plant and Design-Build (**Yellow Book**);
- Conditions of Contract for **EPC**[10]/Turnkey Projects (**Silver Book**).

Their General Conditions (**GCs**) include 20 Clauses with similar wording, except where different risk allocation is stipulated. The users have more certainty with a familiar form. For example, if someone has used the Red Book, he/she feels comfortable with the Yellow Book, learning the main differences related to the design risk and method of payment.

FIDIC considers that one of the main factors in the success of the FIDIC 1999 forms of contract has been the PM procedures and mechanisms that were included in the GCs of Contract. In the 21st century, FIDIC expanded its suite for major works with the following contracts:

- Form of Contract for Dredging and Reclamation Works (**Blue-Green Book**);
- Conditions of Contract for Design, Build and Operate Projects (**Gold Book**);
- Conditions of Contract for Underground Works (**Emerald Book**).

FIDIC launched the second edition of the Red/Yellow/Silver Books in December 2017. Many of the procedures and mechanisms have been enhanced and clarified to reflect modern international best practice. FIDIC explains that with the aim of increasing clarity and reducing potential misunderstandings on interpretation, the contract provisions in the 2017 edition are more prescriptive and introduce step-by-step PM and procedural mechanisms. These set out exactly what is expected from the project participants during the performance of the Contract.[11]

The 1999 Short Form of Contract (**Green Book**) was not so popular. In 2021, FIDIC launched the second edition. It contains the important features of the 2017 books for major works drafted in a concise way and introduces new user-friendly tools for more efficient PM.

Besides the suite for works, FIDIC published a suite of agreements for services led by the Client/Consultant Model Services Agreement (**White Book**).

FIDIC contracts integrate the best international practice, but technology develops rapidly, while the standard forms stay conservative. The usual process of drafting a new standard form followed by rigorous multi-tier reviews and revisions takes several years. After the publication, several years are spent on promotion, training, and integration in the bidding documents before the new standard form is implemented.

Most of the users do not switch to new forms until they have been tested on the market, and disputes have reached the courts, giving legal interpretations resulting in certainty, which takes a decade.[12] The process is facilitated when the International Financial Institutions (**IFIs**) integrate the new form in their standard bidding documents. FIDIC has achieved excellent results in the promotion of the 2017 Edition, signing licence agreements for the use of the new suites for works and services with the most important Multilateral Development Banks (**MDBs**): the World Bank in 2019, followed by the Inter-American

10 Engineering, Procurement and Construction

11 FIDIC, 'FIDIC Rainbow Suite ed. 2017' <https://fidic.org/sites/default/files/press%20release_rainbow%20suite_2018_03_1.pdf> accessed 14 January 2022

12 One third of the participants in the FIDIC International Contracts Users' Conference in December 2021 reported that they were using the 2017 Edition of the FIDIC books for major works

Development Bank (**IADB**), the Caribbean Development Bank (**CDB**), the European Bank for Reconstruction and Development (**EBRD**), the Asian Infrastructure Investment Bank (**AIIB**), the African Development Bank (**AfDB**), the Islamic Development Bank in 2020, the International Fund for Agricultural Development (**IFAD**), and the Asian Development Bank (**ADB**) in 2021. Additionally, FIDIC signed agreements for the use of its standard forms with bilateral financial institutions like the *Kreditanstalt für Wiederaufbau* (**KfW**), the Japan International Cooperation Agency (**JICA**), the *Agence Française de Développement* (**AFD**), and the Millennium Challenge Corporation (**MCC**), and Memorandums of Understanding with many international organisations (including the European Investment Bank (**EIB**), the largest multilateral financial institution in the world).

1.3 Aims of the book

The aim of this book is to demonstrate that the FIDIC standard forms include tools for good PM and to suggest how the drafters of the contract documents can accommodate innovations. At its root, PM is about solving problems. Therefore, the more tools we have for problem solving, the more effective we will be in achieving successful outcomes.[13]

Additionally, special attention shall be paid to the integration of net zero clauses, because it is time for urgent actions in line with the United Nations (**UN**) Sustainable Development Goals (**SDGs**).[14]

The FIDIC forms envisage that the agreement between the parties will consist of a number of documents. Each of the FIDIC books includes GCs together with Guidance for the Preparation of Particular Conditions (**GPPC**) and annexed forms of securities, Contract Agreement, Dispute Avoidance/Adjudication Board (**DAAB**) Agreement, etc.

FIDIC strongly recommends that the drafters of the Particular Conditions take due regard of the five FIDIC Golden Principles (**GPs**):

GP1: The duties, rights, obligations, roles and responsibilities of all the Contract Participants must be generally as implied in the GCs, and appropriate to the requirements of the project.

GP2: The Particular Conditions must be drafted clearly and unambiguously.

GP3: The Particular Conditions must not change the balance of risk/reward allocation provided for in the GCs.

GP4: All time periods specified in the Contract for Contract Participants to perform their obligations must be of reasonable duration.

GP5: Unless there is a conflict with the governing law of the Contract, all formal disputes must be referred to a Dispute Avoidance/Adjudication Board (or a Dispute Adjudication Board, if applicable) for a provisionally binding decision as a condition precedent to arbitration.

13 Devyn Colson, 'How Technology Allows Us to Solve Problems Better and Faster' (*The official PMI Blog*) < https://community.pmi.org/blog-post/70333/How-Technology-Allows-Us-to-Solve-Problems-Better-and-Faster> accessed 9 January 2022

14 <https://sdgs.un.org/goals> accessed 3 February 2023

The FIDIC GPs are described and explained in the freely downloadable publication.[15] They aim to ensure that modifications to the GCs:

- are limited to those necessary for the particular features of the Site and the project, and necessary to comply with the applicable law;
- do not change the essential fair and balanced character of a FIDIC contract; and
- the Contract remains recognisable as a FIDIC contract.[16]

Following the FIDIC guidance, contract drafters are invited to use the recommendations of this book to create Particular Conditions and modernise the Specification/Employer's Requirements/Schedules incentivising the Parties towards innovation and sustainability. The new provisions may use 'the stick and the carrot' (liquidated damages versus bonuses) and stipulations enhancing certainty, collaboration, and value engineering (**VE**).

The next breakthrough in the construction sector is likely to involve the connection of law, management, and technology.[17] The book will link the FBOK with the PMBOK and give recommendations for integration of smart PM tools for improved value.

The technologies develop quickly, but the tips on how to integrate them will remain valid. FIDIC updated the 1999 contracts for major works after 18 years. The 2017 Edition of the Red/Yellow/Silver Books had been on the market for 5 years. Then, FIDIC integrated all errata found in this period in the Reprint 2022, introducing some changes, bringing more clarity, and further strengthening the role of the contract administrator (**CA**). This book refers mainly to the White Book 2017, Red/Yellow/Silver Books Reprint 2022 with amendments of the 2017 Edition, Emerald Book 2019, and Green Book 2021, but all recommendations will be applicable to the previous editions still in use. Some of the provisions in the GCs of the new editions for better PM and innovations may be integrated in the older editions via the Particular Conditions, if users feel more comfortable and certain with the previous shorter editions.[18]

Recent research shows that fear of uncertainty blocks innovation.[19] Applying the guidance of the PMBOK, FBOK, and the Practical Legal Guides for Construction and Technology Projects edited by Dr. Donald Charrett,[20] FIDIC contract users can integrate with confidence tools and methodology supporting innovations and sustainability.

For brevity, the next chapters use the following references to the FIDIC standard forms:

- Sub-Clause/Clause numbers in the sections dedicated to service agreements where no books are specified refer to the White Book fifth edition 2017;

15 FIDIC, *The FIDIC Golden Principles* (FIDIC 2019) < https://fidic.org/books/fidic-golden-principles-2019> accessed 2 April 2022

16 For further details see also Donald Charrett, *FIDIC Contracts in Europe: A Practical Guide to Application* (Informa Law from Routledge 2023) Chapter 2 'The FIDIC Golden Principles'

17 Alan McNamara, 'Automating the Chaos: Intelligent Construction Contracts' Chapter in *Smart Cities and Construction Technologies* (IntechOpen 2020) 8

18 This approach was followed by Fusion for Energy for the **ITER** (International Thermonuclear Experimental Reactor) project: some 2017 provisions have been integrated in the 1999 forms, source: Karoly Tamas presentations on FIDIC Contract Users' Conferences

19 Laura Furstenthal and others, 'Fear Factor: Overcoming Human Barriers to Innovation' (McKinsey & Company 2022) <https://www.mckinsey.com/business-functions/strategy-and-corporate-finance/our-insights/fear-factor-overcoming-human-barriers-to-innovation> accessed 3 July 2022

20 This book belongs to this series published by Routledge

- Sub-Clause/Clause numbers in the sections for works contracts where no books are specified refer to the Red/Yellow/Silver Books Reprint 2022 with amendments of the 2017 edition and Emerald Book first edition 2019;
- Green Book references are for the second edition 2021;
- Gold Book references are for the first edition 2008;
- 'Specification/Employer's Requirements' to be read 'Specification (Red Book Reprint 2022 with amendments of the 2017 Edition)/Employer's Requirements (Yellow/Silver Books Reprint 2022 with amendments of the 2017 edition and Emerald Book 2019)'.

Capitalised terms not specifically defined in this book are defined in the relevant FIDIC Conditions of Contract. The Glossary contains definitions of capitalised terms specifically defined in this book. The term 'Employer' from works contracts is often used when discussing the PM, instead of Client in service agreements.

CHAPTER 2

Project Management Knowledge Areas in Construction

2.1 Tailoring the project management approach

The sixth edition of the PMBOK® Guide provides more detail than the previous editions about key concepts, emerging trends, considerations for tailoring the PM processes, and information on how tools and techniques are applied to projects.[1] Tailoring is necessary for each project to adapt the PM approach, governance, and processes to make them more suitable for the given environment and work scope.

The PMBOK is constantly evolving. PMI asserts that the PMBOK® Guide sixth edition identifies a subset of the PMBOK that is generally recognised as good practice.

With the development of the information technology (**IT**) sector, the majority of PMI members are presently software developers. Construction project managers are in a minority. PMI revised its documents to support the flexibility needed for IT development by adopting performance domains in the seventh edition of the Guide, in place of the Knowledge Areas.

The project performance domains overlap and interconnect, trying to embrace the full spectrum of project approaches, in order to satisfy all industries. The Knowledge Areas are interrelated similarly to the project performance domains, but they are defined separately from the PM perspective. Tailoring should be adapted to the industry. The seventh edition of the PMBOK® Guide goes too far away from the construction practices, arguably to serve the IT sector better. The PMBOK® Guide sixth edition's Knowledge Areas, as developed in the Construction Extension, can be better matched to the contract provisions than the seventh edition's more general project performance domains. Following the sixth edition's Knowledge Areas, this book recommends the use of some modern tools from the seventh edition, like the key performance indicators (**KPIs**) and tracking baseline performance from the Measurement Performance Domain.

The 2016 Construction Extension upgraded the PMBOK® Guide's ten Knowledge Areas with several additional, specific for the industry. The authors assert that the Construction Extension is not tied to a specific edition of the PMBOK® Guide, so it can be used with the sixth edition. Chapters 5–17 in this book link the PMBOK with the FBOK in each Knowledge Area. They point out how FIDIC contract provisions can underpin the PM. Those chapters follow the flow of the processes but focus on the principles behind efficient PM. They demonstrate how users of the FIDIC contracts can innovate PM, integrating modern tools and methodologies.

1 PMI Inc., *PMBOK® Guide* (PMI sixth edition 2017) 2

DOI: 10.4324/9781003272892-2

2.2 Knowledge Areas for the analysis

The Knowledge Areas in the PMBOK® Guide are applicable to construction projects, albeit with modifications to address the unique attributes, practices, and applications.[2] The sixth edition significantly modernises the Knowledge Areas, including key concepts, trends, and emerging practices, tailoring considerations, and for the first time, considerations for an agile/adaptive environment. The ten Knowledge Areas described in the sixth edition of the PMBOK® Guide are:

- **'Project Integration Management'**: includes the processes and activities to identify, define, combine, unify, and coordinate the various processes and PM activities;
- **'Project Scope Management'**: includes the processes required to ensure that the project includes all the work required, and only the work required, to complete the project successfully;
- **'Project Schedule Management'**: includes the processes required to manage the timely completion of the project;
- **'Project Cost Management'**: includes the processes involved in planning, estimating, budgeting, managing, and controlling costs so that the project can be completed within the approved budget;
- **'Project Quality Management'**: includes the processes for incorporating the organisation's quality policy regarding planning, managing, and controlling project and product quality requirements in order to meet stakeholders' expectations;
- **'Project Resource Management'**: includes the processes to identify, acquire, and manage the resources needed for the successful completion of the project;
- **'Project Communications Management'**: includes the processes required to ensure timely and appropriate planning, collection, creation, distribution, storage, retrieval, management, control, monitoring, and ultimate disposition of project information;
- **'Project Risk Management'**: includes the processes of conducting risk management planning, identification, analysis, response planning, response implementation, and monitoring risk on a project;
- **'Project Procurement Management'**: includes the processes necessary to purchase or acquire products, services, or results needed from outside the project team;
- **'Project Stakeholder Management'**: includes the processes required to identify the people, groups, or organisations that could impact or be impacted by the project, to analyse stakeholder expectations and their impact on the project, and to develop appropriate management strategies for effectively engaging stakeholders in project decisions and execution.

The PMBOK® Guide clarifies that the needs of a specific project may require one or more additional Knowledge Areas. The Construction Extension introduces several additional Knowledge Areas applicable to all construction projects:

2 PMI Inc., *Construction Extension to the PMBOK® Guide* (PMI 2016) 19

- **'Project Health, Safety, Security, and Environmental (HSSE) Management'**: addresses both the management of the project and the product of the project; and
- **'Project Financial Management'**: covers the processes to acquire and manage the financial resources for the project. It is more concerned with revenue sources and monitoring net cash flows for the construction project than with managing the day-to-day costs covered in the Project Cost Management Knowledge Area. Since both deal with the PM of money and the procedures for verification and payment, they will be discussed in the same chapter, starting with the Project Financial Management Knowledge Area, which is of key importance for the project outcome.

Annex A1 of the 2016 Construction Extension 'Managing Claims in Construction' is devoted to project claims, including dispute prevention and resolution. It was developed as part of the 'Project Procurement Management Knowledge Area'. However, the previous edition of the Construction Extension included 'Project Claims Management' as a separate Knowledge Area. Due to the importance of timely claims management, this book will treat it as a separate Knowledge Area:

- **'Project Claims Management'**: covers the processes for claims management.

The Construction Extension addresses dispute avoidance and resolution within the Project Claims Management. Following the new FIDIC approach to separate Claims from Disputes, the book will allocate a separate chapter to dispute management. The additional Knowledge Area is named:

- **'Project Dispute Management'**: it deals with the procedures for dispute avoidance and multi-tier dispute resolution under the FIDIC contracts.

Currently, information management is becoming of utmost importance. Chapter 5 discusses it within the 'Project Procurement Management' Knowledge Area in the light of the relevant ISO standards.

The Construction Extension recognises that it does not address the 'green' construction trends that are influencing the way projects are conceptualised, designed, and constructed. Chapter 15 integrates the sustainability and net zero tools within the 'Project HSSE Management' Knowledge Area as part of the environmental management.

Matching the PMBOK with the FBOK will start in the 'Project Procurement Management' Knowledge Area, since it discusses the tender strategy, the choice of conditions of contract, the preparation of tender documents, the launching of tenders, and the selection of consultants/ contractors.

2.3 Tailoring the contract documents

The tailoring involves understanding the project context, goals, and operating environment. Projects operate in complex environments that need to balance potentially competing demands, which include, but are not limited to:

- delivering as quickly as possible;
- minimising project costs;
- optimising the delivered value;
- creating high-quality deliverables and outcomes;
- providing compliance with regulations and contract requirements;
- satisfying diverse stakeholder expectations;
- adapting to change.

The tailoring is even more complex in construction, where every project is unique and faces uncertainty due to:

- split design and construction responsibilities;
- long duration and aggressive schedules;
- changing environment, including physical, political, cultural, and social;
- complex technical processes;
- open locations exposed to climatic changes;
- unskilled workforce;
- material shortage;
- many stakeholders with competing goals;
- public interest;
- changing regulatory requirements and complicated building control procedures;
- volatility of prices of fuel, energy, labour, materials, etc.

The contract documents may be tailored in line with the modern trends in PM. They cannot change human behaviour, but they can influence it, providing tools to incentivise the project team to achieve improved value.

Prior to tailoring, the project environment needs to be analysed and understood. Depending on the time/budget constraints and risks, the PM approach has to be selected. This will be discussed in Chapter 3.

CHAPTER 3

New Trends in Construction Project Management

DOI: 10.4324/9781003272892-3

3.1 New challenges for PM in construction

New market developments and emerging disruptions drive changes in the construction industry and transform ways of working. McKinsey's 2020 survey found that 90 per cent of respondents strongly believe that the industry needs to change.[1] Construction project managers have to adapt their practices to address the new challenges. The following are some of the current challenges to be faced in construction projects.

3.1.1 Increasingly complex projects

Higher skyscrapers and deeper earthscrapers are being constructed due to the lack of space in the megacities. The infrastructure has to satisfy competing needs complying with many restrictions. Megaprojects with budgets above $1 billion have more challenges and uncertainties and a poorer track record than smaller projects and need new methodologies to solve complex issues and satisfy many stakeholders.[2]

3.1.2 More rigorous regulations

Project implementation and outputs have to comply with more rules. The essential requirements for construction products gradually increase; the building control procedures are

1 Maria Joro Ribeirinho and others, 'The Next Normal in Construction: How Disruption Is Reshaping the World's Largest Ecosystem' (McKinsey & Company 2020) <https://www.mckinsey.com/business-functions/operations/our-insights/the-next-normal-in-construction-how-disruption-is-reshaping-the-worlds-largest-ecosystem> accessed 13 January 2022

2 e.g. see Donald Charrett, *Construction Contracts for Construction and Engineering Projects* (2nd edn, Informa Law from Routledge, 2022) Chapter 13 'Challenges in Achieving Successful Megaprojects'

often bureaucratic. Additional regulatory and sustainability considerations require a more adaptive approach.

3.1.3 Internationalisation

International teams have to adapt to a new hostile environment, different legal system, lack of infrastructure, and lack of skilled workforce. Disagreements between partners in multinational joint ventures often escalate into disputes, impacting the project implementation.

3.1.4 Climate changes

Resilient design and methods of construction become important. The PM approach has to deal better with changes.

The SDGs drive the implementation of innovations during the design and construction towards whole life net zero and net resource outcomes.

3.1.5 Lack of resources

The increasing lack of natural resources and skilled workers drives changes and innovations in materials and technology. Modern supply change management, the lean approach, reuse of materials, and smart resource management disrupt traditional project resource management.

3.1.6 Impact of COVID-19

The pandemic caused continuous impediments to movement of people, disruption of the supply chain, and increased costs. The new normal needs flexibility in the PM approach to deal with the uncertainty.

3.1.7 Volatility of prices

The volatility of fuel, energy, and materials prices causes uncertainty in project budgets and increasingly drives contractors to insolvency and projects to termination. Only adaptive scope, resource, and cost management can enable project resilience.

3.1.8 Competitive market for public procurement

Public employers usually do not have enough funds and try to find the cheapest solutions. They plan inadequate contingencies for proper change management. They try to transfer the risk to the contractors. The low prices result in hiring an unskilled workforce, poor quality, and dispute escalation. When the risk allocation is unbalanced in the contract, termination or insolvency may end the project before any results have been achieved. A flexible PM is more likely to save the project than the rigid traditional PM methods.

3.1.9 Construction industry underperformance

According to McKinsey, annual productivity growth in the construction sector over the past 20 years was only a third of total economy averages. Risk aversion and fragmentation as well as difficulties in attracting digital talent slow down innovation. Digitalisation is lower than in nearly any other industry. Profitability is low. High unpredictability and cyclicality have led construction firms to rely on temporary staff and subcontractors, which hampers productivity, limits economies of scale, and reduces output quality.[3] Modern PM has to address those problems.

3.1.10 Increasing client dissatisfaction

Customer satisfaction is hampered by regular time and budget overruns, low quality, and lengthy claims procedures. A 2016 McKinsey's analysis found that construction projects typically take 20 per cent longer to finish than scheduled and are up to 80 per cent over budget, frequently resulting in litigation.[4] Project stakeholders have become more sophisticated, and the constantly changing competing requirements are difficult to satisfy. The traditional PM did not address those issues, so a relatively new Knowledge Area is used to manage stakeholders' satisfaction.

3.2 From predictive to hybrid PM

3.2.1 Predictive approach

The **predictive approach** is a development approach in which the project scope, time, and cost are clear and justify the detailed upfront planning. It is the traditional PM approach, also known as 'waterfall'. **Waterfall development** has been defined as 'A breakdown of project activities into linear sequential phases, meaning they are passed down onto each other, where each phase depends on the deliverables of the previous one and corresponds to a specialisation of tasks'.[5] It allows the project team to reduce the level of uncertainty early and do much of the planning upfront. It may explore options at concept level, but the majority of project work follows the plans developed near the start. Templates from previous similar projects are used. This approach is traditionally used in construction, because:

- significant investment is involved;
- regulations and rigid building control procedures impede changes during construction;

3 Maria Joro Ribeirinho and others, 'The Next Normal in Construction: How Disruption Is Reshaping the World's Largest Ecosystem' (McKinsey & Company 2020) <https://www.mckinsey.com/business-functions/operations/our-insights/the-next-normal-in-construction-how-disruption-is-reshaping-the-worlds-largest-ecosystem> accessed 13 January 2022

4 Rajat Agarwal, Shankar Chandrasekaran, and Mukund Sridhar, 'Imagining Construction's Digital Future' (McKinsey 2016) <https://www.mckinsey.com/business-functions/operations/our-insights/imagining-constructions-digital-future> accessed 20 January 2022

5 Kai Peterson, Claes Wohlin, and Dejan Baca, 'The Waterfall Model in Large-Scale Development' in Frank Bomarius, Markku Oivo, Päivi Jaring, and Pekka Abrahamsson (eds) *Product-Focused Software Process Improvement* (Springer 2009) 386 https://link.springer.com/chapter/10.1007/978-3-642-02152-7_29, accessed 21 April 2023

- the scope, schedule, cost, resources, and risks can be well defined in the early phases in most cases and are relatively stable;
- the design phase may significantly reduce the level of uncertainty;
- the majority of the project work follows the plan developed during the project preparation;
- the participants in the process have experience and templates from similar projects;
- the construction industry uses a lot of unskilled labour and its own equipment, and there are no incentives for innovation;
- construction has many stakeholders and complex governance;
- the community is monitoring how the infrastructure is changing the environment and the taxpayers' money is spent;
- the project participants try to follow the plan, focusing on their interests.

3.2.2 Adaptive approach

The **adaptive approach** is a development approach facilitating change. It prioritises learning from previous outcomes and decisions to improve future practices.

The adaptive approach may use iterative and incremental approaches. The **iterative approach** is a development approach setting requirements at the start of each iteration, mapping them to overall project objectives. These iterations give the flexibility to learn, adjust, and manage risks while making continuous progress. The **incremental approach** is an adaptive development approach breaking down the project development into small, manageable portions known as increments. While agility is a broader concept, **agile approaches**[6] can be considered adaptive. An iterative approach is useful for clarifying requirements and investigating different options. An incremental approach is used to produce a deliverable throughout a series of iterations.

In construction, the adaptive approach is often applied in the initial stages. During the design development, iterations are necessary to address the information gathered through site surveys. Utility operators and authorities lay down specific requirements during the planning process. Public discussions of the project values may be held as part of the mandatory permitting procedures. New legislation may come into force and new interfering projects might be constructed during the project preparation, changing the built environment.

According to the Royal Institute of British Architects (**RIBA**) Plan of Work 2020, the pre-design stages include Stage 0 'Strategic Definition' and Stage 1 'Preparation and Brief'. The design work begins at the start of Stage 2, when the initial ideas are created, and continues to the end of Stage 4 'Technical Design', when all of the design information required for manufacturing and constructing the building is completed. Although the information might seem complete after the Stage 2 'Concept Design', it is likely to require

6 'A project management methodology that uses short development cycles called "sprints" to focus on continuous improvement in the development of a product or service.' Moira Alexander, 'Agile Project Management: 12 Key Principles, 4 Big Hurdles' (CIO 2018) https://www.cio.com/article/237027/agile-project-management-a-beginners-guide.html accessed 21 April 2023

further design iterations to conclude the engineering aspects, coordinate it with project strategies, and align it with the cost plan.[7]

3.2.3 Hybrid approach

A hybrid approach combining adaptive and predictive approaches should be used for better change management, innovation, and improved value. The iterative and incremental approaches bring flexibility.

For example, an iterative and incremental approach can be used for a road/railway project preparation. Consultants explore different routes to compare the economic and social feasibility of different options. They propose the best route after a multi-criteria analysis. After the approval of the proposed route by the stakeholders, the consultants develop the design in several stages with deliverables that go through a rigorous process of consents and agreements by many stakeholders (municipalities, land owners, public utilities, control authorities, and non-governmental organisations [**NGOs**]). The stage deliverables are completed only once the permitting and agreement procedures have successful outcomes.

A lot of unforeseeable issues need resolving during the design phase. For example, the NGOs want to stop the project because the bears cannot cross the highway. The designers have to change the route or add a tunnel for the bears. They use an iterative approach, developing different options and discussing them with stakeholders (who have conflicting interests) before they reach an agreement with clear conditions for further development. This is the gateway for the next design stage or tender procedure for a design-build contractor. Since the project needs huge funds, iterations are also possible in relation to project financial management. The employers apply the incremental approach, splitting the scope of the initial sections into smaller parts, so they can commence the procurement of those parts that have permits and financing, then continue with other parts as soon as financing is available and the permitting is achieved.

Usually, the predictive approach works well for construction projects, because the financing institutions and public prefer to monitor processes planned from the initiation of the project, and the building control procedures for deviations are rigorous. But the hybrid approach has become very popular, since in many cases, employers would prefer not to use the traditional design-bid-build procurement. This is related to the rising complexity of projects, the need to adapt to the constantly changing building environment, including climate, political, social, and regulatory changes, technology breakthroughs, and recently, the COVID-19 pandemic.

A hybrid approach can be used in construction with:

- early contractor appointment to develop/optimise the design;
- tier-two subcontractors' and tier-three suppliers' involvement to bring further improvements;
- continuous involvement of the client to adapt the requirements to the changing environment;
- early involvement of the end-users to review the deliverables at several gateways and introduce new requirements in case innovations are necessary.

7 RIBA, *RIBA Plan of Works 2020* (RIBA 2020) 99

According to RIBA President and Executive Director Simon Allford, 'successful implementation lies in collaboration at the boundaries between traditional disciplines and in developing hybrid solutions that draw on a wide range of different bodies of knowledge'.[8]

3.2.4 Case study – adaptive approach

A technical assistance consultant for the modernisation of a double track railway line of about 100 km selected the best option and delivered a preliminary design based on existing maps and surveys. The client decided to develop the project with a budget of 800 million Euro split into two sections: one through the valley, and the other through the mountains, where uncertainty prevailed. For the next stage, the client selected a new consultant, responsible for the project preparation, support in the selection and coordination of designers for the two sections, and compliance assessment of the design. During the initial site surveys, the new consultant encountered many issues to be resolved (conflicts with existing infrastructure and buildings, water supply zone, etc.). As a result of the iterations (proposing alternatives to the client, discussing, selecting a solution, design revision, and consent), 50 per cent of the route was changed. Based on the changed route, designers were selected and appointed to elaborate the technical design development. During the design development, the consultant prepared application forms for European co-financing. In order to use European funds from different programmes, the client requested the second section to be split into three lots, changing the design accordingly. There were different iterations due to new requirements from utility operators and authorities, problems with land acquisition, and crossing of existing infrastructure.

3.3 Integration of new PM methods in construction

It is suggested that in a constantly changing world with increasing population and decreasing resources, construction needs hybrid PM. It is important to use the successful PM methods from other industries, since construction is lagging behind. Researchers consider that the lean production methods from the automobile industry and some agile tools from the IT sector can be successfully used in construction PM.

3.3.1 Lean methods from the automobile industry

The 'lean revolution' started with the just-in-time (**JIT**) technique introduced by Taiichi Ohno way back in 1960. In 1988, Toyota introduced lean as a formal approach, publishing the Toyota Production System.[9]

8 Simon Allford RIBA President and Executive Director <https://www.architecture.com/knowledge-and-resources/resources-landing-page/dfma-overlay-to-the-riba-plan-of-work#available-resources> 18 January 2022

9 Suhail Iqbal, *Leading Construction Industry to Lean-Agile (LeAgile) Project Management.* (Paper presented at PMI® Global Congress 2015)

Lean manufacturing has enabled businesses to increase production, reduce costs, improve quality, and increase profits by following five key **lean** principles: identify value, map the value stream, create flow, establish pull, and seek perfection.[10]

Rather than executing predefined tasks over and over again, workers must continually improve their way of working (the so-called 'Kaizen' philosophy).

The Lean Construction Institute aims to extend to the construction industry the lean revolution started by Toyota in manufacturing. This approach maximises the value delivered to the customer while minimising waste.[11]

Research demonstrates that lean practices can be implemented either as an integrated system or as stand-alone practices in construction projects. The Last Planner® System (**LPS**), followed by JIT, are identified as the most often used lean tools. Both help in the reduction of the turnaround time of activities, leading to minimised interruptions and cost and time overruns in construction project delivery.[12]

Glenn Ballard and Greg Howell created LPS in the 1980s to improve the predictability and reliability of construction production. LPS enables issues to be identified and resolved before they become problems on site and increases the chances that work will flow and projects be completed on time. LPS is a lean transformation tool, supporting lean thinking to solve the problem with unreliable and unpredictable production in project-based environments.[13] Initially, LPS consisted only of look-ahead planning, weekly work planning, and learning from breakdowns. In the early 2000s, planning and scheduling project phases (which provide inputs to look-ahead planning) were added to its scope. The 2020 Current Process Benchmark further extends LPS in principle to both production (i.e. striving for targets) and project planning and control (i.e. setting targets).[14]

The LPS® focuses on five conversations around which all planning occurs: *should, can, will, did, learn*[15]:

1. Pull-planning establishes what **should** be done.
2. Make-ready planning gets the upcoming work into such a condition that it **can** be done.
3. Weekly work planning establishes the set of promises from specific people for the work that **will** be done.

10 James Womack and Daniel Jones, *Lean Thinking: Banish Waste and Create Wealth in Your Corporation* (Simon & Shuster, New York 1996)

11 Lean Construction Institute, 'A3-Brief Introduction to Lean Construction' (LCI) <https://leanconstruction.org/uploads/wp/media/library/id29/A3-Brief_Introduction_to_Lean_Construction.pdf> accessed 19 January 2022

12 Oluwatosin Babalola and others, Implementation of Lean Practices in the Construction Industry: A Systematic Review (*Building and Environment* 2018) 148

13 Alan Mossman, 'Last Planner®: 5 + 1 Crucial & Collaborative Conversations for Predictable Design & Construction Delivery'(2018) <https://www.researchgate.net/publication/321804312_Last_Planner_5_1_crucial_collaborative_conversations_for_predictable_design_construction_delivery_Dec_2017> accessed 19 January 2022

14 Glenn Ballard and Iris Tommelein, 2020 Current Process Benchmark for the Last Planner® System of Project Planning and Control (*Lean Construction Journal* 2021), 53–155

15 Lean Construction Institute, 'The Last Planner System®' <https://leanconstruction.org/lean-topics/last-planner-system/#:~:text=The%20Last%20Planner%20System%20focuses,is%20done%20through%20pull%20planning> accessed 9 April 2023

4. Daily commitment management supports the last planners[16] in staying on track with their promises so that the work **did** get done.
5. At least weekly, the last planners take time to **learn** from their performance.

JIT enables prompt delivery of materials, information and drawings, or anything required for a project, only when needed to the point of usage.

Some of the other lean tools used in construction are the following:[17]

- Total quality management (**TQM**): used to recognise and assess possible problems, develop and apply new solutions, and appraise results;
- Kaizen: encourages continuous improvement in every process;
- Kanban: involves the use of inventory control card or sign for stock taking on the project site, helping to promote effective inventory and record keeping;
- Visualisation tools/management;
- Pull scheduling/planning: pulling materials in the entire value chain process to get them ready on time for production;
- 5S on-site management: sorting, straightening, shining, standardising, and sustaining of all site processes and activities;
- 6 Sigma: problem identification, performance measurement, analyses of variations;
- Prefabrication and modularisation;
- Design workshops/big room workshop;
- Integrated project delivery (**IPD**);
- Teamwork and partnering;
- Target value design;
- Concurrent engineering: collaborative sharing of information in the process of executing different tasks in a project with the aim of producing a functional design, a product with good quality, and a productive process;
- Standardisation;
- Benchmarking: dividing construction works into sections with an encouragement package attached to each section assigned to a team of workers as a source of competition and motivation among project teams;
- Value based management (**VBM**) or value stream mapping (**VSM**): used in mapping out non-value-adding activities in the construction process in order to maximise value.

The lean tools used in construction enable elimination of waste and sustainability through collaboration and standardisation.

16 The last planner is the person closest to the work with authority to make decisions regarding the schedule and to make reliable commitments to complete the work of their discipline. <https://leanconstruction.org/lean-topics/last-planner-system/#:~:text=What%20is%20the%20Last%20Planner,they%20slow%20down%20the%20flow> accessed 30 April 2023

17 Oluwatosin Babalola and others, Implementation of Lean Practices in the Construction Industry: A Systematic Review (*Building and Environment* 2018) 148

3.3.2 Agile methods from the IT industry

The traditional PM approach is not always the most appropriate one for complex construction projects subject to successive changes. **Agile** approaches might help to deal with uncertainties. The history of agile PM dates back to the 1930s, with Walter Sherwart's plan-do-check-act (**PDCA**) approach to project quality.[18] PDCA is in the heart of the updated ISO standards,[19] widely used in construction. Agile approaches to management in manufacturing and IT developed independently and with different objectives. Research concludes that agile IT is more suitable for construction PM than agile manufacturing.[20]

In 2001, 17 software developers published the Agile Manifesto, a 'formal proclamation of four key values and 12 principles to guide an iterative and people-centric approach to software development':

1. Customer satisfaction is always the highest priority and is achieved through rapid and continuous delivery.
2. Changing environments are embraced at any stage of the process to provide the customer with a competitive advantage.
3. A product or service is delivered with higher frequency.
4. Stakeholders and developers collaborate closely on a daily basis.
5. All stakeholders and team members remain motivated for optimal project outcomes, while teams are provided with all the necessary tools and support, and are trusted to accomplish project goals.
6. Face-to-face meetings are deemed the most efficient and effective format for project success.
7. A final working product is the ultimate measure of success.
8. Sustainable development is accomplished through agile processes whereby development teams and stakeholders are able to maintain a constant and ongoing pace.
9. Agility is enhanced through a continuous focus on technical excellence and proper design.
10. Simplicity is an essential element.
11. Self-organising teams are most likely to develop the best architectures and designs and to meet requirements.
12. Regular intervals are used by teams to improve efficiency through fine-tuning behaviours.[21]

Agile methods are recommended when:

- projects are ever-evolving and have an undetermined closing;
- organisational teams are quite flexible and adaptable to change;

18 Abdallah Lalmia and others, A Conceptual Hybrid Project Management Model for Construction Projects (*Procedia Computer Science* 2021) 181, 921–930

19 *ISO 9001:2015, Quality Management Systems – Requirements*

20 Selim Demir and others, *Re-conceptualising Agile for Lean Construction: the Case for 'Agilean' Project Management.* (Conference: 48th ASC Annual International Conference Proceedings Birmingham, UK 2012) <https://www.arcom.ac.uk/-docs/proceedings/ar2012-1013-1023_Demir_Bryde_Fearon_Ochieng.pdf> accessed 18 January 2022

21 Moira Alexander, 'Agile Project Management: 12 Key Principles, 4 Big Hurdles' (CIO 2018) <https://www.cio.com/article/237027/agile-project-management-a-beginners-guide.html> accessed 24 January 2022

- the project team is lean and small;
- the timeline is flexible;
- the project manager is well experienced in agile methodology and can lead the team efficiently.

As with the waterfall model, with this methodology you still plan, schedule, and identify key milestones and dependencies. But this model provides flexibility, which accommodates changing requirements along the way. The team works in shorter sprints and with short-term goals. The development cycle is quite rapid and iterative. Adaptive teams are mission-focused and are clear about the work and tasks for the coming weeks, but not for the complete duration of the project. The impetus here is on quality and flexibility rather than speed.[22]

The key principle of incremental delivery of value in the form of working software does not translate well to construction. However, agile concepts such as customer collaboration and responsiveness to change have a place in a construction project.[23] Agile PM in information systems has obvious parallels with the design phase of construction. Commentators consider that agile PM offers considerable potential for application in pre-design and design, but that there are significant hurdles to its adoption in the actual construction phase.[24,25] Others demonstrate how agile facilitates the PM during the construction phase in a complex environment, by decomposing the scope into small manageable parts and then completing these parts in order of greatest value.[26]

It is suggested that digitalisation and collaboration can enable agility during construction, enhancing efficiency, improved value, and client satisfaction.

3.3.3 Leagile or Agilean in the hybrid PM approach

Engaging lean thinking and agile techniques can make any project better, faster, and more cost effective, upgrading the foundation of traditional PM.[27] Where lean focuses on reducing waste, agile focuses on being alert to opportunities to make changes in a fast manner.[28] Agile needs an adaptive PM approach, while lean needs a well-planned predictive approach.

In construction projects, both lean and agile approaches are beginning to create cost and time efficiencies with tools and materials, construction operation sequencing, and

22 Sid, 'Adaptive vs Predictive Project Management' (*Zvolv Blog* 2019) <https://zvolv.com/blog/blog/2019/12/10/adaptive-vs-predictive-project-management/> accessed 15 January 2022

23 Enterprise Agile Planning, 'Agile Construction Management' (*CollabNet VersionOne Blog* 2013) <https://digital.ai/catalyst-blog/agile-construction-management> accessed 24 January 2022

24 Robert Owen, Lauri Koskela, and others, *Is Agile Project Management Applicable to Construction?* (IGLC 2006) 51–66

25 Abdallah Lalmia and others, A Conceptual Hybrid Project Management Model for Construction Projects (*Procedia Computer Science* 2021) 181, 921–930

26 Adrian Smith, 'Agile and Lean for Construction' (ENNOVA 2011) <http://www.slideshare.net/adrianl-smith/agile-and-lean-for-construction-9384237> accessed 28 January 2022

27 Terra Stern, *Lean and Agile Project Management: How to Make Any Project Better, Faster, and More Cost Effective* (Routledge 2020)

28 Suhail Iqbal, *Leading Construction Industry to Lean-Agile (LeAgile) Project Management.* (Paper presented at PMI® Global Congress 2015)

safer job sites.[29] The aim of a hybrid approach combining waterfall with agile and lean approaches is to take the advantage of the benefits of the approaches and eliminate their weaknesses.[30]

Researchers are commenting how agile and lean may be combined to improve traditional PM in the construction industry:

- the use of lean and agile elements can help in coping with project complexity;[31]
- some apply a lean-agile, or '*LeAgile*', method of PM, using the LPS with agile iterations of the weekly work plans;[32]
- others consider that a new methodology '*AgiLean PM*', may be introduced, merging lean and agile into a holistic unit.[33] Waste elimination through lean, and the ability to react to change through agile, are achieved at the operational level, with PM methods providing the necessary strategic oversight.[34]

It seems possible to integrate the agile concepts with the lean construction methods in order to innovate the PM in construction for improved value. In the new normal, it is not enough to reduce the waste in the PM processes: agility is needed to manage changes and prevent disputes.

Figure 3.1 summarises the main factors for the use of a traditional predictive approach, lean and agile, to be integrated in a hybrid model for innovative PM in construction.

3.4 Innovative PM tools to address the challenges

Innovations facilitate PM. Opportunities arise in the fields of digital technology, automation, additive manufacturing, and robotics. The UN SDGs[35] also drive innovations and efficiency in construction. The PM tools have been known for many years, but project teams have to be trained to use them effectively. Contract provisions have an important role to mandate the use of innovative practices.

---

29 PMI Inc., *Construction Extension to the PMBOK® Guide* (PMI 2016) 27

30 Abdallah Lalmia and others, A Conceptual Hybrid Project Management Model for Construction Projects (*Procedia Computer Science* 2021) 181, 921–930

31 Maher Thakkar and Krittika Iyer, *Lean Principles on Agile Project Management* (ResearchGate 2021)

32 Suhail Iqbal, *Leading Construction Industry to Lean-Agile (LeAgile) Project Management.* (Paper presented at PMI® Global Congress 2015)

33 Selim Demir and others, *Introducing AgiLean to Construction Project Management* (2014)

34 Selim Demir and others, *Re-conceptualising Agile for Lean Construction: the Case for 'Agilean' Project Management* (Conference: 48th ASC Annual International Conference Proceedings Birmingham, UK 2012) <https://www.arcom.ac.uk/-docs/proceedings/ar2012-1013-1023_Demir_Bryde_Fearon_Ochieng.pdf> accessed 18 January 2022

35 UN, 'Transforming Our World: the 2030 Agenda for Sustainable Development' (UN Department of Economic and Social Affairs Sustainable Development) <https://sdgs.un.org/2030agenda> accessed 18 January 2022

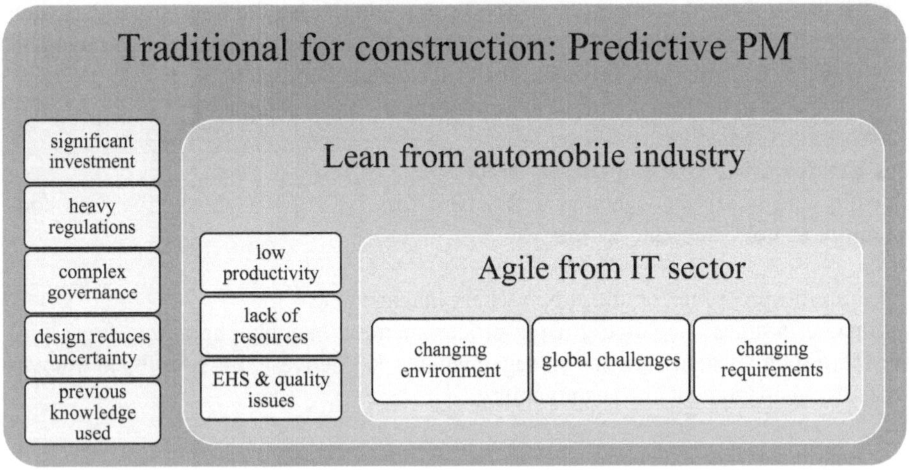

Figure 3.1 Factors for hybrid PM approach in construction

3.4.1 Actions to increase productivity

In 2017, McKinsey's Report[36] recommended the following actions to boost productivity on construction projects by some 50 to 60 per cent, giving important insights for the available innovative PM tools, many of which can be addressed in contracts:

3.4.1.1 Reshape regulation and raise transparency

Permit management will benefit from streamlining permitting and approvals processes. Financing for innovation and training is important.

3.4.1.2 Rewire the contractual framework

These recommendations are very important for the contract tools underpinning innovative PM:

- to move away from the hostile contracting environment to collaboration and problem solving;
- to base tendering processes on best value and past performance rather than cost alone;
- to establish a 'single source of truth' on projects for monitoring progress early, supported by collaborative technology, to minimise misalignments and enable joint corrective action;
- to introduce incentives in traditional contracts that significantly improve performance and alignment not at a trade or package level, but at the project-outcome level;

36 McKinsey Global Institute (MGI's), 'Reinventing Construction: a Route to Higher Productivity, February 2017' <https://www.mckinsey.com/~/media/mckinsey/business%20functions/operations/our%20insights/reinventing%20construction%20through%20a%20productivity%20revolution/mgi-reinventing-construction-a-route-to-higher-productivity-full-report.pdf> accessed 13 January 2022

- to move toward best practices, appropriate alternative contracting models such as IPD help build long-term collaborative relationships; relational contracts will need to become more prevalent than transactional contracts;
- to invest in upfront planning incorporating all parties' input.

3.4.1.3 Rethink design and engineering processes
The quality and safety management will benefit from VE with a greater focus on constructability and off-site manufacture.

3.4.1.4 Improve procurement and supply-chain management
Best practice in areas such as digitising procurement and supply-chain workflows will enable more sophisticated logistics management and JIT delivery. This results in innovative resource management and sustainability.

3.4.1.5 Improve on-site execution
There are four key approaches that are well known in the industry but have not been universally adopted, which can be integrated into contracts:

- a rigorous planning process – LPS is a useful tool;
- KPIs with additional forward-looking plan conformance metrics for better risk management;
- ensuring that all pre-work (for instance, obtaining approvals and developing project milestones) has been completed prior to starting on site;
- careful planning and coordination of different disciplines on site along with the application of lean principles to reduce waste and variability.

3.4.1.6 Infuse digital technology, new materials, and advanced automation
McKinsey suggests that the biggest barriers to innovation by construction companies are underinvestment in IT and technology more broadly, and a lack of research and development processes.

3.4.1.7 Reskill the workforce
Construction firms and workers need to continuously reskill and train to use the latest equipment and digital tools.

3.4.2 Digital tools innovate PM in construction

Digitalisation enables rapid creation and sharing of data. The digital tools listed in the following subsection facilitate and innovate PM in construction. McKinsey reported construction to be lagging behind other industries in 2017.[37] In 2020, McKinsey demonstrated that the call to action was heard, and the change is already here.[38] Commentators consider

37 McKinsey Global Institute (MGI's), 'Reinventing Construction: a Route to Higher Productivity, February 2017'

38 Maria Joro Ribeirinho and others, 'The Next Normal in Construction: how Disruption Is Reshaping the World's Largest Ecosystem' (McKinsey & Company 2020)

that today, no industry seems to be introducing disruptive new technologies faster than construction.[39] Digitalisation has contributed to increasing efficiency, reducing costs and disrupting the traditional PM practices.

3.4.2.1 Mobile technology

Wireless networks, mobile apps, smartphones, and video streaming facilitate project team connectivity and provide real-time project information. Platforms for virtual meetings enable collaboration and remote working. **5G**[40] will boost the application of Internet of Things (**IoT**). 5G opens the possibility of using drones or remote cameras to inspect, sending images, video, and other sensor data to artificial intelligence (**AI**) in the cloud for automated detection of problems. 5G will enable easier connectivity of sensors in buildings to drive new solutions for energy efficiency, occupancy management, and visitor experience. The lower latency of 5G coupled with more bandwidth and more reliable connectivity will enable many kinds of applications in connected vehicles.[41]

3.4.2.2 Cloud platforms

The controlled access to cloud-based digital information brings certainty and transparency. It reduces errors due to using a previous design version for construction. Digital compliance assessment and permit management through online platforms save paper, time, and cost.

Multi-cloud infrastructures allow an IoT solution to live up to its full potential. They allow work in multiple regions, improve network security, and more. Proper multi-cloud service management is essential.[42]

3.4.2.3 Virtual reality (VR) collaboration

Holding meetings in VR offers new benefits. A designer can walk team members through a virtual design from the inside. With features such as the ability to turn walls transparent, the experience creates a better understanding of the layout and how the design fits together.[43]

3.4.2.4 Computer-aided design (CAD)

Digitalisation of design started with two-dimensional CAD. The three-dimensional (**3D**) development of CAD significantly contributes to stakeholder management with the 3D visualisation of the project. It enables better coordination between different disciplines, bringing improved constructability and early detection of conflicts and errors. It may be linked with the off-site manufacturing and on-site processes when used with IoT/AI.

3.4.2.5 Building information modelling (BIM)

BIM is the management of information through the whole life cycle of a built asset, from initial design all the way through to construction, maintaining, and finally

---

39 Viatechnik '14 Technologies Disrupting the Construction Industry' <https://www.viatechnik.com/technologies-disrupting-the-construction-industry/> accessed 18 January 2022

40 Fifth-generation mobile network with increased availability and higher data speed

41 Rishi Vaish and Sky Matthews, '5G Will Accelerate a New Wave of IoT Applications' (IBM) <https://newsroom.ibm.com/5G-accelerate-IOT> accessed 28 January 2022

42 Jacob Schulz, '4 Best Practices to Manage Multi-Cloud Infrastructure'(iot for all 2022) <https://www.iotforall.com/4-best-practices-to-manage-multi-cloud-infrastructure> accessed 23 January 2022

43 Viatechnik, '14 Technologies Disrupting the Construction Industry'

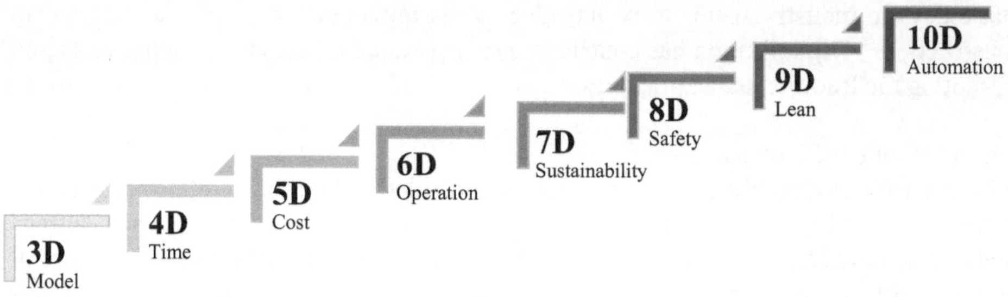

Figure 3.2 BIM development

de-commissioning, through the use of digital modelling.[44] BIM allows collaborative design and construction, thus bringing efficiency and eliminating waste. Figure 3.2 represents one perspective of BIM development.[45] 1D BIM creates a document retrieval system, allowing real-time access to requirements and a comprehensive archive. 2D BIM models the project in two dimensions. 3D BIM enables visualisation with early clash detection within the integration management of different disciplines. BIM involves more than just 3D modelling and is also commonly defined in further dimensions such as 4D (time) and 5D (cost).[46] The sixth dimension introduces as-built operation and facility management. The seventh dimension concerns sustainability. The eighth dimension is about health and safety on site.[47] Some researchers discuss the potential of 9D BIM as integration of lean methods improving resource management and 10D BIM as automation eliminating obstacles to productivity.[48] Others distinguish 10D BIM for adding security, emergency, and disaster management.[49] Software developers advocate that 10D BIM works as a tool that centralises data to optimise PM relying on the use of technology. 10D BIM can be used to align the financial, commercial, environmental, health and safety, and risk analysis sectors, etc.[50]

BIM is being adopted widely across the world as all of those involved in the creation and operations of buildings and infrastructure assets seek ways to bring essential improvements in productivity and performance in response to ever-increasing social, economic, and environmental challenges.[51]

---

44 BSI, <https://www.bsigroup.com/en-GB/Building-Information-Modelling-BIM/> accessed 29 May 2023

45 *Biblus Blog*, 'What Is 10D BIM?' (ACCA Software 2021) <https://biblus.accasoftware.com/en/what-is-10d-bim/> accessed 23 January 2022

46 Peter Smith, BIM & the 5D Project Cost Manager (*Procedia – Social and Behavioral Sciences* 2014) 119, 475–484

47 Benjamin Josseaux, 'The BIM Revolution in Building Management' (*Drawbotics Blog* 2018) <https://blog.drawbotics.com/2018/11/07/the-bim-revolution-in-building-management/> accessed 22 January 2022

48 Mahmoud Ershadi, Peter Davis, and others, Implementation of Building Information Modelling in Infrastructure Construction Projects: a Study of Dimensions and Strategies (*International Journal of Information Systems and Project Management* 2021) 9 (4), 43–59

49 Jordi Battle and others, 'Why It's Time to Adopt Building Information Modelling' (Accenture 2017) <https://www.accenture.com/_acnmedia/PDF-59/Accenture-BIM-POV-Final.pdf> accessed 23 January 2022

50 *Biblus Blog*, 'What Is 10D BIM?' (ACCA Software 2021) <https://biblus.accasoftware.com/en/what-is-10d-bim/> accessed 23 January 2022

51 FIDIC, *Yellow Book* (FIDIC 2022) GPPC 56

3.4.2.6 4-D simulation

Sophisticated animation software allows employers to view the construction of a project virtually. It is used for better planning, time management, claims visualisation, and dispute resolution.

3.4.2.7 Digital fabrication

These processes link data from BIM and CAD sources directly to manufacturing equipment to guide the fabrication process. The most widely known application lies in 3D printing, but its use is growing rapidly in construction, particularly in the fabrication of irregularly shaped objects or linking the BIM to computer numerical control (**CNC**) machine output. CNC machining is a manufacturing process in which preprogrammed computer software dictates the movement of factory tools and machinery. 3D printing can create faster and more accurate 3D components for use in the construction process.[52] Unique architectural forms can be produced, cutting waste, costs and time.

3.4.2.8 Reality capture

This process enables the creation of 3D models from existing building conditions using everything from laser scans to photographs. It can be used to model a site and test the ways the topography, climate, or other factors will affect the building's structure or performance. Experts estimate that reality capture software reduces costs by 50 per cent over traditional 2D scanning, mostly in terms of time saved.[53] 360 photo capture platforms provide uncanny detail quickly with the ability to map an entire site during a quick walkthrough.

3.4.2.9 Digital twin

Producing digital twins of the buildings can provide crucial information on their life performance, enabling improvements. While BIM focuses on a building's design and construction, a digital twin models how people interact with built environment.[54] Leveraging AI and machine learning, the digital twins can improve energy resilience, reduce operating costs, increase resource efficiency, and help decarbonise buildings.

3.4.2.10 Unmanned aerial vehicles

Unmanned aerial vehicles may be used for scanning, monitoring, and mapping. Drones can monitor and record data from multiple angles and elevations to facilitate quality, schedule, resource, and risk management. They can help with the land survey, geomorphology maps, construction monitoring, progress reporting, measurement of quantities, and virtual site visits of the Dispute Board. Drones eliminate health and safety risks during inspections at dangerous heights or in narrow spaces.

---

52 PMI Inc., *Construction Extension to the PMBOK® Guide* (PMI 2016) 26

53 Viatechnik '14 Technologies Disrupting the Construction Industry'

54 ThoughtWire, 'Digital Twins vs. Building Information Modeling (BIM)' (iot for all 2020) <https://www.iotforall.com/digital-twin-vs-bim> accessed 23 January 2022

3.4.2.11 Artificial intelligence (AI)

AI technologies rapidly push innovations. Among the AI applications in construction are computer-vision enhanced safety procedures and 3D design optimising software.[55] AI can facilitate quality and schedule management by early detection of defects and execution of works like bricklaying, plastering, or tiling by robots. AI can be combined with drones or 360-degree image capture tools to extract data and report the progress of the works and non-conformities with the requirements (comparing with the digital twin). Using machine learning technology, AI can learn and adapt based on the data it receives and processes, rather than requiring explicit programming.

3.4.2.12 Distributed ledger technology (DLT) and iContracts

A self-contained and secure database with the ability to share data across a network of geographical locations and users brings transparency and facilitates PM. Distributed ledgers provide security, integrity, and authentication along with ownership resolution. Blockchains[56] implement distributed ledgers with various approaches and functionalities. Smart contracts (also called **iContracts**) allow the creation of self-executing code with the ability to read, evaluate, and append data to the blockchain.[57] Regardless of the automation of contract performance under smart contracts, the use of technologies like blockchain to track and record contractual events will generate a more robust and reliable audit trail of past events if and when a dispute arises.[58] These technologies can remove the discretion of people related to entitlements of the parties in a contract, thus securing timely payments and dispute prevention. DLT can streamline the supply-chain transactions.[59]

3.4.2.13 Supply-chain optimisation tools

Understanding the global supply-chain network connections and using that information to make better purchasing decisions is the job of supply-chain optimisation software, and its use is expanding rapidly.[60]

3.4.2.14 Automated Layout software

The layout of building components and systems is essential for any facility to meet design intent. Automated Layout tools that link the model layout to the field take the guesswork out of the process, from optimising the speed at which teams lay out walls and hangers

55 McKinsey & Company, 'McKinsey Technology Trends Outlook 2022. Applied AI' (2022) <https://www.mckinsey.com/~/media/mckinsey/business%20functions/mckinsey%20digital/our%20insights/the%20top%20trends%20in%20tech%202022/McKinsey-Tech-Trends-Outlook-2022-Applied-AI.pdf> accessed 3 September 2022

56 David L Shirer, *Basic Blockchain* (Robinson Great Britain 2020) 136. A distributed ledger that uses a particular mechanism to ensure the integrity of the database, specifically block transactions with some form of mathematical proof of integrity of the transaction (such as proof of work).

57 Amit Patwardhan and others, Distributed Ledger for Cybersecurity: Issues and Challenges in Railways (*Sustainability* 2021) 13 (18) < https://www.mdpi.com/2071-1050/13/18/10176/htm> accessed 28 January 2022

58 Charlie Morgan and Rebecca Reed, 'Dispute Resolution in the Era of Big Data and AI' (Herbert Smith Freehills Legal Briefings 2019) <https://www.herbertsmithfreehills.com/insight/dispute-resolution-in-the-era-of-big-data-and-ai> accessed 28 January 2022

59 David Mosey, *Collaborative Construction Procurement and Improved Value* (John Wiley and Sons 2019) 262

60 Viatechnik, '14 Technologies Disrupting the Construction Industry' <https://www.viatechnik.com/technologies-disrupting-the-construction-industry/> accessed 18 January 2022

to providing a more precise construction process that follows BIM.[61] Digital tools can help with levelling surfaces, projecting colours on the working surface, or giving voice commands.[62]

3.4.2.15 Augmented reality

Augmented reality allows users to supplement real-world images with contextual information, including overlaying visuals of the model on the current state of a project in various stages of construction. The benefits are many, including allowing on-site verifications based on inconsistencies between the plan and the current reality.[63]

3.4.2.16 Operations Science

In the construction industry, projects that are late and/or over budget are common. Operations Science studies production processes.[64] The results can diagnose and identify solutions to inefficiencies, bottlenecks, and barriers, saving time and money.

3.4.2.17 Data analytics

This software category brings together all the data generated by other technologies and uses it to support better decision-making. Mathematical models developed through data analytics can capture data on projects to assess process and tool improvement opportunities. Extracting data from the numerous project records, the analytics software can create significant time and costs savings for claims management. A data-driven analysis should establish the truth and prevent the escalation of disputes.[65]

3.4.2.18 IoT

IoT allows data to be automatically transmitted through networked computers, intelligent machines, and even wearable devices.[66] Information from BIM is transferred to the construction mechanisation, saving time and reducing waste. Machines equipped with geospatial positioning, sensing, and measurement equipment transmit data for diagnostics, design, BIM, work progress monitoring, and project site security. IoT enables tracking of equipment and materials. Among many benefits, construction site safety is enhanced by wearable devices connected to the IoT. Sensors monitor structural deformations and predict landslides, failure in tunnels and bridges.

---

61 Viatechnik, '14 Technologies Disrupting the Construction Industry'

62 Construction Aid; iLevel (Pelless) <pelles.com> accessed 31 January 2022

63 For the potential of augmented reality throughout the life cycle of a construction project: Hala Nassereddine and others, Augmented Reality in the Construction Industry: Use-cases, benefits, Obstacles, and Future Trends (*Frontiers in Built Environment* 2022) 8 <https://www.frontiersin.org/articles/10.3389/fbuil.2022.730094/full> accessed 1 May 2023

64 Operations Science Institute <https://opscience.org/what-is-operations-science/> accessed 1 May 2023

65 Rhumbix, ' Examples of Data Analytics Sweeping Construction Project Management' (2023) <https://www.rhumbix.com/blog/10-examples-of-data-analytics-sweeping-construction-project-management> accessed 1 May 2023

66 Arka Ghosh and others, 'Use Cases for the Internet of Things (IoT) in the Construction Sector: Lessons from Leading Industries' <https://www.open-access.bcu.ac.uk/9900/1/Use%20cases%20for%20the%20Internet%20of%20Things%20in%20the%20construction%20sector.pdf> accessed 1 May 2023

3.4.2.19 Schedule automation
Software can automatically generate optimised future schedules based on automated updates to project status in real time. Automating this process has the potential to improve outcomes while reducing costs.

3.4.3 Standardisation brings efficiency and improved value in construction

Standardisation guarantees quality and compliance with requirements. Every construction project is unique, but higher efficiency will be achieved using:

3.4.3.1 Standard contract forms
They are balanced, the risk allocation is clear, and the procedures for change and claims management are well thought out. Standard forms bring certainty. Common understanding is developed through training. Books and court decisions/ arbitration awards give reliable legal interpretation. The standard forms have arguably become part of ***lex constructionis***;[67,68]

3.4.3.2 Standards for design and construction
Standards are voluntary (unless mandated in legislation or regulations) but become mandatory when integrated in the contract documents. A list of standards may cover many more requirements than a specification, thus reducing the gaps leading to disagreements. The integration of green certification, quality, sustainability, health and safety, earned value, risk management, information management, and other available standards in the contracts facilitate and innovate the PM. For example, the *ISO 20887:2020 (en) Sustainability in buildings and civil engineering works – Design for disassembly and adaptability – Principles, requirements and guidance* standard for DfD/A (Design for disassembly and adaptability) is a strategy to optimise both the service life and the design life in a sustainable way. Incorporating DfD/A concepts early in the planning and design phase will increase the likelihood that activities during the stages of use, maintenance, and end-of-life (e.g. disassembly, reuse, recycling, and disposal) will be conducted more efficiently from a total resource perspective.[69]

3.4.3.3 Supply of standard components
Components for installations and façade systems, modular built-in furniture, and fixtures are becoming increasingly popular. The producers guarantee the compliance with essential requirements. The components may be delivered just in time with clear instructions for assembly and use. The clients can choose from catalogues and approve samples before procurement, which brings certainty and higher satisfaction.

---

67 Philip Loots and Donald Charrett, *Construction Contracts for Infrastructure Projects* (Informa Law from Routledge, 2022) Chapter 4 *Lex constructionis*

68 Lucas Klee, *International Construction Contract Law* (Wiley 2018) 223. 'These days, FIDIC forms (as a standardized law applicable to construction projects) are without doubt part of **lex mercatoria**' (from Latin 'merchant law': supranational customs and rules of international trade)

69 ISO, *ISO 20887:2020 (en) Sustainability in Buildings and Civil Engineering Works – Design for Disassembly and Adaptability – Principles, Requirements and Guidance* <https://www.iso.org/obp/ui#iso:std:iso:20887:ed-1:v1:en> accessed 22 January 2022

3.4.3.4 Off-site manufacturing of standard modules
Modularisation facilities typically follow manufacturing processes under controlled conditions.[70] Prefabricated modules can range from panels to be integrated on site as parts of standardised structures, to volumetric cells equipped with all installations. Besides the savings in time and cost, off-site manufacturing guarantees reduction of waste, work in a safe indoor environment, and characteristics in compliance with the specification. Moreover, buildings can be easily modified, relocated, and disassembled.

3.4.3.5 Case study – standard modules

In 2015, a Chinese construction company built a 57-storey glass-and-concrete skyscraper made of 2,736 rectangular modules in a record-breaking 19 days.[71]

3.5 Conclusion

Construction needs a hybrid PM methodology using the traditional PMBOK Knowledge Areas with lean methods to eliminate the waste and agile to adapt to the changing requirements. The tools are here: the contract can mandate their use for innovative PM.

70 PMI Inc., *Construction Extension to the PMBOK® Guide* (PMI 2016) 26

71 Stephen Johnson, 'Modular Construction: Using Lego-like Blocks to Build Structures of the Future' (The Future 2021) <https://bigthink.com/the-future/modular-construction/> accessed 22 January 2022

CHAPTER 4

FIDIC Contracts and Project Governance

4.1 Background and development of the FIDIC works contracts

FIDIC publishes standard forms for works and services used internationally. Due to the different colours of their covers, they became popular as the 'FIDIC Rainbow Suite'.

4.1.1 The Red Book 1957–1977 and the role of the Engineer

In 1957, FIDIC published its first standard form: Conditions of Contract (International) for Works of Civil Engineering Construction. Due to its long name and red cover, it became

DOI: 10.4324/9781003272892-4

popular as the Red Book. New editions followed in 1969 and 1977. The FIDIC Red Book early editions were recognisably derived from the Institution of Civil Engineers (**ICE**) forms.[1] The provisions used specific terminologies and concepts referable to English standard forms. The FIDIC form followed the English basis of re-measurement with quantities treated as approximate together with the system of nomination of subcontractors. Commentators point out that no attempt was made to depart from the principles of the **common law**, under which the UK forms were drafted, or recognise that there could be a conflict of laws between the common law system and any other system of law to which the applicable law of the contract belonged.[2]

Dr. Bunni explains the introduction of the Engineer in the Red Book:

> As in the ICE Form of Contract for civil engineering works, the Red Book is drafted on the basis that when a project is initiated by a developer or a promoter, the duties related to feasibility, design and supervision during construction of the project are entrusted to an independent consulting engineer who is referred to as the 'Engineer' in the contractual arrangements between the developer and the contractor. The developer of such a project is referred to in these forms as the 'Employer'.[3]

The role of the Engineer, including certification and other determinations, inherently presupposed the impartiality of the Engineer in relation to decision making.

4.1.2 The Yellow Book 1963–1980

The standard terms in the Red Book were less than satisfactory for contracts where major items of plant and the like were manufactured away from Site. Focusing on this aspect led to the first edition of the FIDIC Yellow Book for mechanical and electrical works in 1963, with its emphasis on testing and commissioning, more suitable for the manufacture and installation of plant. The second edition was published in 1980.[4] The Engineer had a similar role under the Yellow Book, with some differences related to the design supervision.

4.1.3 1987 Edition of the Red and Yellow Books: express impartiality of the Engineer

In 1987, FIDIC published new editions of the Red and Yellow Books. The 1987 fourth edition of the Red Book introduced an express term requiring the Engineer to act impartially when giving a decision or taking any action that might affect the rights and obligations of the Parties, whereas the previous editions had assumed this implicitly. In broad terms, there was greater concentration on the allocation of risk, with the significant addition of

1 Ellis Baker and others, *FIDIC Contracts: Law and Practice* (Informa Law from Routledge 2009) 1.17

2 Nael Bunni, *The FIDIC Forms of Contract* (Blackwell Publishing 2005) 10

3 Nael Bunni, *The FIDIC Forms of Contract* (Blackwell Publishing 2005) 73

4 Jeremy Glover, 'FIDIC an Overview: the Latest Developments, Comparisons, Claims and a Look into the Future' (Fenwick Elliot 2008) <https://www.fenwickelliott.com/sites/default/files/FIDIC.pdf> accessed 19 March 2022

a ground for extension of time (**EOT**) in the event of Employer's delay, impediment, or prevention.[5]

4.1.4 The 1992 Reprint of the Red Book with a Supplement introducing the Dispute Adjudication Board (DAB)

The 1987 Red Book was reprinted in 1992 with modifications. A Supplement was published in 1996, which provided the user with the ability to incorporate alternative arrangements, comprising an option for a DAB with model terms of appointment and procedural rules, and an option for payment on a lump-sum basis rather than by reference to bills of quantities (**BQs**). Thus, the somewhat archaic provision for the submission of matters to the Engineer for his/her 'Decision' prior to an ability to pursue a dispute was eliminated. In its place, an independent DAB was introduced, consisting of either one or three members appointed jointly by the Employer and the Contractor at the commencement of the contract, with the cost being shared by the Parties.

4.1.5 The 1995 Orange Book

FIDIC responded to the increasing popularity of projects being procured on a design and build or turnkey premise. This resulted in the 1995 **Orange Book**, namely, the Conditions of Contract for Design-Build and Turnkey. The 1995 Orange Book reflected a significant move away from the FIDIC forms, which had adopted the traditional role of the Engineer. While the 1987 Red Book had introduced the obligation upon the Engineer to consult with the Employer and Contractor, the 1995 Orange Book dispensed with the Engineer entirely and provided for the 'Employer's Representative'. The express requirement to be impartial was also relinquished, albeit when determining value, costs, or EOT, the Employer's Representative had to '*determine the matter fairly, reasonably and in accordance with the Contract*'.[6]

4.1.6 The 1999 Rainbow Suite

In 1994, FIDIC established a task group (**TG**) to update both the Red and the Yellow Book in the light of developments in the international construction industry, following the new format established by the Orange Book. The TG had to standardise the terminology in the different forms and to make the documents as user-friendly as possible. The publication of the first editions in 1999 included the Red/Yellow/Silver Books for major projects and the Short Form of Contract, popularly known as the Green Book.

The 1999 forms are not revisions of the 1987 editions of the Red and Yellow Books. While the previous editions were distinguished from each other based on the type of project for which they are used, the 1999 forms are distinguished on the basis of the risk allocation. The number of clauses has been drastically reduced (from 72 in the 1987 Red Book to 20 in the 1999 books, albeit each clause has a number of sub-clauses), while the number

5 Jeremy Glover and Simon Hughes, *Understanding the FIDIC Red Book: A Clause by Clause Commentary* (Sweet & Maxwell 2011) xiii

6 Jeremy Glover and Simon Hughes, *Understanding the FIDIC Red Book: A Clause by Clause Commentary* (Sweet & Maxwell 2011) xiii

of words has largely increased. Dr. Bunni identifies important changes in concept, some in favour of the Employer and others in favour of the Contractor, including the following:[7]

- new provisions are added, giving certainty to the Contractor, like:
  - Sub-Clause 2.4 [Employer's Financial Arrangements];
  - an indemnity specified under Sub-Clause 4.2 if the Employer claims under the Performance Security amounts to which it is not entitled;
- 'Limitation of Liability' is introduced in Sub-Clause 17.6, excluding liability for loss of use, loss of profit, loss of contract, or any indirect or consequential loss or damage and limiting the total liability of the Contractor;
- new requirement for Progress Reports;
- the 1999 Edition of the Red and Yellow Books substitutes the duty of the Engineer to '*exercise discretion impartially*' with the duty the Engineer to '*make a fair determination*'.
- FIDIC took away the adjudication of disputes role from the Engineer and handed it to the DAB, recognising that the Engineer belongs to the Employer's Personnel;[8]
- the Employer is entitled to terminate for convenience under Sub-Clause 15.5;
- the 1987 concept of 'frustration' is substituted with '*Force Majeure*' in the 1999 books, with clearly defined criteria to be met;
- the claims provisions for the Employer (Sub-Clause 2.5) and Contractor (Sub-Clause 20.1) have been changed, introducing strict time limits, but only for the Contractor.

So, in the 1999 Red and Yellow Books, the Engineer is no longer formally required to be impartial, but whenever required to make a determination in respect of a value-, cost-, or time-related matter, he/she has 'to make his/her determination fairly, and in accordance with the Contract, having taken into consideration all relevant circumstances'.[9] But before that, the Engineer shall consult with each Party in an endeavour to reach agreement. If agreement is not achieved, the determination process can start.

4.1.7 The Pink Book

In 2005, FIDIC licensed MDBs to use the GCs of a MDB Harmonised Edition of the 1999 Red Book, popular as the **Pink Book**. New editions followed in 2006 and 2010. The participating MDBs agreed in 2007 that the contract could be used by bilateral aid agencies and in 2011 by national procurement agencies.[10]

---

7 Details of the changes may be found in chapters 23 to 27 of Nael Bunni, *The FIDIC Forms of Contract* (Blackwell Publishing 2005)

8 FIDIC, *Yellow Book* 1999 Sub-Clause 1.1.2.6

9 Paul Battrick and Phil Duggan, 'The Rainbow Suite. The 1999 FIDIC Suite' <https://www.fidic.org/sites/default/files/FIDIC-rainbow-suite-2012.pdf> accessed 4 March 2022

10 For the differences from the Red Book 1999, see Frederic Gillion, 'FIDIC Pink Book: The MDB Harmonised Edition of the Red Book' (Fenwick Elliott) <https://www.fenwickelliott.com/sites/default/files/FIDIC%20Pink%20Book%20The%20MDB%20Harmonised%20Edition%20of%20the%20Red%20Book.pdf> accessed 1 May 2023

4.1.8 The 2008 Gold Book

In 2008, FIDIC published a first edition of the Gold Book: Conditions of Contract for Design, Build and Operate Projects. FIDIC adopted the green-field design-build-operate scenario, with a 20-year operation period, and has opted for a single contract awarded to a single contracting entity (which will almost certainly be a consortium or joint venture (**JV**)) to optimise the coordination of innovation, quality, and performance, rather than award separate contracts for design-build and for operation. The Contractor has no responsibility either for financing the project or for its ultimate commercial success.[11] The provisions of the Gold Book upgrade the 1999 Yellow Book with the operation service. It contains a lot of improvements, related to the risk allocation and strengthening the role of the DAB.

4.1.9 The Blue-Green Book

FIDIC established a joint TG with the International Association of Dredging Companies to prepare a second edition of the Form of Contract for Dredging and Reclamation Works. The first edition, published in 2002, was based closely on the FIDIC Green Book. FIDIC published the second edition of the Blue-Green Book in 2016. It may be used for all types of dredging and reclamation work and ancillary construction with a variety of administrative arrangements. Under the usual arrangements for this type of contract, the Contractor constructs the works in accordance with design provided by the Employer or by its Engineer. However, this form may also be suitable for contracts that include, or wholly comprise, contractor-designed works. In addition, the Employer has a choice of valuation methods.[12]

4.1.10 The Red/Yellow/Silver Books 2017 Edition

In 2017, FIDIC published the second editions of the 1999 books for major projects (Red, Yellow, and Silver). FIDIC asserts that the

> three contracts have been significantly amended with the main goal to increase clarity and certainty. These amendments should lead to fewer disputes and more successful projects. Underlying this overarching aim, is the recognition of a need to reflect current best practice and the issues raised by users of the 1999 editions over the past 18 years.[13]

The 2017 FIDIC contracts update many of the PM procedures. For example, they include:

- new sub-clauses supporting collaboration for advance warning, management meetings, and dispute avoidance;
- new detailed Claim procedures identical for both Parties with a 42-day period for consultations managed by the Contract Administrator (**CA**) focusing on the collaborative behaviour;

11 FIDIC, <https://fidic.org/books/dbo-contract-1st-ed-2008-gold-book> accessed 8 March 2022

12 <https://fidic.org/books/dredgers-contract-2nd-ed-2016-blue-green-book> accessed 8 March 2022

13 FIDIC, 'FIDIC Rainbow Suite ed. 2017' <https://fidic.org/sites/default/files/press%20release_rainbow%20suite_2018_03_1.pdf> accessed 14 January 2022

- detailed Variation procedures, which can be used in relation to more provisions, thus enabling serious reduction of Claims;
- detailed requirements for the programme of works, records, and review of the Contractor's submissions;
- additional quality management provisions, introducing a Compliance Verification System;
- GPPC for introducing milestones and bonuses to incentivise the Contractor;
- new reasonable time periods for action by the Parties and the CA and 'deeming' provisions if the time period is not met.

In the 2017 Red and Yellow Books, there is an express provision in Sub-Clause 3.2 [Engineer's Duties and Authority] that the Engineer is not required to obtain the Employer's consent before the Engineer exercises its authority under Sub-Clause 3.7 [Agreement or Determination] and that the Engineer must act '*neutrally*' when exercising its duties under Sub-Clause 3.7 [Agreement or Determination].[14] The intention is that '*the Engineer treats both Parties even-handedly, in a fair minded and unbiased manner*'.[15]

4.1.11 The 2019 Emerald Book[16]

In 2019, FIDIC published, in collaboration with the International Tunnelling and Underground Space Association ITA-AITES, the Conditions of Contract for Underground Works, well known as the *Emerald Book.* It uses 90 per cent of the Yellow Book provisions, introducing a new model for managing the risks that are peculiar to tunnelling or other subsurface construction works. The risks deriving from expected subsurface conditions are assigned to the Contractor, who proposes the production rates and the general cost of performing the Works under unchanged conditions. The Employer takes the risks for subsurface conditions deviating from the Geotechnical Baseline Report (**GBR**) and for changes in the quantities of the excavation and lining works. For unexpected subsurface conditions, a new provision, unique to the FIDIC Books, was introduced,[17] enabling reduction of the Time for Completion in the case of more favourable conditions.

Besides the duties and authorities similar to those under the *Yellow Book*, the Engineer has specific duties and authority to monitor and record progress of the execution of the excavation and lining works (Sub-Clause 3.2.2) and in relation to the adjustments of the Time for Completion and Contract Price under Sub-Clauses 13.8.3/13.8.4.

4.1.12 The 2021 Green Book introduces the Engineer

In 2021, FIDIC published the second edition of the Green Book. The new Short Form is much more advanced than the 1999 Edition. It may serve as an alternative to the 2017 Red and Yellow Books for projects where the perceived level of risk is low and/or the Parties

14 Victoria Tyson, 'Clause 3: The Engineer' (Corbett & Co Knowledge Hub)

15 FIDIC, *Yellow Book* GPPC, Sub-Clause 3.7 (FIDIC 2022) 21

16 The Emerald Book is discussed in more detail in: The Hon. Peter Vickery QC, 'Chapter 4 The FIDIC Conditions of Contract for Underground Works (the Emerald Book)' in Donald Charrett (ed), *FIDIC Contracts in Asia Pacific: A Practical Guide to Application* (Informa Law from Routledge 2022) 71.

17 Fabian Bonke, Provisions of the new FIDIC 'Emerald Book' (*Tunnelling Journal* July 2019) 25

wish to use a form that does not require significant PM resources. It captures the essential rights and obligations of the Parties in 20,000 words (in comparison, the 2017 Red and Yellow Books contain more than 90,000 words). It has user-friendly features and provides different contract solutions: Employer's or Contractor's design, and a wide range of valuation methods. A formula for calculation of Prolongation Costs agreed in the Contract Data (Sub-Clause 1.1.35), tabulation of the Contractor's entitlements in case of Employer's risk events (Sub-Clause 11.1) and of insurance requirements (Contract Data Sub-Clause 12.1), and early appointment of adjudicator (Sub-Clause 14.1) facilitate the PM and prevent disputes. Unlike the 1999 edition, the 2021 Short Form introduces the Engineer, similarly to the Red and Yellow Books.

4.1.13 The 2022 Reprint amends the Red/Yellow/Silver Books 2017 Edition

In 2022, FIDIC reprinted the 2017 Edition of the Red/Yellow/Silver Books with combined amendments, arising from three sets of errata published in 2018, 2019, and 2022. Besides correcting errata and integrating some amendments to improve the publications, FIDIC updated the Advisory Notes to Users of FIDIC Contracts Where the Project Uses Building Information Modelling Systems to reflect recent developments in digital technology.[18]

The 2022 Reprints provide welcome clarifications to the meanings of 'matter[s] to be agreed or determined', 'Claim', and 'Dispute'. By specifying the provisions referring to 'matters to be agreed or determined', the 2022 Reprints eliminate a major uncertainty under the 2017 forms distinguishing this category from 'Claim'. The new definition of Dispute, by requiring the giving of a Notice of Dissatisfaction (**NOD**) with respect to a CA's determination, reinforces the role of the Engineer/Employer's Representative in resolving Claims and 'matters' between the Parties. Parties can no longer sidestep the CA in its agreement and determination role. The Reprints acknowledge and give effect to work cultures in a post-COVID world by providing for the possibility of holding meetings online.[19]

4.1.14 Subcontracts for the works contracts

The following **Subcontracts** are drafted back-to-back to the most often used FIDIC main contracts:

- Subcontract for Construction (2011) back-to-back with the 1999 Red Book;
- Subcontract for Design-Build (2019) back-to-back with the 1999 Yellow Book.

4.1.15 New forms in the pipeline

In 2023, FIDIC is preparing for release the **Bronze Book**, Operate-Design-Build-Operate Contract for brownfield projects.

18 FIDIC, *Yellow Book* 'Notes on Reprinting in 2022' (FIDIC 2022)

19 Christopher Seppälä and Vanessa Pitassi, 'Important Changes to FIDIC's Rainbow Suite' (White & Case 2023) <https://www.whitecase.com/insight-alert/important-changes-fidics-rainbow-suite?s=Christopher%20Sepp%C3%A41%C3%A4,%20Vanessa%20Pitassi,%20%E2%80%98Important%20Changes%20to%20FIDIC%E2%80%99s%20Rainbow%20Suite%E2%80%99> accessed 18 March 2023

In 2021, TG17 commenced its activities for the development of a collaborative contract.

In 2022, FIDIC established TG18 for a Public–Private Partnership (**PPP**) contract.

In 2022, FIDIC restarted TG9 with new members divided into three Sub-TGs to draft six new Subcontracts for the 1999 Silver Book, Reprint 2022 of the 2017 Red/Yellow/Silver Books, the 2019 Emerald Book, and the 2021 Green Book.

In 2023, TG14 started drafting a new form for offshore windfarm projects, and TG20 was established to develop a JV Agreement for Contractors.

4.2 Background and development of the FIDIC suite for services

Besides the Suite for Works, FIDIC published a suite of agreements for services:

- Client/Consultant Model Services Agreement, popular as the White Book (first edition 1990, fifth edition 2017);
- Sub-Consultancy Agreement (first edition 1992, second edition 2017);
- Model Joint Venture Agreement (first edition 1992, second edition 2017);
- Model Representative Agreement, known as the **Purple Book** (first edition 2013).

Any further mentioning of the White Book refers to the 2017 edition.

The White Book is suitable for general use for the purposes of pre-investment and feasibility studies, detail design, and administration of construction and PM, both for Employer-led design teams and for Contractor-led design teams on design and build commissions.

The overall allocation of risk between the parties under the White Book would appear to be similar to the previous fourth edition; however, the provisions of the contract have been developed and strengthened.[20] The fifth edition has enhanced the Consultant's obligation to the reasonable skill and care to be expected from a consultant experienced in the provision of services for projects of similar size and complexity. Without extending this obligation, the services must satisfy the function and purpose described in the Agreement.[21]

The White Book includes provisions under Sub-Clause 3.9 (Construction Administration) designed to apply where the Consultant is performing the function of the Engineer, Employer's Representative, project manager, or similar under a works contract. In the specified capacity, the Consultant shall act on behalf of the Client to the extent provided in the works contract. If the Consultant is authorised under the works contract to certify, determine, or exercise discretion in the discharge of its duties, then the Consultant shall act fairly, exercising independent professional judgement and using reasonable skill, care, and diligence.

The Sub-Consultancy Agreement has been developed back-to-back with the White Book and is intended for use where a Consultant appointed under the White Book desires to engage a Sub-Consultant to undertake part of the services. The Sub-Consultancy Agreement therefore incorporates many of the terms and conditions of the White Book,

20 Julian Bailey and Michael Turrini, 'The New FIDIC White Book for Consultancy Services' (White & Case 2017) <https://www.whitecase.com/publications/alert/new-fidic-white-book-consultancy-services> accessed 4 March 2022

21 <https://fidic.org/books/clientconsultant-model-services-agreement-5th-ed-2017-white-book> accessed 8 March 2022

such that the obligations and rights of the Sub-Consultant are largely consistent with those of the Consultant under the White Book.[22]

The Model Joint Venture Agreement is an alternative arrangement for the Consultant performing the services to constitute several legal persons. Usually, consulting companies unite their efforts to meet the tender criteria. The standard form allows the parties to agree, on a project-specific basis, their obligations, services, and rewards within the JV created among them by the effect of the Agreement. It is not intended to create a legal entity or to be used to create a more permanent non-project-specific legal association. The JV's action, capacities, and internal decision-making processes have been clearly structured. The authors explain how the agreement supports the JV's horizontal governance:

- one purpose is to raise the awareness of the JV Members as to what should be in the Agreement to mitigate their individual risks and avoid disputes between themselves;
- a second purpose is to provide the Members with a manageable agreement that establishes clear responsibilities and legal capacities to act.[23]

The terms of the Model Representative Agreement are intended for Consultants wishing to enter into a contract with a Representative for the provision of Representative Services.[24]

Due to the market demand, FIDIC established TG22 to draft an Engineering, Procurement, and Construction Management (**EPCM**) contract.

4.3 The FIDIC Rainbow Suite and project governance

4.3.1 Project governance in the light of FIDIC contracts

The PMBOK® Guide defines the **project manager** as 'the person assigned by the performing organisation to lead the team that is responsible for achieving the project objectives'.[25] Its Construction Extension explains that this position may reside with the contractor (seller) or owner (buyer).[26] So, each participant in a construction project has its own project manager: the owner, designer, contractor, supervisor, subcontractor, supplier, future operator, etc. The project manager shall have significant general management, technical, and interpersonal skill for leading, communicating, negotiating, and problem solving.

Project governance comprises the framework, functions, and processes that guide PM activities. In construction, project governance establishes an overall integrated view of how the project should be executed in light of the relationship among the owner (FIDIC Employer), the contractor (FIDIC Contractor), and other stakeholders.[27] It influences the timely risk management and the performance optimisation. Recent PMI research suggests that the concept of governance is multidimensional and:

22 FIDIC, *Sub-consultancy Agreement* (FIDIC 2017) Foreword
23 FIDIC, *Model Joint Venture (Consortium) Agreement* (FIDIC 2017) Foreword
24 FIDIC, *FIDIC Model Representative Agreement* (FIDIC 2013) Foreword
25 PMI Inc., *PMBOK® Guide* (PMI sixth edition 2017) 52
26 PMI Inc., *Construction Extension to the PMBOK® Guide* (PMI 2016) 6
27 PMI Inc., *Construction Extension to the PMBOK® Guide* (PMI 2016) 14

- includes consideration of people, roles, structures, and policies; also
- requires providing direction and oversight through data and feedback.[28]

The FIDIC contracts are not only concerned with the rights and obligations of the Parties but also contain detailed contract administration rules and PM procedures. FIDIC developed more detailed provisions for instructions, communications, and records in the 2017 editions of the suite for Works.

A paper by Anthony Morgan and Sena Gbedemah uses research and case studies to conclude that effective project governance:

- would ensure that costly changes were identified at an early stage (with communication of the foreseeable knock-on consequences) and seek to mitigate their effects;
- is not about the apportionment of blame or risk via the contract, it is about identifying risk and uncertainties and putting in place a control environment that ensures that the risks and uncertainties are proactively managed to enhance *success* in delivery of the project objectives.[29]

FIDIC contracts support project governance with distinct Variation and Claims management procedures. They are designed not only to regulate the Parties' entitlements but also to incentivise them for early warning and risk management. For example, the following provisions under the 2017 Yellow Book enhance collaboration for timely solving of issues:

- Sub-Clause 3.7 consultations to reach an agreement;
- Sub-Clause 3.8 management meetings;
- Sub-Clause 4.12 procedure in case of Unforeseeable physical conditions;
- Sub-Clause 8.4 early warning;
- Sub-Clause 17.2 actions in case of damages to the Works;
- Sub-Clause 21.4 Dispute avoidance;
- Clause 13 Variation and Clause 21 Claims step-by-step procedures with strict time limits and deemed provisions in case of failure to respect them.

A project governance framework varies depending on the scale, complexity, and other factors of the project. An effective project governance framework identifies who has decision-making responsibility and authority. Guidelines recommend to public agencies:

- to establish clear governance roles and responsibilities that are communicated to everyone involved in the project, with clear communication lines;
- to appoint a project senior responsible owner (**SRO**) with appropriate seniority;
- project governance boards to be a small group of key individuals who are empowered to support the SRO in making the decisions required to drive the project.[30]

---

28 PMI, PMI Inc., *PMBOK® Guide* (PMI sixth edition 2017) 43

29 Anthony Morgan and Sena Gbedemah, 'How Poor Project Governance Causes Delays' (SCL 2010) <https://www.pwc.co.uk/assets/pdf/society-of-construction-law-oct10.pdf> accessed 13 March 2022

30 New Zealand Government Procurement, *Construction Project Governance. Construction Procurement Guidelines* (2019) <https://www.procurement.govt.nz/assets/procurement-property/documents/construction

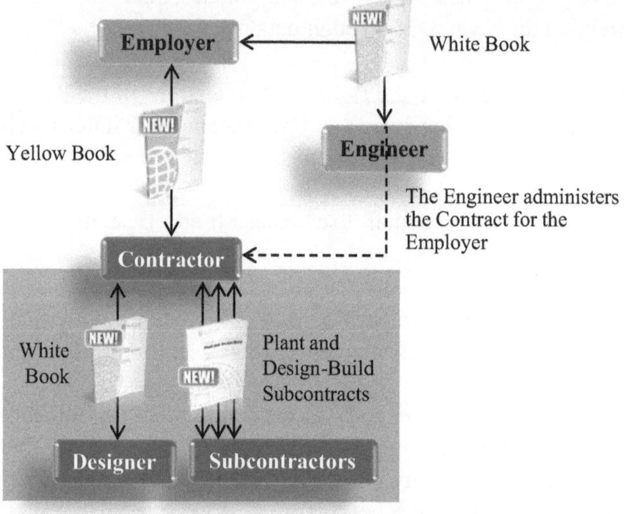

Figure 4.1 Project governance with a FIDIC Yellow Book contract

Researchers suggest that the governance structure of construction projects is divided into:

- vertical governance, focused on the client and first-tier supplier relationship;
- horizontal governance, involving series of contracts between the first-tier supplier and the supply chain;
- project manager-related governance – the relationship between the project manager and the Contractor.[31]

Each relationship should be regulated by a certain contract. Figure 4.1 presents the project governance structure with a FIDIC Yellow Book contract:

- the vertical governance is regulated by a Yellow Book contract;
- the horizontal governance of the Employer is regulated by the structural rules of the relevant organisation (it may be a public entity forming a steering committee with the project beneficiary; it may be a special project vehicle, including the owner of the land, developer, etc.);
- the horizontal governance of the Engineer is regulated by the structural rules of the relevant organisation – a consulting company or JV; in the case of a JV, the FIDIC JV Agreement may be used; the Model Representative Agreement may be used when an international consultant appoints a local Engineer's Representative; the Sub-Consultancy Agreement may be used when the Engineer outsources expertise (e.g. FIDIC expert);

-project-governance-construction-procurement.pdf> accessed 13 March 2022

31 Kaixun Sha, Understanding Construction Project Governance: an Inter-organisational Perspective (*International Journal of Architecture, Engineering and Construction* 2016) 5

- the horizontal governance of the Contractor may be regulated by the White Book Agreement with the designer and the Design-Build Subcontract with other Subcontractors;
- the project manager-related governance is regulated by a White Book Agreement for services for appointment of the FIDIC Engineer; the duties and authority of the Engineer are stipulated in the Yellow Book Contract between the Employer and the Contractor, where the Engineer has a dual role: to act on behalf of the Employer, but to act fairly certifying and determining payments and also '*neutrally*' when exercising its duties under Sub-Clause 3.7 [Agreement or Determination].

Figure 4.2 presents the project governance structure with a FIDIC Red Book contract:

- the vertical governance is regulated by a Red Book contract;
- the horizontal governance of the Contractor is regulated by FIDIC Construction Subcontracts; a JV agreement, if any, regulates the relationship between several companies forming the Contractor;
- the project manager-related governance is regulated by a White Book Agreement for Services for appointment of the FIDIC Engineer; the duties and authority of the Engineer are stipulated in the Red Book Contract between the Employer and the Contractor, where the Engineer has a dual role, similarly to the Yellow Book; the agreements for services with the Engineer and designer may empower the Engineer to interact with the designer, keeping the Employer informed.

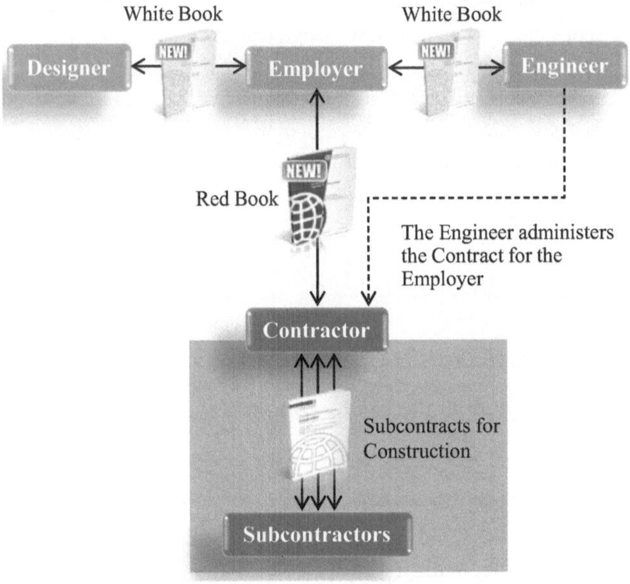

Figure 4.2 Project governance with a FIDIC Red Book contract

Under the Red Book, the Employer is responsible for the design. FIDIC created the old Red Book following the English tradition, where the Engineer was the designer and project manager. Projects are more complex now. Financing is organised from different sources during the project life cycle. Usually, different consultants are responsible for the design and PM.

Depending on the local law, the designers in Red Book projects may have continuing obligations to:

- monitor the construction for compliance with the design;
- authorise design changes related to Engineer's instructions or Contractor's Value Engineering (VE);
- develop detailed design during the construction;
- participate in the commissioning and taking-over procedures.

The designer's agreement with the Employer regulates their relationship. The lack of Engineer's power and authority under the agreements for services to request timely involvement of the designer may result in delays, claims for EOT/prolongation costs, and disputes under the Works contract.

4.3.2 Case study – project governance

> Two coastal protection projects (one at the seaside, the other along a river) financed by a bilateral bank with experienced FIDIC experts is a good example of Red Book project governance. The Employer selected a consultant (a JV in this case study, led by an international consultant with two local partners, one of them responsible for the mandatory design and construction supervision, and the other one for the PM). The scope of services included checking of the design; proposal for optimisation; preparation of tender documents for both Red Book Contracts, including Particular Conditions of Contract, Specifications, and BQs; compliance assessment of the optimised detailed design; permit management; support to the Employer during the tender procedures; and acting as the FIDIC Engineer under both Red Book Contracts.

4.3.3 Interaction of the project managers in the FIDIC works contracts

Discussing the project governance in the light of the FIDIC contracts for works, <u>the overall PM role</u> is assigned to:

- the <u>Engineer</u> in the Red/Yellow/Blue-Green/Emerald and 2021 Green Books;
- the <u>Employer's Representative</u> in the Gold Book, to be appointed by the Employer prior to the signing of the Contract with a similar role to that of the Engineer

under the Yellow Book,[32] with duties extending after completion during the Operation Service Period;

- the Employer's Representative in the Silver Book, to be appointed by the Employer with limited PM power and authority, since the EPC/Turnkey Contractor takes the overall responsibility.

Often, the Engineer/Employer's Representative is called for brevity the 'Contract Administrator' (**CA**). The Construction Extension describes contract administration as 'negotiating and implementing the contractual terms and conditions of project contracts by following established policies, processes, and procedures.'[33] Under the FIDIC contracts, these duties and authorities are related not only to leading the communications, schedule, scope, and cost control, but also to claims management, supervision of the design and construction, including compliance assessment of the Contractor's Documents,[34] and giving technical instructions, sometimes with time and price impact.

The project manager of the Contractor is usually the Contractor's Representative, to be appointed before any interim payment is certified and before the Commencement Date. He/she, by definition, is the natural person named by the Contractor in the Contract or appointed by the Contractor under Sub-Clause 4.3, who acts on behalf of the Contractor. Given his/her importance for the PM, as well as the close working relationship that is envisaged with the CA, the Contractor must have the consent of the CA before the appointment. Alternatively, tenderers may be required to name the Contractor's Representative in the Tender, and so he/she is named in the Contract.[35] The Contractor's Representative shall be 'qualified, experienced and competent in the main engineering discipline applicable to the Works and fluent in the language for communications defined in Sub-Clause 1.4' (Sub-Clause 4.3).

Under the Red/Yellow/Emerald Books Sub-Clause 3.2:

> Any acceptance, agreement, approval, check, certificate, comment, consent, disapproval, examination, inspection, instruction, Notice, No-objection, record(s) of meeting, permission, proposal, record, reply, report, request, Review, test, valuation, or similar act (including the absence of any such act) by the Engineer, the Engineer's Representative or any assistant shall not relieve the Contractor from any duty, obligation or responsibility the Contractor has under or in connection with the Contract.

In relation to the project governance, this means that the Contractor's Representative has a leading role in many PM Knowledge Areas (assuming that he/she is also the Contractor's project manager), while the Engineer has a controlling role. Figure 4.3 maps the roles of the Engineer and Contractor's Representative in 11 PM Knowledge Areas. The Contractor's Representative has to manage the supply chain in order to achieve the planned project outcomes related to time, quality, and price. The Contractor's role is also leading in the Health

32 Christopher Wade, 'FIDIC Introduces the DBO Form of Contract – the New Gold Book for Design Build and Operate Projects' (2008) 23(1) ICLR 14 at 20

33 PMI Inc., *Construction Extension to the PMBOK® Guide* (PMI 2016) 130

34 By definition, the Contractor's Documents include the documents prepared by the Contractor (specified in the Employer's Requirements, required to satisfy permits and regulatory approvals, including design, As-Built Records, and Operation and Maintenance Manuals), including calculations, digital files, computer programs and other software, drawings, manuals, models, specifications, and other documents of a technical nature (*Yellow Book* 2017 Sub-Clause 1.1.14)

35 FIDIC, *The FIDIC 2017 Contracts Guide* (FIDIC 2022) 17

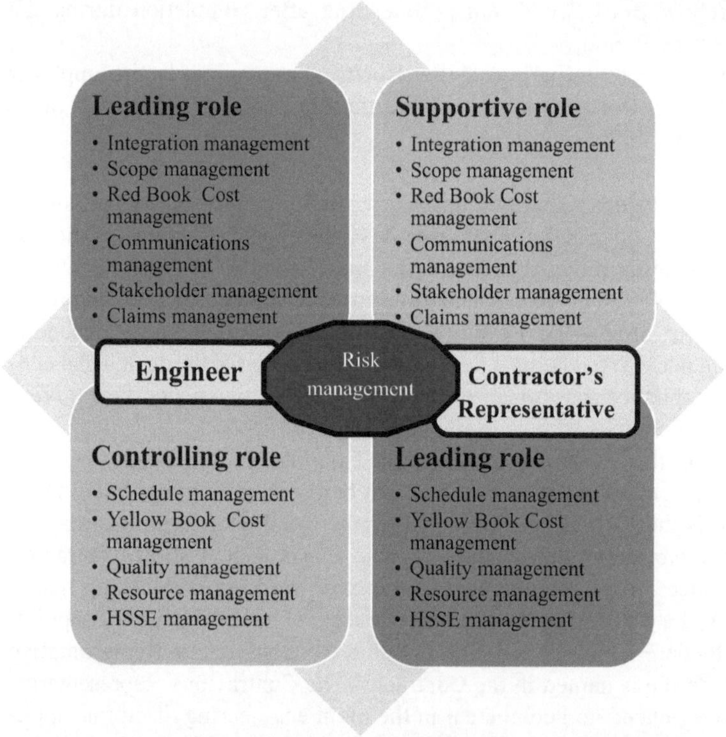

Figure 4.3 PM roles of the Engineer and Contractor's Representative in the Red/Yellow Books

and Safety (HS) management (Sub-Clause 4.8), security management (Sub-Clause 4.21), and environmental protection management (Sub-Clause 4.18). The Contractor is leading in the resource management, being fully responsible for its Subcontractors (Sub-Clause 4.4/Red Book: 5.1), the Contractor's Personnel (Clause 6), the Contractor's Equipment (Sub-Clause 4.17), transport and storage of Goods (Sub-Clause 4.16), and supply of proper Materials (Sub-Clause 7.1).

The Engineer and the Contractor's Representative have to manage the risks in collaboration.[36] Usually, the Contractor's Representative is leading the risk management with an early warning and planning of risk mitigation measures. In other situations, the Engineer may have a leading role: for example, issuing instructions when Exceptional Events described in Sub-Clause 18.1 impact the project, or for mitigation of other Employer's risks listed in Sub-Clause 17.2.

The Engineer may support the Employer's project manager on demand in the Project Procurement Management and Project Financial Management, where the Employer has a leading role. For example, the Engineer may prepare special reports with justification in the case of Variations under Clause 13 and a draft application for additional financing. Or, the Engineer may prepare a draft Addendum to the Works Contract upon request. If those

36 For example, see the FIDIC COVID-19 Memorandum at <https://fidic.org/sites/default/files/COVID%2019%20Guidance%20Memorandum%20-%20PDF.pdf> accessed 18 May 2023

activities are not in the Scope of Services (Appendix 1 to the White Book Agreement), the Client may initiate a Variation to the services under Sub-Clause 5.1.

4.3.4 Appointment and identity of the CA

The Employer is obliged to appoint the Engineer under the Red/Yellow/Emerald Books and similarly the Employer's Representative under the Silver Book, to be named in the Contract Data. Although the role of the Employer's Representative is not as wide-reaching as that of the Engineer, it is important that the Employer's Representative is identified and named in the Contract Data of the tender documents that are issued by the Employer to tenderers.

The Employer under a works contract appoints the CA by signing an agreement with the consulting company or JV (the White Book may be used). This agreement governs the Consultant's PM duties and authorities. Under the works contract, the Employer has a duty to appoint 'a professional engineer having suitable qualifications, experience and competence to act as the Engineer under the Contract', fluent in the ruling language of the Contract (Red/Yellow Book: Sub-Clause 3.1). Since different qualification is necessary for the quality control of different works (e.g. civil, mechanical, electrical), usually the Engineer is a consulting company with experience in the PM of similar projects and the use of FIDIC contracts. Where the Engineer is a legal entity, the Engineer shall give a Notice to the Parties of the natural person appointed and authorised to act on its behalf (Red/Yellow/Emerald Book: Sub-Clause 3.1). This first Engineer's Notice is very important, since the authority shall not take effect until the Notice has been received by both Parties. The 1999 Edition of the Red/Yellow Book does not discuss this duty of the legal person appointed as the Engineer to notify the Parties, which may create misunderstanding, since the senior person acting for the Engineer may be, for example:

- the representative of the JV appointed as Engineer;
- the Chief Executive Officer (**CEO**) of the JV leading partner;
- the CEO of the consulting company appointed as Engineer;
- the natural person nominated for Team Leader/ Project Manager in the winning proposal of the company/JV appointed as Engineer;
- a project manager from the consulting company, appointed as Engineer.

A similar requirement is included in the Silver Book Sub-Clause 3.1 for the Employer's Representative when a legal person must give a Notice to the Parties of the natural person appointed and authorised to act on its behalf.

In some cases, the Employer appoints a natural person from its management staff. Usually, this appointment is governed by an employment contract and an administrative order for the assignment of this role. Some FIDIC users argue that this is wrong practice, but similar project governance is possible if the arrangements do not infringe the FIDIC GPs. The authors of the new Green Book explain that 'what is of paramount importance is that the person administering the Contract, whether being the Engineer or the Employer's Representative, acts according to the duties assigned to it under the Contract'.[37]

37 FIDIC, *Short Form of Contract* (2021) Guidance Notes 8

4.3.5 Case study – Engineer from Employer's organisation

The author of this book has been involved in energy projects using the FIDIC Red/Pink Book, where a top manager in the Employer's organisation was appointed to act as the Engineer:

- Seismic upgrading of nuclear power plant (**NPP**) units under operation – the Engineer was the NPP Executive Director; the Engineer's Representative was a JV, including an international consulting company with local partners;
- Low Level Radioactive Treatment Plant within the boundaries of an operating NPP – the Engineer was the NPP Investment Department Director; the Engineer's Representative was a JV, including an international consulting company with a local partner, licensed for mandatory third-party supervision;
- Reconstruction of 400-kV power line – the Engineer was the Energy System Operator's Chief Engineer, supported by an in-house PM team and a local consulting company appointed for mandatory third-party supervision (authorised under the Particular Conditions of Contract for quality and permit management).

In all these FIDIC Contracts, the Employer was an energy operator and appointed a decision-maker from its organisation with management and technical skills to act as the Engineer, supported by an in-house PM team and consulting companies with engineers qualified in different disciplines. The Employer and Contractor worked in collaboration for the benefit of the project, solving issues before they escalated into Disputes. Specific rules and risk management procedures regulated the access to different parts of the Site and the permits to commence any works. Construction in an NPP under operation had many challenges, and the involvement of the top managers facilitated the PM. The third project was executed during the COVID-19 pandemic, but even with different issues arising, the Contractor completed the Works within the initial Time for Completion with several Variations and zero Claims.

In these examples, the PM functioned much better than in other projects from the author's experience using the FIDIC Red/Yellow Book contracts where the Employer was a public entity, appointing an independent consulting company/JV to act for the Engineer. The Engineer had many constraints on its authority requiring the Employer's approval (e.g. before acting under more than 20 sub-clauses). The delay or refusal of the Employer's specific approval resulted in delays, uncertain cash flow, suspensions, and disputes, and in some cases, even led to termination and arbitration.

The FIDIC 2017 Contracts Guide advises that the Engineer under the Red/Yellow Book or the Employer's Representative under the Silver Book may be anybody of the Employer's choosing, but he/she must be qualified and experienced in the subject matter of the Contract. Besides the tender procedure for selection of consultant for those services, the Employer has various options when making the appointment:

- if the Employer's organisation contains an appropriate engineering department, which has possibly been involved in the planning and development of the project, the Employer may select a senior employee to act as the Red/Yellow Book Engineer/Silver Book Employer's Representative under the Contract;

- alternatively, if the Employer has engaged an international or domestic firm of consulting engineers in the planning and/or conceptual design of the project, he/she may choose that firm as the Engineer/Employer's Representative under the Contract, which then must notify both Parties of the name of the natural person who is to act on its behalf.[38]

The CA's appointment is required to continue until he/she has fulfilled all the duties assigned to him/her under the Contract: in particular, the duties to issue the Performance Certificate under Sub-Clause 11.9 and to issue the Final Payment Certificate under Sub-Clause 14.13. While the Employer's Representative does not certify under the Silver Book, there might be issues pending agreement/determination at the time of the final payment, when the Employer's Representative shall act under Sub-Clause 3.5.

After the CA has been appointed by the Employer, it is not intended that he/she should be changed or replaced, but if a replacement does become necessary, then a replacement must be appointed subject to the Contractor's right to raise a 'reasonable objection' to the intended replacement (Red/Yellow Book: Sub-Clause 3.6, Silver Book: Sub-Clause 3.1).

4.4 Conclusion

The FIDIC Engineer/Employer's Representative is the CA, responsible for the overall PM. Its timely appointment, qualification, experience, and competence is of paramount importance for the success of the project. Recommendations for the selection and appointment process are included in Chapter 5. The transparent allocation of duties and authority and clear lines of communication are crucial for the PM. The relevant procedures are further discussed in Chapters 5 and 12.

38 FIDIC, *The FIDIC 2017 Contracts Guide* (FIDIC 2022) 30, 34

CHAPTER 5

Project Procurement Management with FIDIC Contracts

DOI: 10.4324/9781003272892-5

5.1 Procurement in construction

According to the PMBOK® Guide, Project Procurement Management includes the management and control processes required to develop and administer agreements such as contracts, purchase orders, etc.[1]

Agreements can be as simple as the purchase of a defined quantity of labour hours, or they can be complex multiyear international construction contracts. The contract should reflect the simplicity or complexity of the deliverables or effort and should be written in a manner that complies with local and international laws. A complex project may involve managing multiple contracts simultaneously or in sequence. Each contract life cycle may begin/end during any phase of the project life cycle. The PMBOK® Guide discusses procurement within the context of the buyer–seller relationship, which may exist at many levels. The Construction Extension highlights that procurement in construction is conducted to better transfer the performance risk to sellers who specialise and are skilled in a particular scope of work.[2]

The **Owner** may be a government, a public company, a corporation, a private body, or an individual developer. The Owner is usually called:

1 PMI Inc., *PMBOK Guide* (PMI sixth edition 2017) 459
2 PMI Inc., *Construction Extension to the PMBOK® Guide* (PMI 2016) 113

- the 'Client' employing a 'Consultant';
- the 'Employer' employing a 'Contractor';
- the 'Buyer' buying goods from the 'Seller'.

Construction projects are mostly based on agreement either by choice or by the requirements to satisfy contract obligations.[3] Beyond the typical designer, consultant, and contractor procurements, the Construction Extension lists other provider procurement types specific to construction, like securities; insurances; inspection services (including surveys and third-party inspections); agreements related to permits (e.g. waste disposal, land use, road access); utility relocation and service connections, etc.

The project financing is usually secured by credit agreements or public funds, delivered under a financing agreement. If some conditions of those agreements must be followed by the project participants, they have to be incorporated in the relevant contracts for works and services.

The Make-or-Buy Analysis is important for the early development of the project:

- for the Owner to decide whether to appoint a consultant and a local lawyer or use its in-house team;
- for the Consultant to decide which services to outsource;
- for the Contractor to decide which works to subcontract.

With internationalisation, disruption, and rising complexity of construction projects, the supply-chain management becomes crucial.

5.2 Local legal advice

Construction procurement is heavily dependent on the local law in relation to permits to design, build, and operate. The owner may have in-house lawyers or legal counsel, but local legal advice must be procured when a project is in another jurisdiction to consider laws[4] and regulations dealing with:

- public/private procurement;
- options and restrictions for modifications under public procurement law, if applicable;
- value engineering;
- a specific industry/business;
- property law, land purpose, zoning plans;
- connection to utilities;
- permits, approvals, agreements for design;
- environmental matters;

3 PMI Inc., *Construction Extension to the PMBOK® Guide* (PMI 2016) 118

4 Local laws applicable to the execution of construction are summarised for a number of jurisdictions in the following books edited by Donald Charrett and published by Informa Law from Routledge: *The International Application of FIDIC Contracts* (2020); *FIDIC Contracts in Asia Pacific* (2022); *FIDIC Contracts in Europe* (2023); *FIDIC Contracts in the Americas* (2023); *FIDIC Contracts in Africa and the Middle East* (2023). The tables of legislation in those books highlight the extent of local law that impacts construction in every jurisdiction.

- licences/authorisations for designers, builders, supervisors, different trades;
- site opening and commencement of construction;
- building control;
- taking over or acceptance of the works;
- commissioning and licence/permit to operate;
- labour laws, including welfare, health and safety;
- subcontracting;
- protection of fossils and antiquities;
- taxes, customs duties, foreign currency;
- right to suspend the works;
- *force majeure*;
- right of the parties to terminate;
- intellectual and industrial property rights;
- limitations on rates of interest;
- liquidated damages for delay or non-performance;
- limitations of liability;
- corruption, embargo;
- alternative dispute resolution;
- defects liability, insurance, securities.

5.3 The role of the consultant in Project Procurement Management

The *PMBOK® Guide* includes in the Project Procurement Management Knowledge Area the processes of planning, conducting, and controlling procurement. Only the first two processes, from identification of a project up to receipt of tenders and award of contracts for its implementation, are within the scope of the FIDIC Procurement Procedures Guide.[5] Those processes include procurement activities to:

- establish what is to be procured;
- decide on procurement strategy;
- prepare tender documentation;
- launch tender procedures;
- evaluate tender offers;
- award contracts.

During the development stages of the project from initial and feasibility studies, through conceptual and preliminary design, up to final design, the Employer will normally employ a consulting engineering firm to carry out or manage the studies and design work. Large employers may have engineering staff of their own who can carry out some of the work in-house, but normally, the employer will engage a consultant specialising in the actual type of project. The consultant usually will continue to assist the employer to conduct the procurements with the preparation of tender documents, support during the tender procedures, and evaluation of proposals. The consultant may continue its services as a FIDIC Engineer/ Employer's Representative to control procurements.

5 FIDIC, *FIDIC Procurement Procedures Guide* (2011) 5

The project preparation is the key to the project success. The early appointment of a qualified consultant experienced in similar projects is extremely important. In the initial stage, the employer has not secured project financing and prefers to spend as little as possible before the studies confirm that the project is viable and bankable. Usually, a consultant will be appointed for the initial and feasibility studies only. The consultant also may support the employer to develop the procurement strategy. When the project strategy is clear, the scope of the other consulting services may be defined. At this stage, it may be decided how to package the consulting services, depending on the overall contract packaging. For example, one consultant may be appointed early as project manager, responsible for preparing tender documents and coordinating the procurement of works and services. Another consultant may be selected as a designer.

If the strategy assumes contract packaging needing different specialisation, or in the case of time constraints, several designers may prepare designs in parallel for different works contracts. The project manager will prepare the tender documents for selection of the designers and support the employer during the tender procedures. After the appointment of the designers, the project manager will coordinate the design packages, control the design development, seek optimisations, and incorporate the design into the tender documents for works. The preparation of the employer's requirements for the tender documents may be either in the designer's scope or in the project manager's scope. The preparation of the BQs and budget is usually in the designer's scope, but the project manager has the controlling and directing role.

Where the local building control regulations require third-party design assessment, the mandatory design approval may be organised in different ways:

- the project manager may be responsible for the permit management, interacting with the authorities;
- if the building control is privatised, the project manager may be authorised for technical control services or may subcontract a licenced (authorised) consultant.

In all cases, it is recommended that the permit management should be in the scope of the project manager. If some of the activities for agreements and consents are transferred to designers or contractors, the role of the project manager will be controlling and directing. The main tool for timely permit management is a detailed Road Map for mandatory regulatory procedures with timelines and responsibilities. The Road Map must be prepared in the inception phase of the PM services and detailed during the next phases with inputs from the other participants.

5.4 Consultancy appointments

FIDIC highlights that selecting the consultant is one of the most important decisions. 'The success of any project often depends upon obtaining the most able, experienced and reputable expertise available.'[6] FIDIC strongly recommends prequalification of applicants and advocates for quality-based selection (**QBS**). FIDIC has recognised that quality- and cost-based selection (QCBS) preferred by MDBs is also a good method, if used properly with the

6 FIDIC, *FIDIC Procurement Procedures Guide* (2011) 87

focus on quality (e.g. ADB recommends 90:10 quality-cost ratio). The FIDIC Procurement Procedures Guide describes procedures for selection of consultants,[7] which have to be adjusted to the legal framework and the requirements of the financing institution.

Prequalification ensures that the short-listed candidates have the necessary resources and experience to perform the services and that they meet the eligibility criteria. The pre-selection is based on the examination of the expressions of interest submitted by the consultants against the announced criteria.

The cost of consultancy services, including design, is usually in the range of 1–2 per cent of the life-cycle costs of a project (construction, operation, maintenance, and disposal costs).[8] Lower consultancy fees have a bad impact on the project value:

- important considerations of alternatives are skipped, including more sustainable solutions;
- inadequate investigations lead to expensive and time-consuming modifications during construction;
- minimum remuneration leads to poor design, often with low constructability, lack of coordination between design disciplines, design errors, lack of details, and waste of materials;
- limited budget for PM leads to inadequate project preparation, delays due to missing permits, deficiencies in the tender documents for selection of project participants, minimum monitoring and control of design and construction, bad stakeholder management, and poor risk management.[9]

The design of a facility has a pervading influence, not just on the cost of construction, but even more importantly, on the life-cycle costs.[10]

Due to the impact of the consultancy services on the project success, FIDIC recommends:[11]

- QBS as the only method for those projects where there is significant complexity, or where the consequences of failure are significant, or where the competing candidates will all be national consultancy firms;
- if the Quality and Cost Based Selection method is considered, then the maximum weight given to the financial component should be within the range of 0 and 10 per cent, with 20 per cent as the absolute maximum in exceptionally simple or straightforward cases;
- FIDIC strongly recommends that the selection of a Consultant only on the basis of price should never be used for consultancy services.

7 Chapter 7: Prequalification: Consultancy Appointments; Chapter 9: Consultancy Appointments; Chapter 16: Evaluation and Recommendations for Award: Consultancy Appointments; Chapter 18: Award of Contracts: Consultancy Appointments

8 FIDIC, *FIDIC Guidelines for the Selection of Consultants* (Third Edition 2019) 7 <https://fidic.org/sites/default/files/selection%20of%20consultants%20guide_2019.pdf> accessed 20 March 2022

9 Procurement of design is discussed in more detail in Donald Charrett, *Contracts for Construction and Engineering Projects* (second edition, Informa Law from Routledge, 2022) Chapter 5 'Managing Design Risk' and Chapter 6 'Contracts for Design Services'.

10 Philip Loots and Donald Charrett, 'Reconsidering Procurement Criteria and Procedures for Major Infrastructure Projects; Why the Use of "Price" as the Determining Factor in Construction Procurement Decisions is Wrong' (1 ICLR 2017) 71

11 FIDIC, *FIDIC Guidelines for the Selection of Consultants* (third edition 2019) 8

In QBS, the short-listed applicants submit a technical proposal only or both technical and financial proposals at the same time but in separate envelopes. If technical proposals alone are invited, after the evaluation of the technical proposals, the financial proposal and the contract shall be negotiated with the winner.

The FIDIC publication Guidelines for the Selection of Consultants offers the best practice. The short-listed firms or JVs are invited to submit a proposal. The Request for Proposal usually includes:

a. Letter of invitation
b. Terms of Reference (Scope of Services in the White Book 2017)
c. Information/instructions to Consultants
d. The proposed Agreement (recommended White Book 2017)

When the winner is selected, the following documents will form the Agreement for Services, based on the White Book 2017:

- The Agreement
- The Conditions of the Agreement:
  - Particular Conditions
    - Part A. References from Clauses in the GCs
    - Part B. Additional or Amended Clauses
  - White Book GCs
- Appendices
  1. Scope of Services (typically where Terms of Reference will sit)
  2. Personnel, Equipment, Facilities, and Services of Others to be Provided by the Client, if any
  3. Remuneration and Payment
  4. Programme (Client's requirements)
  5. Rules for Adjudication
- Any letters of offer and acceptance attached to any of the above.

The FIDIC Definition of Services Guidelines Infrastructure Construction/Civil Works[12] contain examples of detailed task descriptions and sample coordination checklists. They are a useful tool for the preparation of Appendix 1 (Scope of Services).

5.5 Plan Procurement Management: initial studies

The results of the initial studies will influence the decision on whether to proceed to full-scale feasibility studies to verify the overall viability of the project. The tasks are:

- identification of options;
- preparation of conceptual designs;
- consideration of suitable sites;
- assessment of approximate capital and life-cycle costs;

12 FIDIC, *FIDIC Definition of Services Guidelines. Infrastructure Construction / Civil Works* (2016) <https://fidic.org/sites/default/files/_DOS_guide_civil_works_1.pdf> accessed 2 May 2022

- estimation of the time and cost constraints;
- identification of potential sources of finance;
- initial reviews of social and economic viability, sustainability, and environmental impact;
- planning of financing and management responsibilities;
- examination of options for contract packaging.

5.6 Plan Procurement Management: feasibility studies

In order to justify the investment and select the best option and method of implementation, feasibility studies consider:

- location alternatives: planning, geophysical studies;
- laws and regulations, permits, licence if needed;
- financing: project viability usually with a Cost Benefit Analysis (**CBA**);
- programmes and time constraints;
- scope and design responsibility;
- operation and maintenance;
- preliminary cost estimates;
- employer involvement during implementation;
- project risks;
- environmental impact and sustainability.

At the end of the feasibility studies, a detailed report should be elaborated with clear recommendations based on the recorded data and their interpretation and analysis. The FIDIC Procurement Procedures Guide includes a typical format for a Feasibility Study Report. The Guide advises that the quality and usefulness of the feasibility studies will depend upon the time available to complete them, the validity and sufficiency of the information used, and in particular, the soundness of the experience and expertise of those carrying out the work.[13]

The author's experience proves that it is also important to identify the main stakeholders at this stage and manage their concerns and expectations.

5.7 Plan Procurement Management: procurement strategy

5.7.1 Procurement strategy considerations

The procurement strategy is developed based on the recommendations of the feasibility studies. Depending on the size and complexity of the project, the procurement strategy considers:

- financing schemes – loans, own fund, PPP, Build-Operate-Transfer (**BOT**), etc.;
- contract packaging;
- time programme, phasing;

13 FIDIC, *FIDIC Procurement Procedures Guide* (2011) 19

- budget planning, cost estimates;
- design responsibility;
- role of the parties during implementation;
- role of the CA (Engineer/ Employer's Representative);
- risk allocation, risk transfer to insurers, risk planning;
- contract price and payment: risks cost money;
- decision on whether to proceed.

The considerations for tailoring include:

- complexity of procurement;
- options for financing;
- physical location of the project and key stakeholders;
- governance and regulatory environment;
- availability of expertise, facilities, and supplies.

FIDIC highlights the importance of correct strategic decisions. 'The success or failure of the project will rest upon the strategy chosen.'[14]

This section elaborates further on some of these considerations. Chapter 9 discusses considerations related to the financial arrangements.

5.7.2 Levels of control

The project manager-related governance arrangements discussed in Chapter 4 should be clearly defined. The procurement strategy should decide whether the Employer wishes to be fully involved in the PM through the Engineer, or to have limited involvement using a turnkey delivery method. In the case of the turnkey method, the Contractor's project manager has a leading role in the risk, cost, and scope management, while the Employer's Representative has a supportive and controlling role. The design control provisions under Sub-Clause 5.2 are similar in the FIDIC Yellow/Emerald/Silver Book contracts.

The Green Book 2021 introduces the Engineer, replacing the Employer's Representative of the 1999 Green Book. With its flexible innovative provisions, the new Green Book may be used as a smart, concise alternative to the 2017 Red and Yellow Books.

The mandatory third-party building control regulations should be considered as well. If technical control (independent supervisor) authorised by the state has to supervise all the works, having a permanent presence on the site, this function may be integrated within the role of the Engineer. If a turnkey model is used, this third party may fit better within the Contractor's horizontal governance, if legally possible. Where the laws require the Employer to sign an agreement with an authorised third party in the case of a Silver Book contract:

- the Contractor may claim that the Employer is constantly interfering and preventing timely completion;

14 FIDIC, *FIDIC Procurement Procedures Guide* (2011) 82

- the Contractor may claim that the supervising third party has joint liability with the Contractor in the case of poor quality, where the applicable law stipulates similar provisions.[15]

5.7.3 *Case study – integrating privatised building control and mandatory designer's control*

The difficulties of integrating the privatised building control and mandatory designer's control during the construction of a gas interconnector with the Site in Bulgaria and Greece led to the introduction of additional participants into a Silver Book contract, amended and reproduced under a licence agreement with FIDIC. Under Clause 3 [Employer's Administration]:[16]:

- the Employer shall appoint an Employer's Representative to act on his behalf under the Contract;
- the Employer shall appoint an Employer's engineer as the 'Owner's Engineer' to manage the Contract on behalf of the Employer in accordance with the terms of a services agreement entered into between the Owner's Engineer and the Employer;
- the Owner's Engineer shall appoint a Third-Party Inspector (whose scope falls within the scope of the Owner's Engineer), who shall be responsible for conducting field inspections in respect of the performance of the Works in the Republic of Greece;
- the Employer shall appoint a Construction Supervision Entity to perform its obligations under the Bulgarian Spatial Development Act and its contract with the Employer. The Construction Supervision Entity may issue instructions from time to time to the Contractor.

With similar arrangements, the role of the overall project manager (Engineer or Employer's Representative) becomes even more challenging: to coordinate all those representatives of the Employer and manage the risks for delays and poor communications. The project manager should check and integrate the comments from the relevant participants within the 21-day time limit for Review of the Contractor's Documents under Sub-Clause 5.2. The delegation of authority for instructions potentially leading to Variations should be carefully considered and announced to the Parties if possible before the Commencement Date.

5.7.4 *Design responsibility*

The PMBOK Construction Extension differentiates between the following general project delivery approaches, which can be further modified and enhanced by the contracts:

---

15 Article 168 of the Bulgarian Spatial Development Act (SG 1 2001) <https://cpcp.mrrb.government.bg/cms/assets/Laws/SPATIAL%20DEVELOPMENT%20ACT.pdf> accessed 22 March 2022: 'The persons exercising construction supervision shall be liable for damages caused to the employer and to the other participants in the construction, and jointly liable with the contractor for damages caused due to not observing the technical rules and regulations and the approved designs.'

16 Contract for Design, Procurement and Construction of a Natural Gas Interconnector Greece-Bulgaria Dated 10 October 2019 < https://www.icgb.eu/downloads/profile/p5-o42.pdf> accessed 22 March 2022

- Traditional: design-bid-build: the design is ready before the tender procedure for selection of contractor;
- IPD and turnkey: the processes for the procurement of plant, design, and construction are integrated to optimise the efficiency in project cost and schedule.

Sub-Clause 4.1 [General Obligations of the Contractor] of the Red Book and Sub-Clause 4.7 [Contractor's Design] in the Green Book stipulate that if the Contract specifies that the Contractor shall design any part of the Permanent Works, the Contractor shall be responsible for this part, and it shall, when the corresponding Works are completed, be fit for such purpose(s) for which the part is intended as are specified in the Contract (or, where no purpose(s) are so defined and described, fit for their ordinary purpose(s)). This fitness for purpose obligation is relevant for all Works (or Section or Part or major item of Plant) under the Yellow/Emerald/Silver Books.

The Red Book uses the traditional project delivery approach: the design is the Employer's responsibility. The general obligations of the Contractor are limited to execution of the Works in accordance with the Contract and the instructions of the Engineer. The fitness for purpose obligation will be limited only to those specified parts for which the Contractor has to prepare the design. Such design is usually related to some cases when Plant is supplied by the Contractor, since the characteristics of the equipment must be considered by the designers. Even when inserted in the Specification, the requirement for the Contractor to design the majority of the Works under a Red Book contract infringes GP 3.[17]

In addition to the traditional approach where the Contractor executes the Works in accordance with the Employer's design, the Green Book 2021 enables an alternative where the Contractor is required to design the Works to various possible degrees[18] set out in the Specification.

It is up to the Employer to define and describe the purposes of the Works/Section/major item of Plant in the Specification/Employer's Requirements, to the extent that the Contractor is responsible for the design of the Permanent Works. Where there is no such definition and description by the Employer, the 'fall-back' required standard is to be '*fit for their ordinary purpose(s)*'. Generally, the ordinary purpose(s) should be obvious from the scope of the relevant works, but matters such as ease of maintenance and expected life will not be so obvious and so should also be specified by the Employer – even if it could be said that they should be implied by law.[19]

Under the Gold Book, the Works shall be 'fit for the purposes for which the Works are intended as defined in the Contract, and the Contractor shall be responsible for ensuring that the Works remain fit for such purposes during the Operation Service Period'.

The Employer usually includes conceptual, preliminary, or more developed technical design in the Employer's Requirements for a Yellow/Emerald/Gold/Silver Book Contract. The Contractor becomes responsible for the final design but has some remedies in the case of timely notified errors in the Employer's Requirements in those books, except under the Silver Book.

17 FIDIC, *The FIDIC Golden Principles* (2019) 9
18 FIDIC, *Short Form of Contract* (2021) Foreword
19 FIDIC, *The FIDIC 2017 Contracts Guide* (FIDIC 2022) 151

The design risk should not be fully transferred to the Contractor if there is insufficient time or information for tenderers to check the Employer's Requirements or carry out their designs, risk assessment, and estimating.

5.7.5 Contract price

According to the allocation of the price risk in the contract, the most common types of construction contracts are:

- **fixed price or lump sum contracts**: the Contractor takes the risk; suitable for projects allowing for an estimate of the total cost;
- **unit-rate/price or remeasurement contracts**: the Contractor takes the risk for the unit prices, while the Employer takes the risk for the quantities; suitable for projects where the types of items are known, but not the quantities;
- **cost-reimbursable contracts**: the Employer takes the risk and pays the Contractor the incurred costs plus a percentage for overheads and profit; suitable for emergency works or where the Employer nominates subcontractors, or for innovations where the scope is uncertain.

Often, **hybrid contracts** are used, combining these pricing methods, for example:

- remeasurement for site preparation and lump sum payment for a thermal power plant;
- remeasurement for reconstructions and lump sum for new build in a complex with many buildings;
- remeasurement for the roads and lump sum for a bridge in an infrastructure project.

The following combinations are possible to incentivise the Contractor for innovative projects:

- Cost plus fixed fee with guaranteed maximum price (**GMP**);
- Cost plus fixed fee with bonus;
- Cost plus fixed fee with target price and pain-gain share.

The Red Book has been drafted by FIDIC as a remeasurement contract. According to the FIDIC definition, this means the form of Contract under which the tenderer prices a BQ, and under which the payment due to the Contractor is based on measurement of the actual physical quantities of the completed work, and then applying the rates and prices in the BQ to those measured quantities.[20]

The Yellow, Gold, and Silver Books are drafted as lump sum contracts. According to the FIDIC definition, this is

20 FIDIC, Glossary of Terminology <fidicterms.org> accessed 26 March 2022

> the form of contract under which the tenderer offers lump sum price(s) for providing the Works, thereby taking the risk that the physical quantities of the constructed Works may differ from those calculated in order to finalise the tender sum. It should be noted that a lump sum contract is not a fixed price contract, because the lump sum offered by the tenderer is generally adjustable under the conditions of the Contract (for example, for Variations instructed by or on behalf of the Employer).[21]

The Emerald Book is a hybrid contract, where the Excavation and Lining Works are measured, and the other works are paid as lump sum.

The Red/Yellow/Silver Book Sub-Clause 14.1 starting statement 'Unless otherwise stated in the Particular Conditions …' means that a different method of determining the Contract Price may be defined in the Special Provisions. Where a different method is used for part of the Works, that part must be clearly identified and specifically defined in the Contract.

5.7.6 Ground risk allocation

The ground risk should not be fully transferred to the Contractor if construction involves substantial work underground or work in areas that tenderers cannot inspect. If the Contractor has to accept high ground risk, the lowest tender is the one submitted by the least knowledgeable tenderer or most reckless gambler, rather than the best tenderer.

The Emerald Book is the most suitable FIDIC form for underground infrastructure. It allocates the foreseeable ground risk referring to the conditions stated in the GBR to the Contractor, and leaves the risk for the Unforeseeable conditions to the Employer.

The Employers who prefer to sign a turnkey contract may either split the Site preparation and connection to utilities in separate contracts, or if they prefer to leave them to the EPC turnkey Contractor, they may introduce measurement provisions for those parts of the Works.

FIDIC considers that the transfer of the risk of Unforeseeable physical conditions to the Contractor under a Red/Yellow Book Contract does not comply with GP1 and GP3.[22]

5.7.7 Selecting a FIDIC contract for works

FIDIC gives detailed answers to the question of how to select a FIDIC Contract for Works on its website[23] and in the FIDIC Procurement Procedures Guide.[24]

Figure 5.1 presents a concise algorithm for the selection of FIDIC Conditions of Contract for Works, taking into account the latest FIDIC publications, but excluding the Blue-Green Book, drafted specifically for dredging and reclamation works. The selection process includes important choices during the tender strategy preparation. The Employer advised by the Consultant answers the following questions:

1. Is the perceived level of risk low, and/or would you like to use a simple form with lean PM? If yes, use the Green Book 2021.

---

21 FIDIC, Glossary of Terminology <fidicterms.org> accessed 26 March 2022

22 FIDIC, *The FIDIC Golden Principles* (2019) 8

23 FIDIC, <https://fidic.org/bookshop/about-bookshop/which-fidic-contract-should-i-use> accessed 29 March 2022

24 FIDIC, *FIDIC Procurement Procedures Guide* (2011) See figure 5.10 for a summary overview

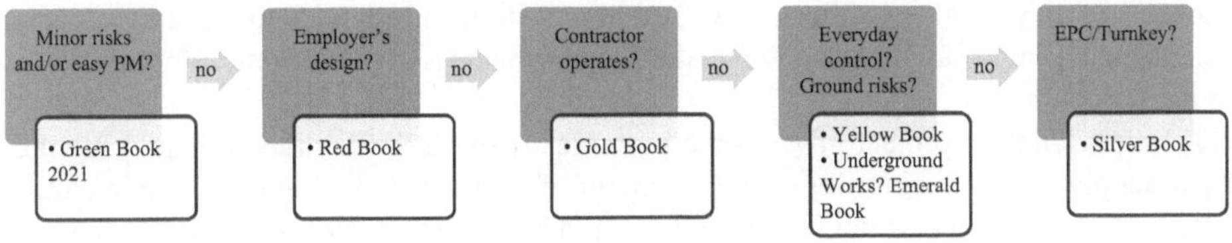

Figure 5.1 Algorithm for selection of a FIDIC works contract

2. Would you like to retain responsibility for most of the design? If yes, use the Red Book.
3. Would you like the Contractor to design, build, and also operate the facility for a certain period? If yes, select the Gold Book.
4. Would you like the Contractor to design, build, and hand over the project to you, and would you like to fully control the project implementation? Are there major risks involved? In the case of predominantly underground works like tunnels, better select the Emerald Book. In the case of energy and industrial projects, use the Yellow Book.
5. Would you like the Contractor to have total responsibility and you to take over a fully equipped project fit for the purpose without everyday control? Are you ready to pay more for allocating the design, ground, price, and climate risks to the Contractor? If yes, then select the Silver Book.

5.7.8 Contract packaging

The project may be constructed using one or several Works contracts. The reasons for contract packaging may be:

- financing from different sources with different timelines;
- the Employer prefers to purchase plant and equipment separately from the construction contract;
- the Employer considers that it can better control the project and make savings in the case of multiple contracts, rather than with a single contract;
- small and medium contractors to be eligible to tender for separate buildings/sections with budget 1–10 million EUR;
- specialised contractors to tender for specific works (e.g. signalling in railway, lining works for infrastructure, or civil, electrical, and mechanical works in a plant);
- frustration from the adversarial behaviour of big contractors, using underpaid subcontractors for most of the works.

When deciding on contract packaging, the first consideration is to determine who will have the responsibility for procurement management.[25] One option is for the Employer to retain this responsibility, with the risk that the failure of a supplier/contractor to deliver as agreed under the contract may delay the other contractors. So, the initial saving from the

25 FIDIC, *FIDIC Procurement Procedures Guide* (2011) 27

in-house PM may result in project delay and higher losses for delay and disruption compensations. The alternative is to allocate the procurement risks to the contractor, resulting in a higher contract price.

The importance of project governance increases with the difficulties of clarifying the project scope from the beginning and the number of contracts to be implemented in parallel. This leads to tender strategies like EPCM and Management Contracting. These strategic models enable the project to begin before the full scope is developed and to use specialised contractors for different works packages.

5.7.8.1 Construction management

Figure 5.2 presents the construction management strategy PM governance. The Client appoints a Designer and a Construction Management Contractor under the FIDIC White Book. The project is divided into works packages. The Construction Manager is responsible for the overall PM, including procurement management. The Works Package Contractors sign contracts with the client, who becomes the Employer under the FIDIC works contracts. The contractual risks that are not passed on to the individual Works Package Contractors will be carried by the Employer. The Construction Management Contractor is paid a management fee, which may be a fixed amount or a percentage of the budget.

5.7.8.2 EPCM

After publishing this model with the use of the White Book for the appointment of the Construction Management Contractor,[26] FIDIC considered that the widely used EPCM

26 FIDIC, *FIDIC Procurement Procedures Guide* (2011) 69

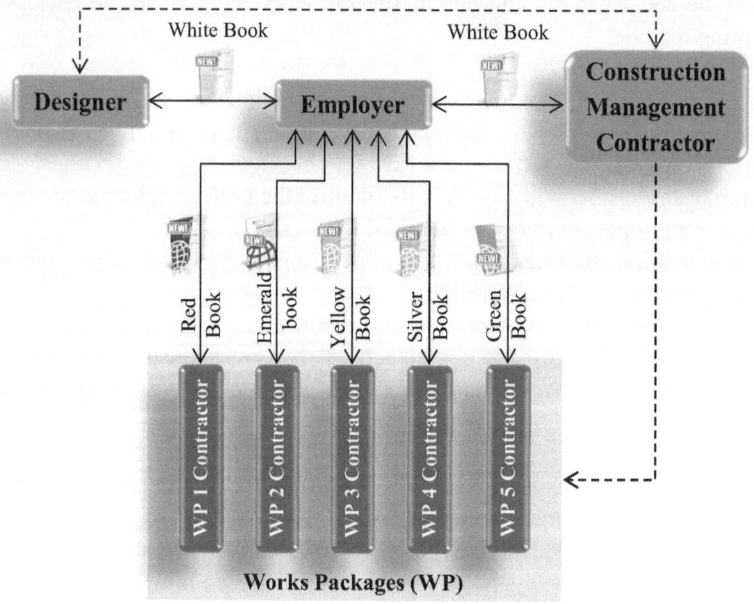

Figure 5.2 Construction management PM governance

model significantly differed from the delivery model contemplated by the existing FIDIC forms, so an EPCM-specific book was needed. TG 22 was formed under the direction of the FIDIC Contracts Committee tasked to develop an EPCM form of contract, guide, and training materials.

The EPCM Contractor has a specific scope of services, requiring besides the PM capability also in-house engineering capability and expertise in procurement management. The EPCM Contractor's engagement may start as early as the feasibility stage of the project. Unlike the Construction Management Contractor, who is mainly involved with the PM and procurement, the EPCM Contractor is typically responsible for:

- basic and detailed design and engineering;
- establishing, implementing, and managing tendering processes for procurement of all equipment and materials and awarding and managing works package contracts;
- overall PM and administration of work package contracts, including during warranty periods.[27]

The EPCM procurement model is recommended when:

- the full scope cannot be defined during the procurement management planning; or
- the project is complex or innovative, and significant changes may be expected; or
- the Employer wants to have input in the design, selection of plant and equipment, choice of contractors, and the construction process.

In the EPCM model presented in Figure 5.3, the Employer retains the control through the EPCM Contractor. The Employer signs the works contracts and retains the risks which are not allocated to the Works Package Contractors.

Besides the traditional EPCM structure presented in Figure 5.3, the EPCM Contractor may enter into contracts directly with Contractors and suppliers, as agent for the Employer (with the EPCM Contractor assuming no or limited liability under such contracts).[28]

5.7.8.3 Management contracting

Figure 5.4 presents the management contracting PM governance. The Client appoints a Designer and a Management Contractor under the FIDIC White Book. The project is divided into works packages. The Management Contractor organises the procurement and then becomes the Employer signing FIDIC contracts with the Works Package Contractors. The responsibility lies with the Management Contractor for the overall PM and the risks not transferred to the Works Package Contractors. The Management Contractor is paid a management fee on top of the payment for the Works. [29]

27 Damian McNair, 'EPCM Contracts: Project Delivery through Engineering, Procurement and Construction Management Contracts' (PricewaterhouseCoopers 2016) <https://www.pwc.com.au/legal/assets/investing-in-infrastructure/iif-8-epcm-contracts-feb16-3.pdf> accessed 30 March 2022

28 Damian McNair, 'EPCM Contracts: Project Delivery through Engineering, Procurement and Construction Management Contracts' (PricewaterhouseCoopers 2016) 5

29 FIDIC, *FIDIC Procurement Procedures Guide* (2011) 70

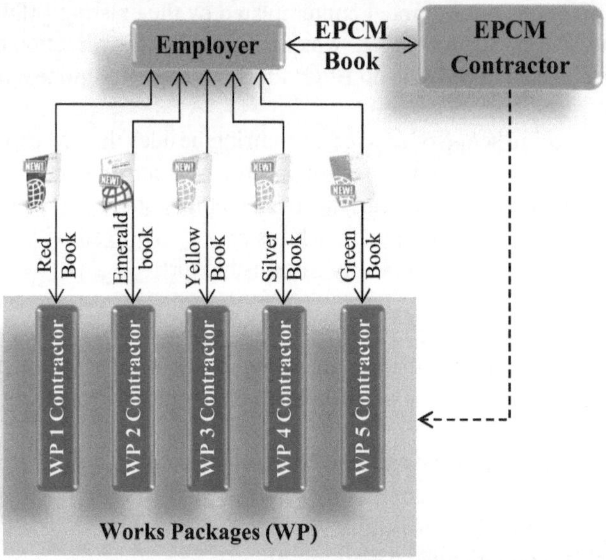

Figure 5.3 EPCM PM governance

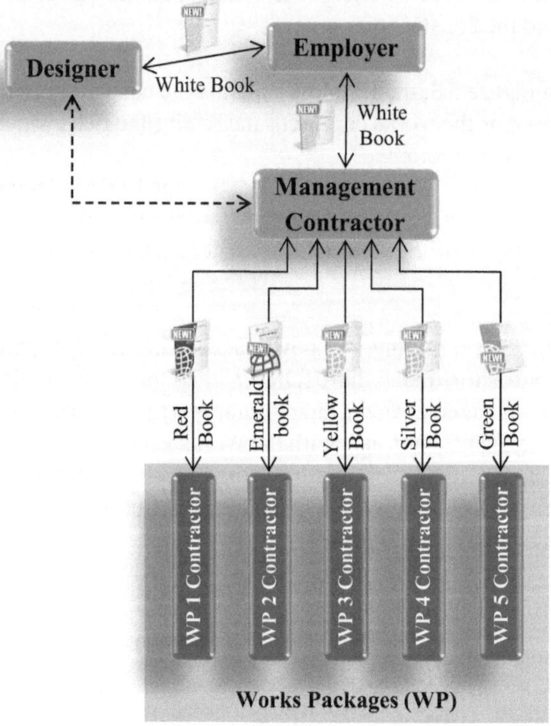

Figure 5.4 Management contracting PM governance

5.8 Conduct Procurements (works)

5.8.1 Procedures for tendering

The method to be adopted for tendering depends on the size, complexity and budget of the project. The certainty about the scope and readiness of the design are also important considerations.

FIDIC describes four usual tendering methods:[30]

- **Open** – a public call for tenders; the prequalification questionnaire is submitted with the tender;
- **Selective** (restricted) – the first stage is prequalification; the second stage is the competition between the short-listed four to eight firms or JVs, which receive the tender documentation and submit tenders;
- **Negotiated** – in case of urgency, secrecy, or lack of skilled competitors for specific works;
- **Competitive dialogue** – used for large and complex projects where the Employer prefers clarification of the scope and discussion of the requirements with the selected tenderers; the contract is awarded on the basis of the Most Economically Advantageous Tender (**MEAT**) in accordance with the announced award criteria.[31]

Figure 5.5 shows the main procurement activities until the award of a Works contract.

The FIDIC Procurement Procedures Guide recommends a prequalification procedure and contains a set of Standard Prequalification Forms for Contractors,[32] which can be used in a two-stage tendering process for award of a FIDIC Works Contract. An open procedure example can be found in the World Bank Standard Procurement Document 'Request

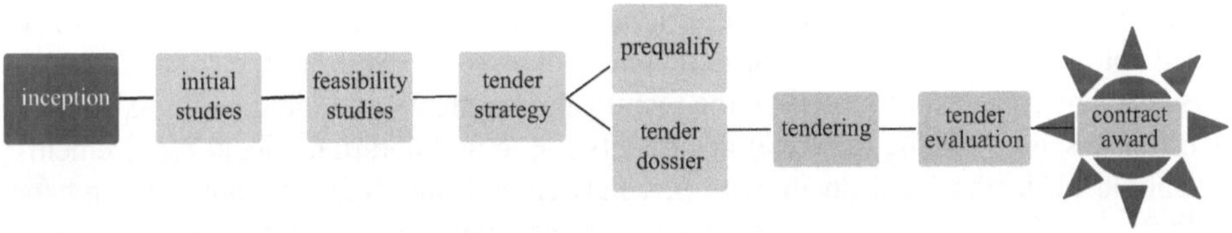

Figure 5.5 Procurement activities

30 FIDIC, *FIDIC Procurement Procedures Guide* (2011) 97–99

31 The European Union (**EU**) recognises that best-value public procurement is not achieved by solely focussing on lowest cost. Directive 2014/24/EU of the European Parliament and of the Council of 26 November 2014 on public procurement Article 67 requires public contracts to be based on the most economically advantageous tender (MEAT), which requires the use of a cost-effectiveness approach such as life-cycle costing. Criteria for determination of the MEAT may include aspects such as quality, including technical merit, aesthetic and functional characteristics, accessibility, design for all users, social, environmental, and innovative characteristics, etc.

32 FIDIC, *FIDIC Procurement Procedures Guide* (2011) Appendix 8.3.1

for Bids – Works (Without Prequalification)'[33] drafted for a one-envelope Bidding process with evaluation and qualification criteria specified to determine the Most Advantageous Bid for the award of a Red Book Contract.

5.8.2 Preparation of tender dossier

The Employer prepares the tender dossier, usually with the help of a consultant. The tender dossier for Works normally contains the following parts:

1. **Tender procedures**
   - Letter of Invitation to Tender
   - Instructions to Tenderers
   - Forms to be filled in by the tenderers (include Schedules)
2. **Conditions of Contract**
   - General Conditions (for FIDIC Contracts only a reference to the relevant Book)[34]
   - Particular Conditions (Part A – Contract Data and Part B – Special Provisions)
   - Forms (Contract Agreement, Letter of Tender, Letter of Acceptance, DAAB Agreement, Forms of Securities)
3. **Technical documents**, defining the scope of the Works and the requirements:
   - Specification and Drawings for the Red Book
   - Employer's Requirements for the Yellow/Silver/Emerald Books (they usually include some drawings)
   - Other information and data for the Site (it is recommended besides the survey reports to attach all agreements, consents, and permits, since they often include conditions resulting in additional works).

The FIDIC Procurement Procedures Guide and The FIDIC Contracts Guide (2000) include example forms for the Letter of Invitation and Instructions to Tenderers. The financing institutions often have their own standard documents available on their websites.[35] The Instructions to Tenderers convey information about the preparation, submission, and evaluation of tenders. They should not form part of the contract, so the drafters have to be careful not to include there any information needed after the award of the contract.

Usually, different experts prepare the technical documents and the conditions of contract. It is essential for the technical documents not to add additional legal requirements that should be in the legal documents (i.e. Particular Conditions of Contract), and for the Particular Conditions not to add technicalities related to the procedures that may be accommodated in the technical documents. It is important that FIDIC experts with technical and legal expertise review the tender dossier for consistency and completeness

33 The World Bank Standard Procurement Document 'Request for Bids – Works (without Prequalification) for Projects with Project Concept Notes (PCN) Decision Notes Dated after October 1, 2018' (World Bank 2019) <https://thedocs.worldbank.org/en/doc/426371562944917302-0290022019/render/SPDRequestforBidsWORKSwithoutprequalification.pdf> accessed 2 May 2022

34 Due to FIDIC copyright restrictions, the full text of the GCs should not be included in the tender dossier. An example for the reference may be found in the Introduction to the GPPC Part B – Special Provisions, *Red/Yellow/Silver Book* 2017

35 For example, World Bank Standard Procurement Documents

before distribution. During the tender procedure, if the tenderers discover inconsistencies between the tender documents, they may ask questions, and the Employer distributes the answers to all tenderers. The questions and answers become part of the Contract. Sometimes when major errors or inconsistencies are found, the Employer may issue an addendum amending a tender document. In such cases, the time for submission of the tender may need to be extended. In other cases, the tender procedure may be cancelled, announcing a new one after a proper revision of the tender documents.

The FIDIC Contracts Guide recommends that the Contractor's obligations be stated in clear unambiguous terms, and not in such a way that the application of opinion or judgement is necessary.[36]

The FIDIC 2017 Contracts Guide recommends when preparing the Particular Conditions for each individual contract to give regard to the comments in this Guide, The FIDIC Golden Principles, and the GPPC in the relevant Book:

> It is essential that all of these drafting tasks, and the entire preparation of the tender documents to be issued by the Employer to tenderers, are entrusted to personnel with the relevant expertise, including the contractual, technical and procurement aspects of such documents ... Expert professional/legal advice should be obtained and carefully considered whenever appropriate, and particularly before entering into or terminating a contract.[37]

Part A of the Particular Conditions-Contract Data includes essential information, presented in a table.[38] The Employer completes most of it. In case the Employer wishes any Contract Data to be completed by the tenderers, the Employer should insert 'Tenderer to Complete' in the space next to any Sub-Clause. Besides the address of the Contractor for communications and nomination of DAAB members, the tenderers usually fill in the Time for Completion. In some cases of urgency, the Employer fills in the Time for Completion, and those tenderers that cannot comply with it do not submit tenders. Failure by the Employer to provide the information and details required in the Contract Data could mean either that the documents forming the Contract are incomplete with vital information missing, or that the GCs' fall-back provisions will automatically take effect, if any.[39] The Contract Data reflects the tender strategy for early taking over of some parts of the Works (if applicable) that should be designated as Sections with different Times for Completion connected with Delay Damages.

The GCs of the FIDIC Books are protected by copyright and trademark and may not be changed without specific written consent, usually in the form of a licence to amend, from FIDIC. If drafters wish to change the GCs, the place for doing this is in the Special Provisions and not by making changes in the GCs.[40]

The Particular Conditions Part B-Special Provisions include modifications to the GCs required to account for local mandatory regulations and in rare cases, to reflect the

36 FIDIC, *The FIDIC Contracts Guide* (2000) 19

37 FIDIC, *The FIDIC 2017 Contracts Guide* (FIDIC 2022) ii–iii

38 In the 1999 Edition of the *Red* and *Yellow Book*, this important information was in the document named 'Appendix to Tender', which was often lost among the numerous appendices submitted with the tender (including evidence of technical and financial capability).

39 FIDIC, *Yellow Book* (FIDIC 2022) GPPC 2

40 FIDIC, *The FIDIC 2017 Contracts Guide* (FIDIC 2022) 39. Further recommendations on the preparation of Particular Conditions are contained in the books edited by Donald Charrett and published by Informa Law from Routledge: *FIDIC Contracts in Asia Pacific* (2022); *FIDIC Contracts in Europe* (2023); *FIDIC Contracts in the Americas* (2023); *FIDIC Contracts in Africa and the Middle East* (2023).

specific project arrangements. The GPPC is intended to assist drafters of the Special Provisions by giving options where appropriate. In some cases, example wording is included between lines, while in other instances, only an *aide-memoire* is given.

JICA published the 'Check List for One-Sided Contracts'[41] to educate Employers that if modifications of the GCs for any particular project alter the originally contemplated risk distribution to a large extent, and the risks allocated to the Contractor become excessively high, the following problems may occur:

- higher bid price;
- bid failure and disruption of project implementation;
- non-participation in the bid of conscientious and capable contractors;
- contract award to a bidder not capable of estimating the risks properly;
- poor construction quality and delay due to lack of risk contingency;
- undermining the relationship of mutual trust and respect between the Parties;
- repetition of groundless claims from the Contractor;
- frequent disputes between the Employer and the Contractor;
- in an extreme case, eventual termination of the contract.

FIDIC strongly recommends that the drafters of the Special Provisions take due regard of the five GPs, discussed in Chapter 1 of this book.

The FIDIC Procurement Procedures Guide and the GPPC in the books include guidance for the preparation of the technical documents.

Using a Red Book, the Employer takes responsibility for the design. The '**Drawings**' by definition include the drawings of the Works included in the Contract and any further modified drawings. Often, the design approval/building permit procedures run in parallel with the selection of Contractor, and the tender drawings may be modified.

The **Specification** includes the requirements for the execution of the Works not shown on the Drawings and not covered by the Red Book GCs. The Specification and the Drawings included in the tender dossier should provide sufficient detail to enable the tenderers to assess the nature and scope of the Works. They should follow local and international standards.

The 2021 Edition of the Green Book has been drafted on the principle that the Contractor constructs the Works in accordance with design provided by the Employer, although some design by the Contractor might be required. However, it can be used for a full Contractor's design scenario with four short amendments, proposed by the authors.[42] This is reflected in the scope of the Specification defined under Sub-Clause 1.1.40. The **Specification** specifies the Works and includes:

- the Employer's design and/or
- the Employer's requirements if any design is to be carried out by the Contractor.

41 JICA, 'Check List for One Sided Contracts For Use with 'Sample Bidding Documents under Japanese ODA Loans – Procurement of Works' (JICA 2011) https://www.jica.go.jp/activities/schemes/finance_co/procedure/guideline/pdf/check_e.pdf (accessed 23 April 2023)

42 FIDIC, *Green Book* (FIDIC 2021) Guidance Notes, 12

The **Employer's Requirements** in the other FIDIC Books should specify the particular requirements for the completed Works, including functional requirements, quality, and scope. If the Contractor is required to supply certain items, such as consumables, these should be listed in a Schedule. Drafters of the Employer's Requirements must remember that if any matters are not covered, the Contractor may well be relieved of any responsibility in respect of such matters.[43]

It is recommended to include all available information related to the Site and any approvals, consents, and permits in the technical documents, which become part of the contract documents. FIDIC Contracts are internationally used, and the drafters of the tender documents should assume that there will be international tenderers, which do not know the local laws and might receive incorrect or incomplete legal advice. The FIDIC Books include many provisions stipulating that the Contractor has to comply with the local laws, procedures, and labour practices.[44] For clarity and certainty, it would be better to explain the mandatory procedures and as a minimum, include a list of the relevant construction regulations and links to the websites where the updates may be checked. The drafters should include in the Specification/Employer's Requirements the Contractor's obligations related to regulatory consents and permits, Site opening, and building control procedures. The most important of those obligations may be added in the Particular Conditions, but without reference to a dated issue of a specific regulation. A reference to the mandatory procedure under the 'Laws of the Country' (by definition the local laws) is the best practice.

The **Schedules** are the forms included in the tender dossier to be completed by the tenderers to become part of the contract documents. They differ depending on the FIDIC Book and the information that the Employer requires. Table 5.1 includes the Schedules named in the definitions of the Red/Yellow/Silver/Emerald/Gold/Green Books. The Employer decides which of them will be used depending on the method of payment, uncertainties, and other considerations. Schedules of Spare Parts, Schedule of DNP Consumables, Availability Schedule, Schedule of Carbon Emissions, and others may be added.

5.8.3 Site visit

The Employer should arrange a formal visit to the Site for all tenderers. When the access to the Site is not limited, the Instructions to Tenderers usually include a form for a statement to be submitted that the tenderer has visited the Site. Sub-Clause 4.10 of the FIDIC Red/Yellow/Emerald/Silver Books stipulates that 'the Contractor shall be deemed to have inspected and examined the Site'. It is important for the Employer to provide equal opportunities to all tenderers to visit the Site.

5.8.4 Contractor's Proposal

Under the Yellow Book, the Contractor's Proposal may include a preliminary design. The Employer should define in the Instructions to Tenderers what is expected to be prepared

43 FIDIC, *Yellow Book* (FIDIC 2022) GPPC 10

44 FIDIC, *Yellow Book* (FIDIC 2022) Sub-Clauses 1.13, 4.8, 4.10, 4.18, 5.1, 5.2, 5.3, 5.4, 6.2, 6.4, 7.4, 7.7, 13.6

Table 5.1 Schedules named in the definitions

Red Book	*Yellow Book*	*Silver Book*	*Emerald Book*	*Gold Book*	*Green Book*
BQs			BQs for Excavation and Lining Works		BQs
Schedule of Payments	Schedule of Payments	Schedule of Payments	Schedule of Payments	Schedule of Payments	Schedule of Payments
Daywork Schedule	Daywork Schedule	Daywork Schedule	Daywork Schedule		
Schedule of Rates and Prices	Schedule of Rates and Prices	Schedule of Rates and Prices	Schedule of Rates and Prices	Schedule of Rates and Prices	Schedule of Rates and Prices
	Schedule of Performance Guarantees	Schedule of Performance Guarantees	Schedule of Performance Guarantees	Schedule of Guarantees	Schedule of Performance Guarantees
			Schedule of Baselines		
			Completion Schedule		
			Schedule of Contractor's Key Equipment	Asset Replacement Schedule	

and submitted as a responsive tender. Tenderers will be reluctant to develop a costly, detailed design if they have little chance of winning.

The Contractor's Proposal is not defined or used in the Red Book, since this form of contract envisages that the Works will be fully, or for the most part, designed by the Employer.

The Contractor's Proposal is also not defined in the Silver Book 2017, because EPC/turnkey projects generally require negotiations between the Employer and one or more tenderers after the tender submissions but before the Contract is entered into. In practice, these negotiations will include discussion of (and often modifications of, or adjustments to) the technical options proposed in the tenders. Accordingly, for EPC/turnkey projects, the proposal included in the successful tender for the execution of the Works is rarely exactly the same proposal that is included in the Contract.[45] The Employer should plan a longer tender procedure for a Silver Book Contract, where the tenderers have to check the Employer's Requirements carefully for errors in order to take total responsibility for the design and fitness for purpose.

5.8.5 Award of Contract

The FIDIC Procurement Procedures Guide contains good practice for tender evaluation and recommendations for award using the MEAT identification method, combining the technical assessments with assessed tender prices on an 80/20 basis.

The Red/Yellow/Emerald/Gold Books contemplate the possibility of a contract being formed by the exchange of the Letter of Tender and the Letter of Acceptance. A Contract Agreement may, however, be required under certain governing laws or procurement rules.[46] The drafters do not need to amend the GCs when no Letter of Acceptance is used: in this case, the expression will be read as Contract Agreement.[47]

Under the Silver Book, the Contractor takes total fitness-for-purpose responsibility, and the award of the Contract will be completed by the execution by the Parties of the Contract Agreement.

The Green Book envisages a simple procedure, setting out a one-page form incorporating the offer and acceptance within the Contract Agreement. The tenderer will sign the 'Offer' part of the Contract Agreement and send it to the Employer with the rest of the tender. The Employer fills in and signs the 'Acceptance' part and returns it to the winner, who becomes the Contractor.

All books include sample forms for the Contract Agreement, but they may need modifications to reflect the mandatory provisions of the governing law. For example, the object of the Contract outlining the scope is an essential element in civil law jurisdictions. The lawyers preparing the Contract Agreement must be knowledgeable about FIDIC contracts and not add redundant texts. For example, the forms do not include the Time for Completion in calendar days, because the Contract has a mechanism for its extension; if a Completion Date is stated in the Contract Agreement, an amendment must be signed

45 FIDIC, *The FIDIC 2017 Contracts Guide* (FIDIC 2022) 15

46 Ellis Baker and others, *FIDIC Contracts: Law and Practice* (Informa Law from Routledge 2009) para 2.117

47 Sub-Clause 1.1.50 (*Red/Yellow*)/1.1.48 (*Gold*)/1.1.58 (*Emerald Books*)

for every EOT.[48] The form of the Contract Agreement should be included in the tender documents.

5.8.6 Contract documents

Most of the documents included in the tender dossier, excluding the Instructions to Tenderers, form the Contract. The main documents forming the Contract are defined in Sub-Clause 1.1 and listed in Sub-Clause 1.5 [Priority of Documents]. Any further documents may be added, but they have to be listed in the Contract Agreement or in the Letter of Acceptance. Table 5.2 lists the documents forming a Contract in their order of priority under Sub-Clause 1.5 of the Red/Yellow/Silver Books/Emerald/Gold/Green Books. In order to achieve certainty, the Parties usually initial all pages of the Contract documents. After that, they may scan them in the order of precedence in one file. A header with the Contract identification and footer with continuous numbering may be added to facilitate traceability. Where the Contract is in electronic form and signed digitally, all Contract documents should be integrated in the order of priority (the GCs only by reference to the FIDIC publication, using the sample text in the GPPC).[49]

If a Party finds a conflict, ambiguity, or discrepancy in the documents, that Party should promptly give a Notice to the Engineer under the Red/Yellow/Emerald/Green Books or Employer's Representative under the Gold Book or other Party under the Silver Book. Then, the CA (Silver Book: Employer) shall issue the necessary clarifications. The following rules may be applied:[50]

1. All the documents forming the Contract shall be considered together when ascertaining a particular meaning or requirement;
2. Where the same subject-matter is covered in different Contract documents, the document in a higher position on the list is to be given priority;
3. The Contract Agreement has the highest priority;
4. The Particular Conditions amend the GCs;
5. Although the GCs are stated to have a higher priority than the documents listed thereafter, the provisions in these latter documents may have priority in clarifying or changing the risk allocation when the text says that it applies 'unless otherwise stated in the Contract';
6. The priority of the documents listed below the Particular Conditions is based on the principle that the Employer's documents should have priority over the Contractor's documents;

48 But under the Russian law, the Time For Completion must be specified either by a calendar date or through an inevitable event, so the Russian Court of Appeal decided that there was no contract between the Parties, using a FIDIC form where the Commencement Date is to be notified, Aleksei Kuzmin, 'Claims in the St Petersburg Flood Protection Barrier Construction', in Lucas Klee, *International Construction Law* (Wiley-Blackwell 2018) 186–189

49 Given the extensive documentation required for a major construction contract, often running to thousands of pages, contracts may be entirely in electronic form

50 Rules 1–8 are based on the text of *The FIDIC 2017 Contracts Guide* (2022) 100; Rule 9 is commented in the Green Book Guidance Notes, referring to Sub-Clause 1.5

Table 5.2 Contract documents in the Sub-Clause 1.5 order of priority

Red Book	*Yellow Book*	*Silver Book*	*Emerald Book*	*Gold Book*	*Green Book*
Contract Agreement	Contract Agreement	Contract Agreement	Contract Agreement	Contract Agreement	Contract Agreement
Letter of Acceptance	Letter of Acceptance		Letter of Acceptance	Letter of Acceptance	Offer/acceptance incorporated above
Letter of Tender	Letter of Tender		Letter of Tender	Letter of Tender	
PC Part A – Contract Data	PC Part A – Contract Data	PC Part A – Contract Data	PC Part A – Contract Data	PC Part A – Contract Data	PC Part A – Contract Data
PC Part B – Special Provisions	PC Part B – Special Provisions	PC Part B – Special Provisions	PC Part B – Special Provisions	PC Part B – Special Provisions	PC Part B – Special Provisions
			Geotechnical Baseline Report		
General Conditions	General Conditions	General Conditions	General Conditions	General Conditions	General Conditions
Specification	Employer's Requirements	Employer's Requirements	Employer's Requirements	Employer's Requirements	Specification
Drawings					Drawings
Schedules	Schedules	Schedules	Schedules	Schedules	Schedules
	Contractor's Proposal	Tender	Contractor's Proposal	Contractor's Proposal	Contractor's tendered design
JV Undertaking, if applicable	JV Undertaking, if applicable	JV Undertaking, if applicable	JV Undertaking, if applicable		
Any other documents forming part of the Contract	Any other documents forming part of the Contract	Any other documents forming part of the Contract	Any other documents, including Geotechnical Data Report, Contract Risk Register, Contract Risk Management Plan	Any other documents forming part of the Contract	Any other documents forming part of the Contract

7. If the Employer's Requirements contain an imprecise requirement, its higher priority may be offset by the precision of the tenderer's proposal, which would form part of the Contract;
8. It is recommended to add an order of precedence between different documents comprising the Specification/Employer's Requirements;
9. The governing law in most cases will have a *contra proferentem* rule, permitting interpretation against the interests of the Party responsible for the preparation of the Contract documents (the Employer).

The statistics show that a lot of disputes are due to different interpretation of the Contract.[51] The Employer must remember that double checking of the tender documents for clarity and avoiding ambiguities and inconsistencies may save future delays and losses.[52]

5.9 Control Procurements process (works)

The CA monitors and controls the Parties to adhere to the Contract. The following PM tools are used to follow the implementation of a FIDIC Works contract with references to the Yellow Book 2017 provisions:

- progress reports (Sub-Clause 4.20);
- records (Sub-Clauses 6.10, 20.2.3);
- management meetings (Sub-Clause 3.8);
- Variation Procedure (Sub-Clause 13.3);
- Review of Contractor's Documents (Sub-Clause 5.2.2);
- inspections and testing (Sub-Clauses 7.3, 7.4, etc.);
- progress payments (Sub-Clause 14.6);
- Review and revision of Programme (Sub-Clause 8.3);
- tracking of progress and earned value (Sub-Clauses 8.7/14.4);
- Claims (Clause 20).

The DAAB can help the Parties with informal assistance under Sub-Clause 21.3 for interpretation of the Contract provisions upon joint request by the Parties.

5.10 Procurement Management closing (works)

The FIDIC Red/Yellow/Silver/Emerald Book contracts may be closed only after the Parties fulfil their obligations under Clauses 11 and 14. This happens when:

51 In 2020, the main cause of disputes was the owner/contractor/subcontractor failing to understand and/or comply with its obligations under the contract according to Arcadis, '2021 Global Construction Disputes Report. The Road to Early Resolution' (Arcadis Knowledge Hub), <https://www.arcadis.com/en/knowledge-hub/perspectives/global/global-construction-disputes-report> accessed 3 April 2022

52 The United Kingdom (**UK**) case of *MT Hojgaard a/s v E.ON Climate and Renewables Robin Rigg East Ltd* [2017] UKSC 59 highlights that, notwithstanding the well-established canons of contractual construction, construing the meaning of contractual provisions is not a trivial exercise for a complex and detailed construction contract. As this case shows, skilled and experienced judges can have different views on the objective meaning of contract provisions. This case is discussed in: Donald Charrett, *Contracts for Construction and Engineering Projects* (Informa Law from Routledge 2022) Chapter 8 'Fit for Purpose, or Due Skill and Care?'

- the latest Defects Notification Period (**DNP**) has expired;
- the Contractor has supplied all the Contractor's Documents in compliance with the Contract;
- the Contractor has completed and tested all the Works (including remedying any defects) in accordance with the Contract;
- the Performance Certificate has been issued or deemed to be issued on the date 28 days after the expiry of the latest DNP or as soon thereafter as the Contractor has completed its obligations;
- the Contractor has:
  - removed any remaining Contractor's Equipment, surplus material, wreckage, rubbish, and Temporary Works from the Site;
  - reinstated all parts of the Site that were affected by the Contractor's activities during the execution of the Works and are not occupied by the Permanent Works; and
  - left the Site and the Works in the condition stated in the Specification/ Employer's Requirements (if not stated, in a clean and safe condition);
  - submitted a final Statement and discharge confirming the money due;
- the Employer has returned the Performance Security and paid the amount due to the Contractor.

After the issue of the Performance Certificate, each Party remains liable for the fulfilment of any obligation that remains unperformed at that time. For the purposes of determining the nature and extent of unperformed obligations, the Contract shall be deemed to remain in force (Sub-Clause 11.10). There might be pending disputes to be resolved. Under many civil code jurisdictions, the Contractor has strict decennial liability for latent defects.[53]

5.11 Procurement Management early termination (works)

Early terminations in construction are not infrequent and can occur any time during the project.[54]

5.11.1 Termination under the contract or at law

The law of most jurisdictions will grant a party the right to terminate a contract in limited circumstances due to the other party's breach of the contract. The precise circumstances in which a party is entitled to treat itself as being discharged of its future obligations due to breach of the contract by the other party, and the consequences that follow such a

53 Decennial liability is enshrined in the Civil Code of many civil law jurisdictions, such as France, United Arab Emirates, Oman, Qatar, and Saudi Arabia. These statutory provisions impose 'strict liability', arising without the claimant having to prove fault or causation. The Contractor and the supervising Consultant are jointly and severally liable to the Employer for a period of 10 years from the date of delivery of the Works in respect of serious structural defects or collapse. (Niel Coertse, 'Decennial Liability in the Middle East: What Is It and Does Insurance Cover It?' (Charles Russell Speechlys 2022) <https://www.charlesrussellspeechlys.com/en/news-and-insights/insights/constuction-engineering-and-projects/2020/decennial-liability-in-the-middle-east-what-is-it-and-does-insurance-cover-it/#_ftnref2> accessed 2 May 2023

54 PMI Inc., *Construction Extension to the PMBOK® Guide* (PMI 2016) 134

termination, are heavily dependent on the governing law. However, generally, the grounds upon which a party may terminate the contract at law are usually very narrow.

The contract may grant the parties express rights to terminate under the contract. The grounds for termination under the contract often do not replicate the grounds for termination at law; for example:

- they may entitle a party to terminate the contract where the other party has failed to comply with a particular provision, breach of which at law would not be considered sufficiently serious as to give rise to a right to terminate at law;
- where a sufficiently serious event occurs (like insolvency) that does not place the other party in breach of contract, but is considered to attract the same consequences.

Contractual termination provisions may permit the party that exercised the right to greater or different remedies than are available under the governing law. The right to terminate under a contract does not necessarily exclude the right to terminate at law, although this is possible under certain jurisdictions if the same breach is expressly addressed under the contract. Generally, contractual termination provisions will be construed strictly, particularly where they grant a party with greater rights than it would otherwise have at law. They also often require a particular procedure to be followed, compliance with which is a condition precedent to the right to terminate. The clauses referred to in the following give general and specific grounds and PM procedures for termination under the Red/Yellow/Silver/Emerald Books.

FIDIC advises the Employer before inviting tenders to obtain and carefully consider expert professional/legal advice to verify that the wording of the termination provisions is consistent with the law governing the Contract.[55] For example, termination for convenience as provided for in Sub-Clause 15.5 in the Red/Yellow/Silver Books is not permitted under the law of some jurisdictions.

5.11.2 Termination by the Employer for Contractor's default

The Employer is entitled to terminate under Sub-Clause 15.2 if the Contractor:

(a) fails to comply with a Notice to Correct under Sub-Clause 15.1 or a binding agreement, or final and binding determination, under Sub-Clause 3.7 (Red/Yellow/Emerald)/3.5 (Silver Book) or a binding or final and binding decision of the DAAB under Sub-Clause 21.4, when such failure constitutes a material breach of the Contractor's obligations under the Contract;
(b) abandons the Works or otherwise plainly demonstrates an intention not to continue performance of the Contractor's obligations under the Contract;
(c) without reasonable excuse fails to proceed with the Works in accordance with Clause 8 or if the Employer would be entitled to Delay Damages that exceed this maximum amount, if applicable;

55 FIDIC, *The FIDIC 2017 Contracts Guide* (FIDIC 2022) 434

(d) without reasonable excuse fails to comply within 28 days with a Notice of rejection under Sub-Clause 7.5 or instruction under Sub-Clause 7.6;
(e) fails to comply with Sub-Clause 4.2 [Performance Security];
(f) subcontracts the whole, or any part, of the Works in breach of Sub-Clause 4.4, or assigns the Contract without the required agreement under Sub-Clause 1.7;
(g) becomes bankrupt or insolvent; goes into liquidation, administration, reorganisation, winding-up or dissolution, or similar;
(h) is engaged in corrupt, fraudulent, collusive, or coercive practice at any time in relation to the Works or to the Contract.

For grounds (a)–(e), the Employer shall give a Notice to the Contractor of the Employer's intention to terminate the Contract in 14 days. Unless the Contractor remedies its failure within 14 days of receiving the Notice, the Employer may, by giving a second Notice to the Contractor, immediately terminate the Contract. The date of termination shall be the date the Contractor receives this second Notice. In the case of (f)–(h), the Employer may, by giving a Notice, immediately terminate the Contract, and the date of termination shall be the date the Contractor receives this Notice.

After the termination under Sub-Clause 15.2, the Employer may:

- issue a Notice with reasonable instructions for the protection of life or property or for the safety of the Works;
- instruct the Contractor to assign to it the benefit of any subcontract;
- call upon the Performance Security and Advance Payment Guarantee;
- expel the Contractor from the Site;
- complete the Works itself or arrange a new procurement procedure for others to do so;
- use the Contractor's Documents and Goods to effect completion;
- withhold further payments until all the costs of completing the Works and remedying defects, damages, and other losses have been established.

Sub-Clauses 9.4(b)/11.4(d) include different specific grounds and procedures for termination for Contractor's default with immediate effect if the failure to pass the Tests on Completion, defect, or damage deprives the Employer of substantially the whole benefit of the Works.

5.11.3 Termination by the Contractor for Employer's Default

The Contractor is entitled to terminate under Sub-Clause 16.2 if:

(a) the Contractor does not receive the reasonable evidence for the Employer's financial arrangements within 42 days after giving a Notice under Sub-Clause 16.1;
(b) the Engineer fails, within 56 days after receiving a Statement and supporting documents, to issue the relevant Payment Certificate (except Silver Book);
(c) the Contractor does not receive the amount due for payment within 42 days after the expiry of the agreed time under Sub-Clause 14.7;
(d) the Employer fails to comply with a binding agreement, or final and binding determination under Sub-Clause 3.7(Red/Yellow/Emerald)/3.5(Silver Book), or a

decision of the DAAB under 21.4, and such failure constitutes a material breach of the Employer's obligations under the Contract;

(e) the Employer substantially fails to perform, and such failure constitutes a material breach of, the Employer's obligations under the Contract;

(f) the Contractor does not receive a Notice of the Commencement Date under Sub-Clause 8.1 within 84 days after receiving the Letter of Acceptance (Silver Book: after both Parties have signed the Contract Agreement);

(g) the Employer:
   (i) fails to comply with Sub-Clause 1.6 [Contract Agreement], or
   (ii) assigns the Contract without the required agreement under Sub-Clause 1.7;

(h) a prolonged suspension affects the whole of the Works, and the procedure under Sub-Clause 8.12 is followed;

(i) the Employer becomes bankrupt or insolvent or similar; or

(j) the Employer is found, based on reasonable evidence, to have engaged in corrupt, fraudulent, collusive, or coercive practice in relation to the Works or to the Contract.

For grounds (a)–(g)(i) (Silver Book (a)–(f)(i)) under Sub-Clause 16.2, the Contractor shall give a Notice to the Employer of the Contractor's intention to terminate the Contract in 14 days. Unless the Employer remedies its failure within 14 days of receiving the Notice, the Contractor may by giving a second Notice immediately terminate the Contract. The date of termination shall be the date the Employer receives this second Notice. In the case of (g)(ii)–(j) (Silver Book f(ii)–(i)), the Contractor may by giving a Notice immediately terminate the Contract, and the date of termination shall be the date the Employer receives this Notice.

The Contractor's election to terminate the Contract does not prejudice any of its other rights under the Contract. If the Contractor gives notice and then wishes or agrees to withdraw it, the Parties may agree that:

- the Notice shall be of no effect; or
- the Notice to be put on hold.

5.11.4 Termination for Employer's convenience

The Employer shall be entitled to terminate the Contract at any time for the Employer's convenience by giving a Notice of such termination to the Contractor under Sub-Clause 15.5. The purpose is to permit an Employer to terminate the Contract at any time and for any reason, whether financial, political, or otherwise, where the Contractor is not in default. These provisions are, however, sometimes used where an Employer considers that the Contractor is performing badly but does not wish to take the risk of terminating for cause and having the Contractor dispute the validity of that termination.

After giving a Notice to terminate under this Sub-Clause, the Employer shall immediately:

(a) have no right to further use any of the Contractor's Documents, which shall be returned to the Contractor, except those for which the Contractor has received payment or for which payment is due under a Payment Certificate;

(b) if Sub-Clause 4.6 [Co-operation] applies, have no right to allow the continued use (if any) of any Contractor's Equipment, Temporary Works, access arrangements, and/or other of the Contractor's facilities or services; and
(c) make arrangements to return the Performance Security to the Contractor.

Termination shall take effect 28 days after the later of the dates on which the Contractor receives the Notice or the Employer returns the Performance Security. Unless and until the Contractor has received the payment of the amount due under Sub-Clause 15.6, the Employer shall not arrange any part of the Works to be executed.

The relevant law may restrict the Employer's freedom to terminate for convenience.

5.11.5 Termination resulting from Exceptional Event

The Contract may be terminated if a Party is or will be prevented by an Exceptional Event from performing its obligations for a long period. Either Party may give a Notice of termination under Sub-Clause 18.5 if performance is prevented for a continuous period of 84 days or for multiple periods totalling more than 140 days due to the same Exceptional Event. The Notice becomes effective 7 days after receipt.

5.11.6 Release from performance under the law

In construction projects, events may arise outside the control of the Parties that make it impossible or unlawful for a Party to continue its performance under the Contract, or the law governing the Contract may entitle the Parties to be released from further performance of the Contract, for example:

- Exceptional Events like war or tornado cause total destruction of the Works;
- the authorities suspend the project for breach of zoning regulations;
- new legislation prohibits similar projects (e.g. plants using coal or nuclear energy or waste landfills).

Sub-Clause 18.6 encourages the Parties to attempt to agree 'an amendment to the Contract that would permit the continued performance of the Contract'. If agreement cannot be reached, either Party can give a Notice to the other Party. According to The FIDIC 2017 Contracts Guide, the Contract will, in effect, be deemed to have been terminated under Sub-Clause 18.5.[56]

5.11.7 Actions after termination by the Contractor for Employer's default/convenience, optional termination, and release from performance under the law

When the termination takes effect:

- the Contractor:

56 FIDIC, *The FIDIC 2017 Contracts Guide* (FIDIC 2022) 484

  - ceases all further work, except as may be instructed for the safety of life, property, and the Works;
  - hands over to the Engineer (Silver Book: Employer) any Contractor's Documents, Plant, Materials, etc. that have been paid for;
  - removes from the Site all other Goods, except as necessary for safety, and leaves the Site.
- the Employer promptly:
  - returns the Contractor's Performance Security;
  - pays the Contractor.

5.11.8 The role of the CA after termination

The CA has a passive role while the Parties are leading the termination procedure. Usually, this role is limited to monitoring of the remedial actions and compliance with the contractual procedures, more intense reporting, and consultancy for the Employer. The behaviour and professionalism of the Engineer/Employer's Representative is often an important factor in reaching an agreement for the benefit of the project instead of terminating the Contract.

After the termination has proceeded, the CA has an active role. The important activities include:

- Site inspections for checking the status of the Works and need for any additional safety or protection measures;
- Review of the Contractor's Documents that have not been submitted/examined before the termination;[57]
- agreement or determination of any Claims;
- valuation after termination to agree or determine the value of the Permanent Works, Goods and Contractor's Documents, and any other sums due in accordance with the Contract;
- under the Red/Yellow/Emerald Book, certification of the agreed or determined amount due for payment in the case of termination for Employer's default/convenience, optional termination, and release from performance under the Law.

5.11.9 Payments due following termination

Figure 5.6 summarises the payments due following termination under the Red/Yellow/Silver/Emerald Books.

In the case of termination by the Contractor for the Employer's default/convenience, optional termination, and release from performance, the Employer pays to the Contractor (less any entitled deductions):

57 By definition, under Clause 1.1 of the Red/Yellow/Emerald/Silver Books, '*Review*' means examination and consideration by the Engineer (Silver Book: Employer) of a Contractor's submission in order to assess whether (and to what extent) it complies with the Contract and/or with the Contractor's obligations under or in connection with the Contract.

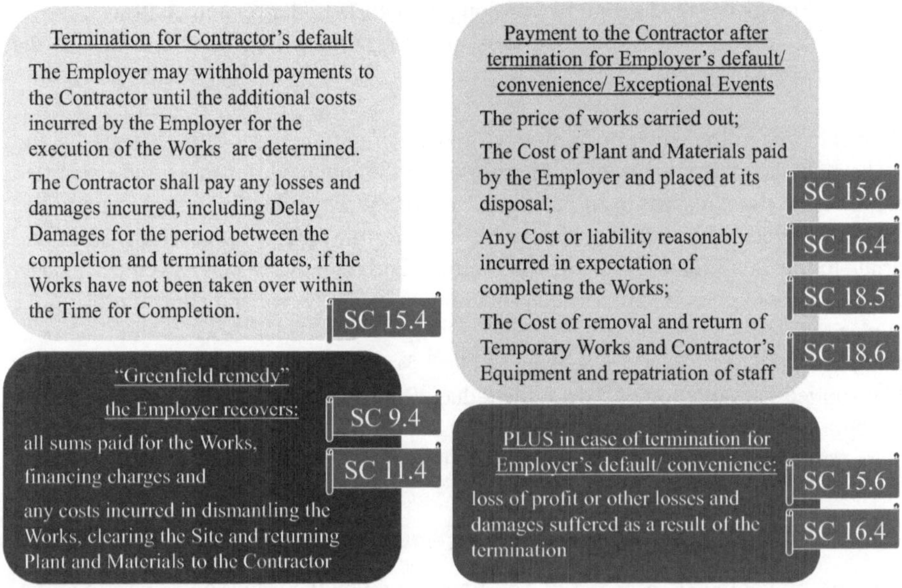

Figure 5.6 Payments after termination

- amounts payable for works complying with the Contract;
- the Cost of Plant and Materials delivered by the Contractor;
- any other Cost or liability reasonably incurred;
- Cost of removal of Temporary Works and Contractor's Equipment from Site and their return to the Contractor's home base;
- Cost of repatriation of the Contractor's staff and labour.

Additionally, the Contractor shall be compensated for any loss of profit or other losses and damages suffered as a result of the termination for Employer's default under Sub-Clause 16.2 or Employer's convenience under Sub-Clause 15.5, as exception to the general rule for limitation of liability under Sub-Clause 1.15 (Silver Book: 1.14).

The Contractor is in a worse position in case of termination for Contractor's default under Sub-Clause 15.2, since the payment due is withheld until all costs, losses, and damages are established. Then, the balance usually changes in favour of the Employer, because:

- the cost for the preparation of a new tender dossier is added;
- the new Accepted Contract Amount (ACA) (Silver Book: Contract Price) for the remaining works usually will be higher due to the price escalation, the smaller scope of the works, and the risk of poor quality: this increase of the price will be due as an ascertained loss after the new tender procedure;
- additional costs for prolonged CA's services may be added;
- additional financing costs may be added;
- deterioration may decrease the works value in the case of inadequate preservation measures;

- the Employer may have to pay prolongation costs under related contracts;
- the Employer may suffer damages under renting/selling contracts due to the delayed handing over.

The Contractor is in the worst position in case of termination under Sub-Clauses 9.4 and 11.4, where the Employer cannot use the Works for the intended purposes and is entitled to recover from the Contractor all sums paid for the Works, plus financing charges and any costs incurred in dismantling the same, clearing the Site, and returning Plant and Materials to the Contractor. This so-called 'Greenfield remedy'[58] is an extreme scenario, which should not happen if appropriate procurement management and control procedures are in place.

In all cases, the Sub-Clause 3.7(Silver Book: 3.5) agreement/determination procedure shall be applied for valuation of the payment due after termination. In case of termination caused by a Party's default, the Sub-Clause 20.2 Claims procedure shall be followed by the other Party (under Sub-Clauses 9.4/11.4/15.4/16.4).

5.12 Innovative Project Procurement Management

The 2017 Edition of the PMBOK® Guide outlines the following trends and emerging practices for Project Procurement Management:[59]

- advances in tools, including online tools providing tenderers with a single source to find procurement documents and complete them directly online and BIM;
- more advanced risk management, an increasing trend to write contracts that accurately allocate specific risks to those entities most capable of managing them;
- changing contracting processes: more megaprojects, involving international contracts with multiple contractors from many countries; the use of internationally recognised standard contract forms is increasing in order to reduce problems and claims during execution.

Obviously, the new trends lead to the increased use of the FIDIC Books, recognised to be the most widespread standard forms for international use. Since innovations are developing faster than the standard forms may be updated, the goal of this book is to demonstrate how procurement may use the FIDIC Books integrating the modern tools.

5.12.1 Digitalisation

Chapter 3 in this book presents a number of digital tools innovating PM in construction. The Project Procurement Management planning must consider how to implement those tools using the FIDIC Contracts.

58 Ellis Baker and others, *FIDIC Contracts: Law and Practice* (Informa Law from Routledge 2009) para. 8.87

59 PMI Inc., *A Guide to the Project Management Body of Knowledge (PMBOK® Guide)* (PMI 2017) 463

5.12.2 BIM with FIDIC Contracts

The Red/Yellow/Silver/Emerald Books include BIM Advisory Notes highlighting the issues to be considered when preparing the tender documents for projects using BIM and FIDIC Contracts. FIDIC recognises that BIM is founded on a team approach and collaboration. FIDIC believes that since FIDIC contracts are designed to be fair to all parties, they may be considered suitable for projects featuring the use of BIM. The parties should recognise the difference in approach and use the contract appropriately. This starts with the proper procurement planning and management, with the help of experienced consultants and legal advisors. Given that there will be many parties involved in the design and construction effort, the Employer must ensure that the information requirements for each contracting party are clearly outlined and there are appropriate limits of liability in place for each participant and for the project as a whole. FIDIC recommends that expectations should focus on the purpose for which the information is required and be connected to the overall outcomes of the project.[60]

Proper procurement planning will help the tenderers be responsive and enable the Employer to select the Consultant/project team best qualified to deliver the desired project outcome.

To manage the digital technology-related risks, FIDIC advises the Employer to set out the requirements with respect to the following matters, and check whether the tenderers address them adequately:

- the specific technologies (including versions/releases) to meet the overall project information requirements;
- access/use rights for the purposes of the project for all information produced by the supplier;
- risk allocation for correctness of external/open data upon which suppliers might rely when developing their own information;
- Service Level Agreements (**SLAs**) for software that might be provided by a supplier upon which the other project participants might be dependent to continue in a timely manner;
- definition of liability in the event of cyber-attack;
- requirements for archiving of all information transactions relating to model-based design to enable audit trail and review of changes.[61]

A recent judgement highlights the impact of these issues on the project information management. In *Trant Engineering Limited v Mott MacDonald Ltd*, the designer hired by the contractor suspended its services and revoked the contractor's access codes to the common data environment (**CDE**). Mrs Justice O'Farrell granted injunction and ordered the designer to make available the design data that had already been procured and completed to date. This would allow Trant to progress the project. Access was permitted only to the public folders that were intended for use by Trant.[62] In circumstances where the BIM coordinator is one of the project participants and not the employer, revocation of access to a CDE could bring the

60 FIDIC, *Yellow Book* (FIDIC 2022) GPPC 56

61 FIDIC, *Yellow Book* (FIDIC 2022) GPPC 57

62 *Trant Engineering Limited v Mott MacDonald Ltd* [2017] EWHC 2061 (TCC)

whole project to a standstill. Commentators recommend tailoring the project procurement to carefully consider which party is best placed to host and control access to the CDE. If, for example, a project participant is best placed to act as the data coordinator and gatekeeper, an employer may wish to regularly extract and store data from the CDE so as to ensure it has the most up-to-date project information. This may help to reduce the risk of a project becoming paralysed in the event access to the CDE is denied.[63]

According to FIDIC, BIM is well suited for integrated project delivery, including design-build and especially design-build-operate projects, where early proactive involvement of the design engineer, Contractor, and Employer are essential. If advanced levels of BIM are anticipated for the project, the possibility of adding operation and maintenance elements might be considered. So, the Yellow/Silver/Emerald/Gold Books may be used in a BIM environment. But, what about the Red Book? Early involvement of the Contractor and its supply chain may be arranged using a multi-party contract, as discussed further in this chapter.

In addition to introducing the general principles for BIM users, the FIDIC Advisory Notes include long non-exhaustive lists of Sub-Clauses of the GCs to be thoroughly reviewed when drafting the Particular Conditions.

5.12.3 Case study – limited use of BIM with a FIDIC Red Book Contract

A case study demonstrates limited use of BIM under a FIDIC Red Book Contract in a Western Balkans country. The Employer included in the Red Book scope BIM, to be developed by a BIM consultant to be appointed by the Contractor under the monitoring of the FIDIC Engineer. In this case, the Employer had not considered BIM before the designer's appointment, but having a detailed design for traditional procurement, decided that it was 'better late than never' to obtain a model of the road section during the construction phase. In this scenario, the full benefits of BIM cannot be used, but the model of the asset can bring benefits as part of a future model of the national road system.

5.12.4 Information management in FIDIC contracts with ISO 19650

5.12.4.1 General requirements for the use of BIM with FIDIC contracts

The FIDIC Contracts do not require particular forms of BIM Protocol and BIM execution or master information delivery plan (**MIDP**). The tender documents for the selection of consultants, contractors, and subcontractors may mandate the use of the ISO 19650 series in order to achieve smooth BIM implementation with FIDIC Contracts. The standard links the information management with the approaches taken to the PM, but ISO 19650 does not require only the ISO 21500[64] PM series to be used; the PMI PM standard will be compatible as well.

The information management process should be applied throughout the delivery phase for each appointment, regardless of project stage. Delivery teams can join and leave at any time.

63 Julian Bailey and others, 'BIM before the English Courts' (White & Case 2017) <https://www.whitecase.com/publications/alert/bim-english-courts> accessed 10 April 2022

64 *ISO 21500:2021 Project, Programme and Portfolio Management – Context and Concepts* (ISO 2021)

ISO 19650-2 supports the effective and efficient procurement, use, and management of information during the delivery phase of assets (design and construction).[65] It is designed to enable an appointing party to establish its requirements for information and to provide the right commercial and collaborative environment within which multiple appointed parties can produce information in an effective and efficient manner. The standard is applicable to projects of all sizes and all levels of complexity. However, its requirements should be applied in a way that is proportionate and appropriate to the scale and complexity of the project.[66] If the FIDIC Employer intends the standard to apply to the project, this should be reflected in the service agreements/works contracts. Then, the selected Consultant should integrate it in the Sub-Consultancy Agreements and the Contractor in the Subcontracts.

The standard requires the Employer as appointing party to establish the project's *information protocol* to be included in each contract. The ISO 19650 series consider all information.[67] In FIDIC works contracts, this includes the programme, records, Contractor's Documents, Employer's design, quality assurance (**QA**) documents, Variations, Notices and other communications under Sub-Clause 1.3, etc. It is essential that the Employer formulates the requirements in relation to the information management during the procurement planning. The requirements for information should be clearly established in the tender documents, including responsibilities, level of information needed, acceptance criteria, supporting information, and dates.[68]

Preparing their proposals, designers, contractors, and subcontractors plan the delivery of the information. The information requirements should be passed down to the most relevant party or the point at which the information can be most easily provided. The requirements may not be identical: only those that are relevant to each appointed party should be passed down. At the start of the delivery phase, relevant information should transfer from the asset information model (**AIM**) to the project information model (**PIM**). The asset information is the starting point in the case of rehabilitations/reconstructions/renovations/upgrading of an existing facility. At the end of the delivery phase, relevant information should transfer from the PIM to the AIM to enable proper operation and facility management.

The information management process during the procurement planning and conducting according to ISO 19650 is linked with the FIDIC contracts in the following subsections.

5.12.4.2 ISO 19650-2 Clause 5.1 Assessment and need and FIDIC contracts

The Employer as Appointing party:

- 5.1.1. **appoints information manager(s)** with appropriate knowledge and skills from its personnel or a third party under agreed scope of services. FIDIC notes that the competence and responsibilities of this position need to be understood and should include experience in data back-up and integrity, continuity planning,

65 *ISO 19650-2:2018(en) Organization and Digitization of Information about Buildings and Civil Engineering Works, Including Building Information Modelling (BIM) – Information Management using Building Information Modelling – Part 2: Delivery Phase of the Assets* (ISO 2018) 0.4 <https://www.iso.org/obp/ui/#iso:std:iso:19650:-2:ed-1:v1:en> accessed 10 April 2022

66 *ISO 19650-2:2018*, 0.1 Purpose

67 UK BIM Framework, *Information Management According to BS EN ISO 19650 Guidance Part 2: Processes for Project Delivery* (fourth edition 2021) https://ukbimframework.org/wp-content/uploads/2020/05/ISO19650-2Edition4.pdf accessed 11 April 2022

68 *ISO 19650-2* Section 5.2.1

and cyber security.[69] The Employer may appoint a BIM consultant as an information manager. Alternatively, the Employer may delegate the responsibility for all or part of the information management function to the Contractor or Designer, but the Employer remains accountable.[70] If, for example, under a FIDIC Silver Book Contract, the information management function is to be delivered by the EPC Contractor as a lead appointed party, the Employer should ensure that the scope of the appointment suitably reflects the activities and responsibilities assigned. For this purpose, the Employer must appoint and upskill an in-house information manager or an advisor for the preparation of the project information requirements and during the project delivery to monitor the compliance with the requirements. It is suggested that the Employer may delegate the responsibility for the review and acceptance of the information model to the Engineer under the Red/Yellow/Emerald Book on its behalf, to be aligned with the Review of the Contractor's Documents under Sub-Clause 5.2/(Red Book: 4.4). It is not recommended the Engineer to be the information manager responsible for the model at the same time, since there might be a conflict of interest.

- 5.1.2. **establishes the project's information requirements**. The considerations are linked with the tender strategy: project scope, intended purpose for which the information will be used, Sections, milestones, stages, contract packages, key decisions based on the delivered information (e.g. Interim Payment Certificate (IPC) including payment for detailed design with the established level of detail).
- 5.1.3. **establishes the project's information delivery milestones.** They may be linked with the dates for delivery of Contractor's Documents, Employer's design/ Site data, key decision points in the Programme; Contract Milestones may be linked to bonuses/damages (Emerald Book Sub-Clause 4.25; Red/Yellow/Silver Book GPPC for additional Sub-Clause 4.22);
- 5.1.4. **establishes the project's information standard.** The considerations are project specific and take into account the future use of information during operation.
- 5.1.5. **establishes the project's information production methods and procedures.** They are project specific and have to be linked with the Contractor's Documents delivery and Review procedures under Sub-Clause 5.2 (Yellow/ Silver/Emerald)/4.4 (Red Book); Sub-Clause 6.10 records; Sub-Clause 8.3 Programme; Red Book Clause 12 measurement and evaluation, etc., considering the security or distribution of information.
- 5.1.6. **establishes the project's reference information and shared resources.** The use of open data standards has to be considered for both reference and shared information to avoid duplication of effort and interoperability issues. This Employer's responsibility should be aligned with its obligations under Sub-Clause 2.5 to deliver to the Contractor information for the Site and items of reference. The Emerald Book GBR and the Employer's requirements (Red Book Specification and Drawings or Employer's Requirements in the other Books) is also important reference information.

---

69 FIDIC, *Yellow Book* (FIDIC 2022) GPPC 57

70 UK BIM Framework, *Information Management According to BS EN ISO 19650. Guidance Part A* (Second Edition 2021) <https://www.ukbimframework.org/wp-content/uploads/2021/02/Guidance-Part-A_The-information-management-function-and-resources_Edition-2.pdf> accessed 12 April 2022

- 5.1.7. **establishes the project's CDE.** CDE should be in place prior to issuing the invitation to tender to enable tender information to be shared with tenderers in a secured manner.
- 5.1.8. **establishes the project's information protocol.** It should be included in the tender documents for the appointment of all designers, Engineer, Contractors, Subcontractors, sub-Subcontractors, and key suppliers, reflecting the differences in each appointment.[71] The Employer should establish it, including associated licence agreements to be incorporated in the contracts. The Employer should consider specific information management/production obligations; any warranties/liabilities associated with the PIM; intellectual property rights of information; the use of existing information/shared resources/information during the project; reuse of information. The information protocol may be incorporated under the Red/Yellow/Emerald/Silver Book Sub-Clause 1.5 with higher priority than the Particular Conditions. In this way, in case of discrepancies between the information protocol and any Conditions of Contract or Specification/Employer's Requirements and other documents with lower priority, the responsibilities and liabilities stipulated by the information protocol will prevail. Thus, the necessity to use incorporation clauses or amend numerous GCs will be minimised. The information protocol may be incorporated in Clause 2 of the Agreement for Services under the White Book with higher priority than the Particular Conditions. The same approach may be followed in the Subcontracts/Sub-Consultancy Agreements.

5.12.4.3 ISO 19650-2 Clause 5.2 Invitation to tender and FIDIC contracts
The Employer as Appointing party:

- 5.2.1. **establishes the exchange information requirements (EIR).** They may be integrated into the Specification/Employer's Requirements (for example, specifying the format for the Contractor's Sub-Clause 6.10 records, Sub-Clause 8.3 Programme, different Contractor's Documents, including As-Built Records/O&M Manuals).
- 5.2.2. **assembles reference information and shared resources.** The Employer may provide access to the project's CDE during the tender procedure and then revoke access for unsuccessful tenderers.
- 5.2.3. **establishes tender response requirements and evaluation criteria.** The Employer should integrate them in the Instruction to Tenderers. Key considerations are likely to be the required content of the (pre-appointment) BIM execution plan; the competency, capability, and capacity of the tenderer to deliver the information requirements; mobilisation plan; and risk assessment.[72]
- 5.2.4. **compiles invitation to tender information;** to be combined with the other tender documents described in this chapter.

71 The UK BIM Framework published an Information Protocol Template supporting *ISO-19650-2* to be used for appointments with English governing law or modified as necessary (2022) <https://www.ukbimframework.org/wp-content/uploads/2021/02/Protocols_Ed3.pdf> accessed 5 February 2023

72 UK BIM Framework, *Information Management According to BS EN ISO 19650. Guidance Part A* (Second Edition 2021) <https://www.ukbimframework.org/wp-content/uploads/2021/02/Guidance-Part-A_The-information-management-function-and-resources_Edition-2.pdf> accessed 12 April 2022

5.12.4.4 ISO 19650-2 Clause 5.3 Tender response

The tenderer as lead appointing party:

- 5.3.1. **nominates individuals to undertake the information management function.**
- 5.3.2. **establishes the delivery team's (pre-appointment) BIM execution plan.**
- 5.3.3. **assesses task team capability and capacity to manage/produce information.**
- 5.3.4. **establishes the delivery team's capability and capacity** (own personnel/ Subcontractors/consultants).
- 5.3.5. **establishes the delivery team's mobilisation plan.** ISO 19650-2 requires the tenderer to consider 11 elements covering testing of information exchange and delivery, CDE, other software and hardware requirements, training, and education.
- 5.3.6. **establishes the delivery team's risk register.** It is suggested that an integrated Risk Register is generated, which includes the information management risks alongside other appointment-related risks[73].
- 5.3.7. **compiles the delivery team's tender response,** including pre-appointment BIM execution plan, capability and assessment summary, mobilisation plan, and information delivery risk assessment.

5.12.4.5 ISO 19650-2 Clause 5.4 Appointment (Agreements for Services, Contracts and Subcontracts, Supply contracts)

The selected Employer's Designer/Engineer/Contractor as lead appointed parties appointing Subcontractors/Sub-Consultants:

- 5.4.1. **confirm the BIM execution plan.** They have to check the (pre-appointment) BIM execution plan and update it after the tender evaluation process with the contribution of the Employer and the Subcontractors. A BIM execution plan is likely to evolve over the life of the delivery team as additional Subcontractors/ Sub-Consultants are appointed.
- 5.4.2. **establish the delivery team's detailed responsibility matrix**, produced from the initial high-level responsibility matrix with the contribution of the Subcontractors/Sub-Consultants.
- 5.4.3 **establish the lead appointed party's EIR.** The level of detail and other requirements in the Subcontracts have to be in line with the EIR in the main Contracts.

The Subcontractors/Sub-Consultants:

- 5.4.4. **establish the Task Information Delivery Plan (TIDP).**

73 UK BIM Framework, *Information Management According to BS EN ISO 19650. Guidance Part A* (Second Edition 2021) <https://www.ukbimframework.org/wp-content/uploads/2021/02/Guidance-Part-A_The-information-management-function-and-resources_Edition-2.pdf> accessed 12 April 2022

The Employer's Designer/Engineer/Contractor as lead appointed parties:

- 5.4.5. **establish the MIDP** as compilation of TIDPs.

The Employer:

- 5.4.6. **completes the main Contracts**, incorporating via the project's information protocol the above applicable information management documents, taking into account that some of the content may evolve.

The Employer's Designer/Engineer/Contractor as lead appointed parties:

- 5.4.7. **complete the Subcontracts**, incorporating via the project's information protocol the above applicable information management documents. The information protocol will usually be the same as that in the main Contract, with any relevant changes to reflect the particular Subcontract.

5.12.4.6 Aligning ISO 19650-2 and ISO 19650-3 with the FIDIC Contracts
Completing the contracts in line with Sections 5.4.6 and 5.4.7 of ISO 19650-2, the parties should ensure that:

- the main contracts and the supply-chain subcontracts/sub-subcontracts incorporate the Information Protocol with higher priority than the Particular Conditions and the rest of the Subcontract documents with lower priority;
- the Information Protocol is properly completed, incorporating all relevant documents detailing the information management process and responsibilities in conformity with ISO 19650-2, applicable for the appointment;
- the change control process is aligned with the Variation procedure (under Clause 13 of the FIDIC Works Contracts/Subcontracts or Clause 5 of the White Book/Sub-Consultancy Agreement) and the entitlement to claim additional time and costs (subject to Clause 20 of the FIDIC Works Contracts/Subcontracts);
- the process of submission, authorisation, rejection, and acceptance is clearly set out in the MIDP and each TIDP, and is aligned with the FIDIC procedures (for example, for Review under Sub-Clause 5.2 of the Yellow/Silver/Emerald/Gold Books/Subcontract 2019);
- the taking-over and project completion procedures under Clause 10 and Clause 11 of the FIDIC Red/Yellow/Silver/Emerald Books/Subcontracts include clear requirements for the Review of the completed project information model, responsibility for archiving, access, security, and continuing integrity/change control during the Tests after Completion under Clause 12 and DNP under Clause 11.

At the end of the delivery phase, relevant information should transfer from the PIM to the AIM. ISO 19650-3 specifies requirements for information management within the context of the operational phase of assets and the exchanges of information within it, using BIM. The appointing party can be the owner of the asset, the operator of the asset (for example, through a long-term concession agreement), or the outsourced asset/facility manager (typically appointed for a period of several years).[74] For the Gold and Bronze Books, the standard may be used for the operation service period, while ISO 19650-2 is used for the design-build period. During the operation with ISO 19650-3, ISO 19650-2 may be used in parallel for the planned maintenance tasks and/or new projects. An ISO 19650-3 compliant Asset Information Protocol[75] should be incorporated in the contracts including operation, besides the ISO 19650-2 compliant Project Information Protocol for the design-build period.

5.12.5 Smart contracts

A **smart contract** is essentially a computer program that incorporates the terms and conditions of a contract into computer code working on a network. The contract self-executes based on the 'if/then' principle, where a situation is monitored and certain terms are executed by the computer program, based on the situation. In reality, it is neither 'smart' (since it operates on an algorithm), nor is it a legal contract.

Blockchain forms the foundation for smart contract applications. The computer code that underpins the contract logic is stored on the blockchain ledger, and consensus algorithms are used to validate the addition of new contract terms and transactions.[76]

A recent study analyses the current state of smart contracts, calling them iContracts. The researchers admit that there are numerous legal ramifications to be overcome.[77]

Prof. Mosey suggests that rather than seeking to automate the whole process, better results may be achieved by collaborative procurement through BIM in tandem with smart contracts, IoT, and AI, to optimise the roles of digital technologies that inform and support:

- human judgement and interactions;
- automatic supply-chain transactions where required decisions have been made, and human judgements and interfaces no longer have the potential to increase efficiency.[78]

---

74 *ISO 19650-3:2020 Organization and Digitization of Information about Buildings and Civil Engineering Works, Including Building Information Modelling (BIM) – Information management Using Building Information Modelling – Part 3: Operational Phase of the Assets* (ISO 2020) <https://www.iso.org/standard/75109.html accessed 17 April 2022

75 The UK BIM Framework published an Information Protocol Template to support the ISO 19650-3 operational phase of assets to be used for appointments with English governing law or modified as necessary (2022) < https://www.ukbimframework.org/wp-content/uploads/2021/11/Information-Protocol-Template-to-support-BS-EN-ISO-19650-3.pdf> accessed 17 April 2022

76 Ernst & Young LLP, 'Smart Contracts Using Blockchain Technology: a Better Way to Deliver Construction Projects' (2018) <https://www.google.com/url?sa=t&rct=j&q=&esrc=s&source=web&cd=&cad=rja&uact=8&ved=2ahUKEwiVvY3xiuiAAxU0VPEDHfL_Cx0QFnoECCYQAQ&url=https%3A%2F%2Fassets.ey.com%2Fcontent%2Fdam%2Fey-sites%2Fey-com%2Fen_ca%2Ftopics%2Fblockchain%2Fey-how-blockchain-can-enable-smarter-contracts-in-infrastructure.pdf%3Fdownload&usg=AOvVaw1dKxKWTxutPtRKmNaPmTHN&opi=89978449> accessed 19 August 2023

77 Alan J. McNamara and Samad Sepasgozar, Intelligent Contract Adoption in the Construction Industry: Concept Development (*Automation in Construction* 2021 122, 103452) <https://www.sciencedirect.com/science/article/abs/pii/S0926580520310323> accessed 4 July 2022

78 David Mosey, *Collaborative Construction Procurement and Improved Value* (John Wiley and Sons 2019) 263

Presently, even for projects using the FIDIC Green Book, where the entitlements are tabulated and Prolongation Costs agreed as liquidated damages, it would be difficult to encode them. In some situations, there might be concurrent delay of the Parties and no entitlement to Costs even in the case of compensable delays. If the Employer wishes to minimise the Engineer's discretion, the Particular Conditions may stipulate further rules and procedures for assessing the Contractor's entitlement to EOT and Costs in the case of concurrency. But in all cases, the Engineer has to analyse the facts, causation, and concurrency in relation to Costs, and criticality in relation to EOT, before agreeing/determining the entitlement, if any. Then, the calculation of the quantum may be done by computer software if the relevant algorithms have been agreed under the Contract and encoded.

A semi-automated solution may be used on this stage of digitalisation[79] with FIDIC contracts. Encoding of repetitive processes that would achieve the greatest cost/time saving with an element of human control may prevent disputes and innovate the PM. The CA will have fewer manual tasks and may allocate more efforts to risk management.

iContracts may bring different benefits to the supply chain. For example, Sub-Clause 14.7 of the 2019 Plant and Design-Build Subcontract has provisions for payments to the Subcontractor within 70 days of receipt by the Contractor of the Subcontractor's statement. The Contractor is entitled to withhold amounts included in the Subcontractor's interim statement not certified by the Engineer or not paid by the Employer. Using the smart contract with the blockchain, the payments to the Subcontractors and their payments to sub-Subcontractors/suppliers may be transferred as soon as the amounts for the relevant Subcontract works are paid by the Employer to the Main Contractor. This will bring trust and certainty and incentivise the supply chain for improved value.

5.12.6 Digital procurement

Digitalisation creates a transparency during the conducting procurement processes and facilitates the tendering and selection of consultants and contractors.

IFIs and the European Commission (**EC**) have implemented digital procurement. The use of electronic tools in public procurement offers a range of important benefits, such as:

- significant savings for all parties;
- simplified and shortened processes;
- reductions in red tape and administrative burdens;
- increased transparency;
- greater innovation;
- new business opportunities by improving the access of enterprises, including small and medium-sized enterprises (**SME**s), to public procurement markets.[80]

The EC digital procurement initiatives include:

79 Alan McNamara, 'Automating the Chaos: Intelligent Construction Contracts' Chapter in *Smart Cities and Construction Technologies* (IntechOpen 2020) 14

80 EC https://ec.europa.eu/growth/single-market/public-procurement/digital-procurement_en accessed 20 April 2022

- Tenders Electronic Daily (**TED**) online portal publishing public procurement notices;
- European single procurement document and eCertis Self-declaration form used by companies to participate in public procurement procedures (only the winner of the tender needs to provide the actual documents);
- eForms used by public buyers to publish procurement notices on TED;
- European Standard for eInvoicing;
- common procurement vocabulary;
- single classification system to standardise the references used to describe procurement contracts;
- emerging technologies in public procurement, enabling better decisions, lower costs, and increased transparency, using AI and machine learning, Big Data and data analytics, blockchain, robotic process automation, augmented and virtual reality, IoT, and drones;
- other initiatives to improve digital procurement in the EU.[81]

The World Bank's e-Procurement online learning course and e-Procurement Toolkit provide practical information for users to quickly and easily take advantage of e-procurement benefits.[82]

5.12.7 Other procurement strategies supporting innovation

In 2020, the UK government launched The Construction Playbook[83] containing 14 key policy reforms to enable 'faster, better, greener' construction by transforming the procurement of public works. Employers are expected to follow those policies on a 'comply or explain' basis, engaging the whole supply chain for improved outcomes:

- Commercial pipelines to be supported with a central government procurement pipeline;
- Market health and capability assessment to enable the use of innovations;
- Portfolios and longer-term contracting where appropriate, to give the certainty required to invest in new technologies to deliver improved productivity and efficiency savings;
- Harmonising, digitising, and rationalising demand;
- Further embedding digital technologies using the UK BIM Framework to standardise the approach to generating and classifying data, data security, and data exchange, and to support the adoption of the Information Management Framework and the creation of the National Digital Twin;
- Early supply-chain involvement;

---

81 EC https://ec.europa.eu/growth/single-market/public-procurement/digital-procurement_en accessed 20 April 2022

82 Sanda Jugo, 'The World Bank e-Procurement Tools' (World Bank 2017) https://collaboration.worldbank.org/content/sites/collaboration-for-development/en/groups/e-procurement/blogs.entry.html/2017/02/14/the_world_bank_e-pro-5pSF.html accessed 20 April 2022

83 UK Cabinet Office, *Construction Playbook* (2020) <https://www.gov.uk/government/publications/the-construction-playbook> accessed 20 April 2022

- Outcome-based approach focusing on outcome with KPIs rather than scope;
- Benchmarking and Should Cost Mode for better investment decisions;
- Delivery model assessments enabling the best outcomes;
- Effective contracting supporting an exchange of data, collaboration, improved value and risk management; adoption of frameworks where appropriate;
- Risk allocation incentivising suppliers to focus on delivering agreed contractual outcomes;
- Payment mechanism and pricing approach to go hand in hand with risk allocation and incentivise desired behaviours or outcomes;
- Assessing the economic and financial standing of suppliers;
- Resolution planning as a new requirement for suppliers of critical works to ensure continuity in case of their insolvency.

Similarly to FIDIC, the UK government recommends tender evaluation to focus on value over cost using MEAT based on published award criteria. Value for money is defined as securing the best mix of quality and effectiveness for the lowest outlay over the life of a project or programme. It is not about minimising initial capital costs. When considering 'outlay', the key factor is whole-life cost, not lowest purchase price. Whole-life cost takes into account the total cost over the life of an asset, including capital, maintenance, management, operation, and exit, and can be very different from the initial price. Paying more for higher quality may be justified if the whole-life cost is advantageous. Contracting authorities should determine whether increased benefits justify higher costs. Affordability will always be a key factor.

FIDIC's 2021 State of the World report, entitled Time to $Tn-vest,[84] recommends a renewed and massive global effort to improve infrastructure spending to meet the investment challenge facing the world, which would amount to a doubling of current infrastructure investment to at least $7 trillion. According to FIDIC, if the investment is not made and/or if the need is not reduced through innovation, meeting the UN's Sustainable Development Goals (SDGs) and making a full economic recovery from the COVID pandemic will be placed in jeopardy. The figures in the report demonstrate that procurement has to enable innovations and sustainability, focusing on the whole-life costs instead of the capital costs.

The ability to influence the project outcome drops with the design development, which makes early supply-chain involvement essential. Spending for better quality and innovations may bring long-term savings.[85]

5.12.8 Green procurement

Green evaluation criteria may be included as eligibility criteria and/or in the technical evaluation to be considered in the selection of the winning tender. The technical evaluation,

84 FIDIC, *State of the World* report, entitled *Time to $Tn-vest* (2021) <https://fidic.org/sites/default/files/SOTW_2021_report01_FINAL_0218.pdf> accessed 20 April 2022

85 FIDIC, *FIDIC Definition of Services Guidelines. Infrastructure Construction / Civil Works* (FIDIC 2016) figure 3 <https://fidic.org/sites/default/files/_DOS_guide_civil_works_1.pdf> accessed 2 May 2022

among other criteria, may rank the proposals for Net Zero actions, environmental action plan, and waste management plan.

The evaluation criteria for the selection of consultants and Contractors may include:

- carbon footprint criteria (carbon footprint question forms to be filled in by tenderers);
- requirement for ISO 14001 certificate;[86]
- requirements for green suppliers, reusable materials, electric equipment, off-Site manufacturing, local personnel, local materials, etc.

The World Bank policy for environmentally responsible procurement envisages applicable evaluation criteria to be included in the technical evaluation and considered in the cost analysis of all proposals received in determining the successful bidders.[87]

Energy performance contracting is a form of innovative financing of energy efficiency investments, where the Contractor gives a financial guarantee for the expected energy savings on a project. The initial investment is covered by the energy service company or by a private partner (usually a financial institution) and repaid through the guaranteed energy savings.[88] The FIDIC 2021 Green Book may be used for this model with payment 'Option B – Lump sum price with stage payments'. The Contract Price may be paid according to the Schedule of Payments, with monthly instalments starting after taking over (for example, 1.19 per cent of the Contract Price for 84 months, with adjustments depending on the monitoring results: in the case of parameters lower than the Performance Guarantees, the payment shall be reduced accordingly; in the case of improved parameters, the saving above the Performance Guarantees shall be split between the Employer and the Contractor). In this model with long-term monitoring and payments, it is better for the Engineer to be an employee of the Employer, fulfilling the Sub-Clause 3.1.2 criteria for suitable qualifications, experience, and competence. Alternatively, a smart contract may be used for monthly payments at an agreed percentage during the 5–10 years after the taking over, to be adjusted with a formula related to the data received from the measuring devices upon validation by the Employer.

5.12.9 Collaborative procurement

Collaborative procurement supports innovation and improved value. One of the recommendations in the 2017 McKinsey's Report[89] referred to in Chapter 3 was to move away from the hostile contracting environment to a system focused on collaboration and prob-

86 *ISO 14001* sets out the criteria for an environmental management system and can be certified to. It maps out a framework that a company or organisation can follow to set up an effective environmental management system. <https://www.iso.org/iso-14001-environmental-management.html> accessed 27 April 2022

87 The World Bank, 'Environmentally Responsible Procurement' <https://www.worldbank.org/en/about/corporate-procurement/vendors/environmentally-responsible-procurement> accessed 17 August 2022

88 European Commission, 'Innovative Financing: Energy Performance Contract' <https://energy.ec.europa.eu/topics/energy-efficiency/financing/innovative-financing_en> accessed 27 April 2022

89 McKinsey Global Institute (MGI's), 'Reinventing Construction: A Route to Higher Productivity, February 2017' <https://www.mckinsey.com/~/media/mckinsey/business%20functions/operations/our%20insights/reinventing%20construction%20through%20a%20productivity%20revolution/mgi-reinventing-construction-a-route-to-higher-productivity-full-report.pdf> accessed 13 January 2022

lem solving. To achieve this, tendering processes can be based on best value and past performance rather than cost alone. Establishing a 'single source of truth' on projects for monitoring progress early helps to minimise misalignments and mitigate risks.

Collaborative procurement comprises a set of processes and relationships through which teams can develop, share, and apply information in ways that improve the design, construction, and operation of their projects. It can be supported by early contractor involvement, by digital technology such as BIM, and by new contractual structures and processes.[90]

The UK platform Construction Excellence, created to improve performance in order to deliver better built environments, defines six critical factors for collaborative working:

- early involvement;
- selection by value;
- aligned commercial arrangements;
- common processes and tools;
- performance measurement;
- long-term relationships.[91]

Partnering commitments may be set down in the form of a charter to be signed by the responsible persons of the partnering organisations. Since the partnering charter/agreement is not legally binding, it does not alter the terms of the contracts between the various parties. The success of the collaboration will depend on mutual trust. When problems arise, litigation may follow.[92] The experience with partnering arrangements not integrated in the contract proves that collaborative procurement cannot thrive separately from commercial expectations and commitments.[93]

Collaborative contracts incorporate features like:

- contractual commitments to co-operate and act in 'good faith';
- early warning mechanisms;
- early involvement of the main contractor and key specialist subcontractors in the design process;
- governance arrangements that facilitate collective problem solving and decision making;
- payment arrangements that financially motivate each participant to act in a manner that is best for the project, rather than best for the participant;
- the agreement of each participant to waive its right to sue any other participant for mistakes, breach, or negligence by another participant (except in the case of wilful default).[94]

---

90 David Mosey, *Collaborative Construction Procurement and Improved Value* (John Wiley and Sons 2019) 2

91 Construction Excellence, Hymn-Sheets (2011) <https://constructingexcellence.org.uk/wp-content/uploads/2015/01/CW-Hymn-Sheet-FIN2.pdf> accessed 29 April 2022

92 FIDIC, *FIDIC Procurement Procedures Guide* (FIDIC 2011) 73

93 David Mosey, *Collaborative Construction Procurement and Improved Value* (John Wiley and Sons 2019) 135

94 Owen Hayford, 'Collaborative Contracting' (PWC 2018) <https://www.pwc.com.au/legal/assets/collaborative-contracting-mar18.pdf> 4, accessed 29 April 2022

For a no blame clause to be effective, it should clarify its scope and impact. Where a badly drafted no blame clause blocks the enforceability of a party's rights, its effect may be more adversarial than collaborative if it transfers risk to the party being deprived of its ability to rely on the contract.[95]

ISO 44001:2017 advises for collaborative procurement:[96]

- contract terms to determine clarity of purpose and encourage appropriate behaviour;
- performance requirements and measurement methods to be mutually agreed to ensure clarity;
- risk and reward models, issue management, exit strategy, knowledge transfer, and sustainability to be considered when developing an agreement;
- what 'value' means to the collaborative partners to be defined;
- a mechanism to be provided for the capture and evaluation against relevant criteria of innovations and ideas for improvement;
- to establish a method for reviewing the success or failure of value creation initiatives and record lessons learned.

FIDIC TG 17 is presently developing a new collaborative form. FIDIC forms like the Yellow/Emerald/Gold/Silver Books involve the Contractor in the design process, but there are no provisions to enable the early involvement of the Contractor under the Red Book. The 1999 and 2017 Editions include the Contractor's duty to co-operate with other contractors (Sub-Clause 4.6). The 2017 Edition of the Red/Yellow/Silver Books and 2019 Emerald Book add several new collaborative provisions in the GCs, such as advance warning (Sub-Clause 8.4); a 42-day consultation of the Engineer/Employer's Representative with the Parties in an endeavour to reach agreement before proceeding to make a determination of a pending matter or Claim (Sub-Clause 3.7/Silver Book: 3.5); management meetings (Sub-Clause 3.8/Silver Book: 3.6); and dispute avoidance (Sub-Clause 21.3). Sub-Clause 13.2 establishes a mechanism for Value Engineering (VE), incentivising the Contractor to submit proposals for improved value, where the Employer may consider sharing of the benefit, costs, and/or delay between the Parties as agreed under the Particular Conditions. More collaborative elements can be introduced with the Particular Conditions, such as:

- target price with pain-gain sharing;
- bonuses for improved value;
- regular risk management meetings.

5.12.10 Frameworks and alliance contracts

A framework agreement is a long-term agreement entered into between an Employer and one or several Contractors that incorporates agreed specifications and rates and prices for

95 David Mosey, *Collaborative Construction Procurement and Improved Value* (John Wiley and Sons 2019) 362

96 *ISO 44001:2017: Collaborative Business Relationship Management Systems – Requirements and Framework* (2017) <https://www.iso.org/standard/72798.html> accessed 29 April 2022

works packages to be carried out as and when instructed by the Employer. In recent years, Employers in the transportation and utilities industries have found framework agreements to be a convenient method of initiating and commencing work at short notice. FIDIC asserts that subject to the introduction of appropriate wording in the Particular Conditions, any of its forms may be suitable for use within a framework agreement.[97]

A recent study by David Mosey recommends going beyond the call-off, measurement, and administration of the framework agreements in order to achieve continuous improvement and the delivery of better, safer, faster, and greener project outcomes. The study provides 24 Gold Standard recommendations with detailed supporting actions designed to improve the outcomes delivered by framework procurement management. It recommends a Gold Standard framework alliance contract that is collaborative, outcome-focused, and flexible, with processes that drive improved value and integrate the work of framework providers, clients, managers, suppliers, and supply-chain members. The contract should support early supply-chain involvement, digital technologies, modern methods of construction, and joint risk management.[98]

The FIDIC Procurement Procedures Guide explains that alliance contracts between Employers and Contractors have been developed with the prime aim of cost and time reduction in the completion of large, complex, multi-discipline, multi-contract projects and the avoidance of conflicts and disputes. Other aims have been to improve the quality of completed works and safety records.[99] FIDIC outlines the following features of the alliance contract strategy:

- the senior persons of the organisations forming the alliance to be committed to its aims and have mutual trust and understanding;
- the alliance to be initiated at the inception of the project to benefit from early Contractor involvement;
- a definition of equitable risk sharing to be included in the alliance contract;
- a reasonable final cost target and time period to be agreed;
- the agreement to include incentives of financial reward for the Contractor to reduce the targeted costs and/or contract period.

FIDIC does not have an alliance form, but the international Framework Alliance Contract **FAC-1**[100] published in 2016 may be used as a multi-party umbrella to embrace two-party FIDIC Contracts. The UK Crown Commercial Service asserts that FAC-1 drives improvements across the construction industry, including:

- ability to influence design through earlier Contractor engagement;
- connecting whole-life considerations through an integrated supply chain;
- focus on newer and safer working practices in construction;
- options for innovative construction methods and technologies;

97 FIDIC, *FIDIC Procurement Procedures Guide* (FIDIC 2011) 73

98 David Mosey, 'Constructing the Gold Standard. An Independent Review of Public Sector Construction Frameworks' (David Mosey 2021) <https://assets.publishing.service.gov.uk/government/uploads/system/uploads/attachment_data/file/1041002/Constructing_The_Gold_Standard_Final.pdf> accessed 30 April 2022

99 FIDIC, *FIDIC Procurement Procedures Guide* (FIDIC 2011) 71

100 ACA, *Framework Alliance Contract FAC-1* (2016)

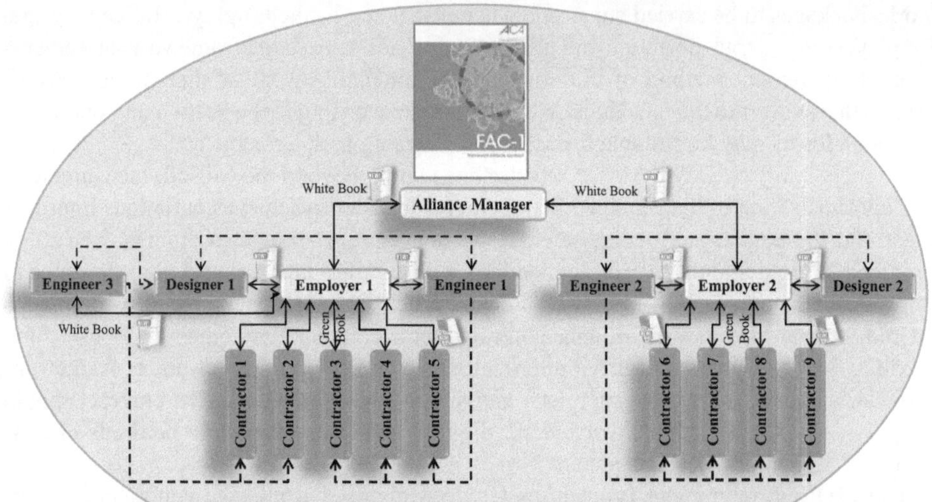

Figure 5.7 An energy efficiency housing framework alliance with FIDIC contracts, integrated by FAC-1

- more opportunities for SMEs, specialists, and local providers;
- sustainable solutions and environmental benefits;
- collaborative risk management and dispute avoidance;
- promoting standardisation in contracting through common 'boilerplate' clauses.

Figure 5.7 presents a practical example for this arrangement for an energy efficiency housing programme. The Employers may be several municipalities, which select several framework Contractors, including FAC-1 in the tender documents as a multi-party integrator. The framework Contractors compete with prices for each project (a high-rise residential building or a lot including several multi-storey houses) and sign Project Contracts with the FIDIC Green Book 2021. The same approach may be followed with the FIDIC Engineers and designers, who sign FAC-1 and the FIDIC White Book 2017 agreements.

5.12.11 Soft landing

The principle of soft landings describes the planned, structured, and graduated handover of an asset from the design and construction team to the operation and maintenance team. The aim of soft landings is to improve operational performance; as a process, it aligns the interests between those who design and construct an asset and those who subsequently use it. BIM and the ISO 19650 series actively support the soft landings process.[101] The FIDIC

101 UK BIM Alliance, Centre for Digital Built Britain and BSI, *Information Management According to BS EN ISO 19650. Guidance Part 1: Concepts* (second edition 2019) Annex B: Soft Landing <https://www.ukbimalliance.org/wp-content/uploads/2019/04/Information-Management-according-to-BS-EN-ISO-19650_-Guidance-Part-1_Concepts_2ndEdition.pdf> accessed 10 April 2022

Gold Book secures the soft landing, since the Contractor continues with the operation of the completed facility.

5.13 Scenario

Appendix 1 includes a scenario inspired by real cases about the procurement management of a complex railway project. It illustrates the risk-based selection of FIDIC Contracts discussed in this chapter with various considerations. The story in Appendix 1 continues with example Notices, related to the relevant PM Knowledge Area, example Issue Log, and Variation Log, helpful for the contract administration of FIDIC Contracts.

CHAPTER 6

Project Integration Management with FIDIC Contracts

DOI: 10.4324/9781003272892-6

6.1 Project Integration Management processes

According to the PMBOK® Guide, Project Integration Management includes the processes and activities to identify, define, combine, unify, and coordinate the various PM processes and activities. These actions should be applied from the start of the project through completion. Project Integration Management includes making choices about:

- resource allocation,
- balancing competing demands,
- examining alternative approaches,
- tailoring the processes to meet the project objectives, and
- managing the interdependencies among the PM Knowledge Areas.[1]

In the construction industry, not only do the processes interact, but the management of these processes is greatly expanded to address multiple performing organisations operating simultaneously.[2] Considering the challenges for PM in construction listed in Chapter 3, the digitalisation and collaboration discussed in Chapter 5 are crucial for successful outcomes. This chapter suggests which innovative tools may support the Project Integration Management with FIDIC Contracts and how to integrate the teams delivering outputs under different FIDIC Contracts in a complex project.

6.2 Project Integration Management initiation

In construction, the Contract typically authorises the project for the Contractors and other service providers. Each entity becomes involved in the project at different stages.[3] Chapter 5 highlighted the benefits of the early appointment of the consultant as a PM integrator, the early appointment of the Contractor, and early supply-chain involvement for improved value. Chapter 4 reviewed the procedure for the appointment of the FIDIC Engineer under Sub-Clause 3.1 of the Red/Yellow/Emerald Books and the Employer's Representative under the Gold/Silver Books.

Chapter 4 discussed the dual role of the Engineer under the Red/Yellow/Emerald/Green Books as Employer's Personnel, which must act 'neutrally' when exercising its duties for agreement/determination. The Employer's Representative under the Gold Book has a similar role as the Engineer, while this role under the Silver Book is more limited, since the turnkey Contractor takes the overall responsibility.

6.2.1 The Engineer's authority

The Engineer should be vested with all the authority necessary to act as the Engineer under the Contract. The Engineer has no authority to amend the Contract or to relieve either Party of any duties, obligations, and responsibilities under the Contract, except as otherwise stated in the Conditions of Contract. Such an exception is, for example, the Engineer's authority to instruct Variations under Clause 13 of the Red/Yellow/Emerald Books.

1 PMI Inc., *PMBOK® Guide* (PMI sixth edition 2017) 69
2 PMI Inc., *Construction Extension to the PMBOK® Guide* (PMI 2016) 29
3 PMI Inc., *Construction Extension to the PMBOK® Guide* (PMI 2016) 31

If the Employer wishes the Engineer to obtain consent to exercise an authority that is not stated in the Particular Conditions, then it will be necessary for the Employer to seek the Contractor's agreement to this amendment of the Contract. When the Employer's consent is required, in accordance with Sub-Clause 1.3, it shall be in writing and shall not be unreasonably withheld or delayed. Furthermore, it must allow the Engineer to exercise his/her duties within the time limits required under the Contract, as any delay caused by the Engineer waiting for the Employer's written consent may entitle the Contractor to EOT and/or to recover the additional costs incurred. Sub-Clause 3.2 expressly states that the Engineer shall not be required to obtain the Employer's consent when he/she is exercising authority under Sub-Clause 3.7 [Agreement or Determination].[4]

6.2.2 The limitations for delegation of authority by the Engineer

After the mandatory appointment of the Engineer and the Notice identifying the natural person authorised to act on behalf of a legal entity is received by the Parties, its authority shall take effect. Then, the Engineer may assign duties and delegate authority to his/her assistants under Sub-Clause 3.4. Since different qualifications are necessary for the monitoring and control of civil, mechanical, electrical, and other works, usually, the assistants have duties and authorities limited within special PM Knowledge Areas, such as Quality Management in the relevant discipline, or overall Cost Management/Schedule Management. The key concept of the PMBOK is that the project manager cannot delegate or transfer to assistants the accountability of the Project Integration Management.[5] Similarly, only the Engineer has the full authorisation under Clause 3 of the FIDIC Red/Yellow/Emerald Books. From the PM point of view, he/she integrates the management in all Knowledge Areas and is in charge of the overall project manager-related governance discussed in Chapter 4.

Figure 6.1 represents the structure of the Engineer and the delegation of authority. The Engineer may appoint an Engineer's Representative under Sub-Clause 3.3 to act on his/her behalf on the Site and assistants under Sub-Clause 3.4. While the Engineer's Representative, like the Engineer, has to be a professional engineer with suitable qualifications, experience. and competence and be fluent in the ruling language, the assistants have to be qualified, experienced, and competent only in the scope of their duties and be fluent in the language for communications. Usually, the Engineer submits to the Parties two separate Notices of Delegation under Sub-Clause 3.4:

- a Notice of full delegation of authority to the Engineer's Representative and Resident Engineers (if any) excluding the authority to act under Sub-Clause 3.7 and to issue a Notice to Correct under Sub-Clause 15.1;
- a Notice of Delegation of limited authority to the assistants under Sub-Clauses 5.2 for Review of Contractor's Documents (Yellow/Emerald Books), inspections under Sub-Clauses 7.3, attendance of tests under Sub-Clauses 7.4, measurement under Sub-Clause 12.1 (Red Book), signing of Daywork forms under Sub-Clause 13.5, etc., in the discipline in which the assistant is qualified.

4 FIDIC, *The FIDIC 2017 Contracts Guide* (FIDIC 2022) 123
5 PMI Inc., *PMBOK Guide* (PMI sixth edition 2017) 72

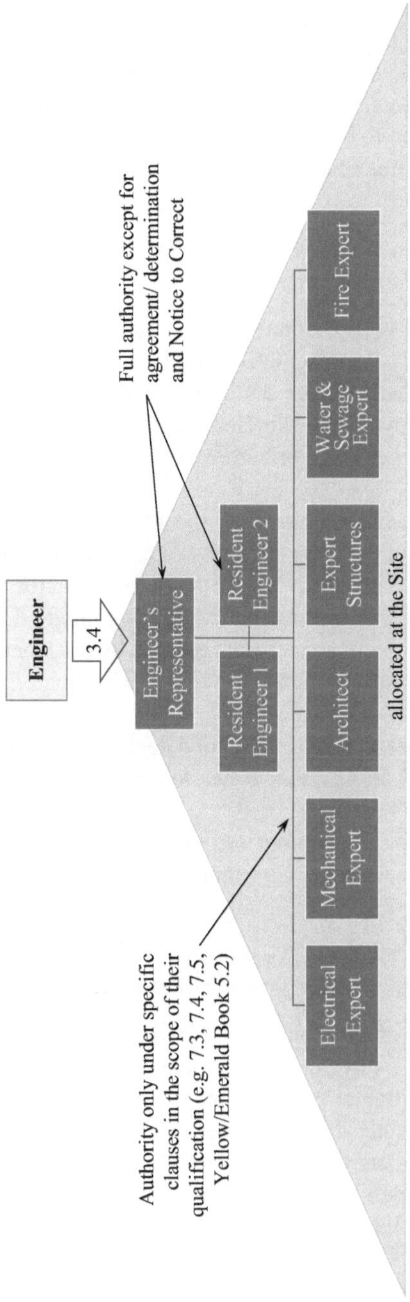

Figure 6.1 Delegation by the Engineer under Sub-Clause 3.4 of the FIDIC Red/Yellow/Emerald Books

The Engineer is usually a senior expert staying in his/her office, often not in the country of the Site. The Engineer will usually visit the Site for the monthly management meetings and other important risk management meetings. The Engineer's Representative and Resident Engineers will be present on the Site all the time and interact with the Contractor's Representative. A lot of issues have to be resolved daily: Unforeseeable events arise; the Contractor sometimes does not manage its suppliers and Subcontractors properly, which leads to continuous rejection of the works and frustration from the Engineer's and Contractor's Site management staff. Due to the difficulties and character clashes, tension often impedes the collaborative solving of problems. By staying away from the Site, the Engineer is in a better position to be fair and unbiased in order to act even-handedly under Sub-Clause 3.7 to agree or determine the entitlements of the Parties.

6.2.3 Silver Book: the Employer's Representative's authority and delegation of authority

Unlike the Engineer, vested with the authority necessary to act as the Engineer under the Contract, the Employer's Representative shall be deemed to have the full authority of the Employer under Sub-Clause 3.1 except for the termination of the Contract by the Employer. The Employer may therefore have no need to require the Employer's Representative to obtain his/her consent before exercising an authority, at the same time being entitled to do so.[6]

Similarly to the Engineer, the Employer or the Employer's Representative may from time to time assign duties and delegate authority to assistants, and may also revoke such assignment or delegation, by giving a Notice to the Contractor of the name, assigned duties, and delegated authority of the assistant. The assignment, delegation, or revocation shall not take effect until this Notice has been received by the Contractor. The limitations for delegation of authority are the same as those for the Engineer:

- the Employer's Representative shall not delegate authority to the assistants to act under Sub-Clause 3.5 [Agreement or Determination]; and/or issue a Notice to Correct under Sub-Clause 15.1;
- the assistants shall only be authorised to issue instructions and communications and/or to give Notices to the Contractor to the extent defined by the Notice of Delegation.

6.2.4 The Contractor's Representative's authority and delegation of authority

Chapter 4 discussed the role of the Contractor's Representative as project manager of the Contractor and the procedure for his/her appointment under Sub-Clause 4.3. The Contractor is obliged to appoint the Contractor's Representative and to give him/her all the authority necessary to act on the Contractor's behalf under the Contract. The Engineer has no obligation to issue any Interim Payment Certificate (**IPC**) under the Red/Yellow/Emerald Books, and the Employer has no obligation to make any interim payment, until the Contractor's Representative has been appointed under the Red/Yellow/Silver/Emerald Books.

6 FIDIC, *The FIDIC 2017 Contracts Guide* (FIDIC 2022) 124

Unlike the natural person authorised to act on behalf of the Engineer, the Contractor's Representative shall be based at the Site during the execution of the Works. His/her whole time shall be given to directing the Contractor's performance of the Contract. If he/she is to be temporarily absent from the Site, a suitable replacement shall be temporarily appointed, subject to the Engineer's (Red/Yellow/Emerald Book)/Employer's (Silver Book) prior consent.

The Contractor's Representative may delegate any powers, functions, and authority to any suitably competent and experienced person, fluent in the language for communications under Sub-Clause 1.4, except the authority to issue and receive Notices and other communications under Sub-Clause 1.3 and to receive instructions under Clause 3. He/she may at any time revoke the delegation. Similarly to the Engineer's delegation, any delegation or revocation shall not take effect until the Engineer (Silver Book: Employer) has received a Notice from the Contractor's Representative, naming the person, specifying the powers, functions, and authority being delegated or revoked, and stating the timing of the delegation or revocation.

The delegation of duties, functions, and authority by the Contractor's Representative is essential for his/her role in the Project Integration Management to direct the performance in compliance with the Contract and the CA's instructions. For this reason, the Contractor should ensure that the Contractor's Representative has sufficient suitably competent and experienced key personnel for whatever duties, functions, and authority have been delegated to them. Clause 6 [Staff and Labour] supplements the Contractor's obligations under Sub-Clause 4.3 regarding the Contractor's Personnel:

- under Sub-Clause 6.8, the Contractor shall provide a sufficient number of competent persons for the superintendence to plan, arrange, direct, manage, inspect, test, and monitor the execution of the Works, with adequate knowledge of the language for communications;
- Sub-Clause 6.12 is a new provision to be used if any Key Personnel are specified under the Contract, with requirements for the CA's consent before the appointment where those persons are not named in the Tender;
- Sub-Clause 6.9 requires all Contractor's Personnel to be appropriately qualified, skilled, experienced, and competent in their respective trades or occupations and empowers the CA to remove any person employed on the Site or Works (including the Contractor's Representative) who:
  - (a) persists in any misconduct or lack of care;
  - (b) carries out duties incompetently or negligently;
  - (c) fails to comply with any provision of the Contract;
  - (d) persists in any conduct that is prejudicial to safety, health, or the protection of the environment;
  - (e) is found, based on reasonable evidence, to have engaged in corrupt, fraudulent, collusive, or coercive practice; or
  - (f) has been recruited from the Employer's Personnel.

The Contractor may readily be persuaded to remove the personnel, since such removal by agreement is preferable to enforcement under Sub-Clause 6.9.[7] Often, the Notice for

7 FIDIC, *The FIDIC 2017 Contracts Guide* (FIDIC 2022) 247

removal of a careless and/or aggressive Contractor's Site Manager persisting in non-compliance with the CA's instructions/Contract/regulations cures the problem: the relevant person changes his/her behaviour under the pressure of the Contractor's top management.

6.2.5 Team building

The first Notices of Delegation may be submitted before the Commencement Date. The Engineer should disclose the project information to his/her assistants and instruct them in this period, transferring the corporative and project knowledge to the team. A FIDIC-accredited in-house training is recommended, tailored for the specific contract documents. It can be delivered for the Employer and Engineer/Employer's Representative, but if the Contractor is invited, this will be an important step for common interpretation of the contract provisions, dispute prevention, and collaboration for the benefit of the project.

The CA should streamline the communication and document control. An assistant should be appointed as a coordinator or document control to follow up the timely submission of documents by the Parties and the CA.

6.2.6 Permit management

The CA should prepare a Road Map or scrutinise the Employer's/Contractor's Road Map, if any, to identify any pending procedures and applicable statutory approval and permits for:

- land acquisition, right of way, expropriation;
- design permits, environmental and zoning plan requirements;
- design agreements/approval and construction permit;
- Site opening and commencement of construction;
- procedures for connections to utilities/crossing or relocation of existing infrastructure;
- traffic control and diversions;
- commissioning and permit to use/licence for operation.

If the provisions of the Contract and the local law are not in conflict, then all requirements should be satisfied. As part of the Project Integration Management, the CA should try to integrate the contractual and legal requirements for the construction documentation in a lean manner. The FIDIC Contracts do not have mandatory forms, so if the local regulations require the compliance of the Works to be documented using certain templates, those forms may be used, referring, for example, to Sub-Clause 7.3 [*Inspections*]. If the ruling language of the Contract is English due to the requirements of the financing institution, the forms may be bilingual: in English and in the local language mandatory for permitting. In this case, the local language should be included in the Contract Data as a language for communications. The Contract Data should clearly state whether all Contractor's Documents shall be in the local official language, or all shall be in English and only those needed for permitting be either bilingual or translated. The lean approach is recommended, since unnecessary translation creates delay and additional costs.

6.3 Project Integration Management planning

6.3.1 PM plan

The PMBOK defines the **PM plan** as the document that describes how the project will be executed, monitored, controlled, and closed. The PM plan integrates and consolidates all of the subsidiary management plans and baselines, and other information necessary to manage the project. The needs of the project determine which components of the PM plan are to be developed.[8]

The CA who is responsible for the overall PM of the FIDIC contracts should develop the PM plan referring to the components included in the Contract documents or other PM documents prepared by its team. The contract outlines progress and completion milestones, contract price, reporting, and inspection and compliance requirements, all of which are integrated for project success.[9] After the Notice to Commence, the CA should update the PM plan, reflecting/integrating components developed by the Contractor.

6.3.2 PM plan components

The PM plan components may include:

Subsidiary management plans:

- **Prepared during the project preparation phase**:
  - **Scope management plan.** Establishes how the scope will be defined, developed, monitored, controlled, and validated. May be updated depending on the available financing sources and the requirements of the financing institutions.
  - **Procurement management plan.** Establishes the tender strategy.
  - **Stakeholder engagement plan.** Establishes how stakeholders will be engaged in project decisions and execution according to their needs, interests, and impact.
- **Integrated in the tender/Contract documents for works**:
  - **Requirements management plan.** Establishes how the requirements will be analysed, documented, and managed.
  - **Schedule management plan.** Establishes the criteria and the activities for developing, monitoring, and controlling the schedule.
  - **Cost management plan.** Establishes how the costs will be planned, structured, and controlled.
  - **Change management plan.** Describes how the change requests throughout the project will be formally authorised and incorporated.
  - **Quality management plan.** Establishes how the quality policies, methodologies, and standards will be implemented in the project.
  - **Communications management plan.** Establishes how, when, and by whom information about the project will be administered and disseminated.
  - **Risk management plan.** Establishes how the risk management activities will be structured and performed.

8 PMI Inc., *PMBOK® Guide* (PMI sixth edition 2017) 86

9 PMI Inc., *Construction Extension to the PMBOK® Guide* (PMI 2016) 34

- **Drafted by the Contractor in the tender proposal and developed after the Notice to Commence**:
  - **Environmental, Health & Safety (EHS) plan.** The scope may be established by the local legislation and the requirements of the financing institution.
  - **Resource management plan.** Provides guidance on how project resources should be categorised, allocated, managed, and released.
  - **Security management plan.** Demonstrates compliance with the Contract requirements.

6.3.3 Baselines to be used as components of the PM plan in FIDIC Contracts

- **Scope baseline** approved by the Employer and integrated in the tender documents. The Specification/Employer's Requirements include the approved version of a scope statement and work breakdown structure (**WBS**), which is used as a basis for comparison. The Red Book includes a detailed WBS in the BQs. The Emerald Book GBR clarifies the Employer's Requirements' scope baseline.
- **Schedule baseline.** The Programme developed by the Contractor in compliance with Sub-Clause 8.3 should be used as a basis for comparison to the actual results.
- **Cost baseline.** The ACA in the Red/Yellow/Emerald Books and the agreed amount stated in the Contract Agreement in the Silver/Green Book with its breakdown may be used as cost baseline.
- **Performance measurement baseline.** An integrated scope-schedule-cost plan for the project work, against which project execution is compared, may be included in the Contract by the Employer in the form of the Schedule of Payment and the Schedule of Performance Guarantees (Yellow Book Sub-Clauses 1.1.73 and 1.1.74) stating the applicable Performance Damages payable in the event of failure to attain any of the guaranteed performances. Also, the Programme developed by the Contractor in compliance with Sub-Clause 8.3 may include the schedule and cost baseline, enabling tracking of the schedule and measurement of the earned value. The Emerald Book 2019 introduces Schedules of Baselines for the excavation and lining works, where the Contractor inserts production rates for quantities estimated by the Employer, to be used for future adjustment of the construction time; the items are based on the subsurface physical conditions described in the GBR.

6.3.4 Elaboration of the PM plan

In some agreements for services, the Engineer has an obligation to elaborate the project PM plan, to be agreed by the Employer. In many cases, this plan remains on paper, since it is a deliverable under the consultancy agreement, which is not incorporated in the works contract and not updated to reflect the quality system and other components drafted by the Contractor. The PM plan may help the integrated PM if:

- the CA is appointed during the project preparation phase, and is responsible for elaborating the works tender documents, and may integrate important requirements for the components of the PM plan in the Specification/

Employer's Requirements (e.g. cost/resource allocation in the Sub-Clause 8.3 programme);

- the agreement for services and the works contract contemplate an integrated PM plan to be agreed within 56 days after the Notice to Commence with detailed verification plans, procedures for approval of Contractor's Documents and works, and risk management, taking into account the Contractor's Quality Management and Compliance Verification Systems and the local legal requirements.

6.3.5 Contract requirements

It would be difficult to ask the Contractor to load resources and costs on the programme and disclose breakdown of costs if those requirements were not incorporated in the Contract, but it may be still possible during the 'honeymoon' period.

In all cases, a formal attitude will create redundant paperwork: FIDIC contracts have many provisions with detailed PM procedures (e.g. updating the Programme under Sub-Clause 8.3, inspections and testing under Clause 7). FIDIC contracts have no requirements for the templates of the documents and the contemporary records, and this has to be planned, harmonising the FIDIC provisions with the local regulations and the contractual technical standards (e.g. Acts and protocols for inspection of the works, non-conformity reports, compliance assessment of design by authorised third parties, etc.).

The following chapters discuss components of the PM plan in relation to the relevant Knowledge Areas.

6.3.6 Pre-commencement meeting

Preferably prior to the issuance of the Notice of the Commencement Date, it is recommended that the Parties and the CA convene a meeting to discuss and record matters including:

- introductions, project governance, and financing;
- appointment of the CA under Sub-Clause 3.1/ delegation of authority under Sub-Clause 3.4 (Silver Book: 3.1 and 3.2);
- appointment of the Contractor's Representative/ delegation of authority under Sub-Clause 4.3;
- status of the availability of the right of access to, and possession of, the Site, as agreed under Sub-Clause 2.1;
- status of the procedures for compliance with the applicable Laws under Sub-Clause 1.13 (Silver Book: 1.12) for the Site opening and commencement of construction;
- status of Drawings under the Red Book/ design authorisation (Sub-Clause 5.1: Yellow Book/Emerald/Silver Book) and arrangements for design compliance assessment under the applicable Laws;
- status of the Contractor's Sub-Clause 4.2 Performance Security and Sub-Clause 14.2.1 Advance Payment Guarantee;
- status of the Contractor's Sub-Clause 19.2 insurances;

- agreed Commencement Date under Sub-Clause 8.1;
- requirements for green procurement, environmental protection under Sub-Clause 4.18, health and safety under Sub-Clause 4.8/6.7, security of the Site under Sub-Clause 4.21;
- requirements for Quality Management and Compliance Verification Systems under Sub-Clause 4.9;
- status of the initial programme under Sub-Clause 8.3, programming software, key dates for information and submissions, delivery of Plant and Materials, periods for Review/third-party approval, sequence and timing of inspections/tests, coordination with utility operators/third parties and specific issues;
- requirements for Subcontractors named, nominated, and to be proposed by the Contractor under Sub-Clause 4.4 (Red Book: 5.1)/ nominated by the Employer under Sub-Clause 4.5 (Red Book: 5.2);
- Employer-Supplied Materials/ Employer's Equipment under Sub-Clause 2.6;
- procedures for requests for inspection under Sub-Clause 7.3 and the applicable Laws, measurement, Notices, instructions under Sub-Clause 3.5 (Silver Book: 3.4), submissions, Review of Contractor's Documents;
- agreement on the templates for documents and Site Diary, Risk Register;
- establishment of the Project Management Information System (**PMIS**)/ review of key roles and responsibilities for information management;
- procedures for interim valuations, certifications, and payments under Clause 14;
- procedures for monitoring the progress of the Works;
- planning of monthly meetings under Sub-Clause 3.8 (Silver Book: 3.6) and weekly risk management meetings.

6.3.7 Commencement of the Works

The Commencement Date marks the start of the Time for Completion, and therefore, determines the date by which the Contractor must complete the Works. Sub-Clause 8.1 includes the procedure for the commencement of the Works: the Engineer (Red/Yellow/ Emerald Books)/ Employer (Silver Book) shall give a Notice to the Contractor stating the Commencement Date, minimum 14 days in advance. These 2 weeks are needed for the Contractor to mobilise its resources and organise the Site offices/workshops in order to proceed after the Commencement Date with due expedition and without delay. This is a period of intensive PM planning.

Unless otherwise stated in the Particular Conditions, the Commencement Date shall be within 42 days after the Contractor receives the Red/Yellow/Emerald Book Letter of Acceptance or 42 days after the date on which the Contract comes into full force and effect under the Silver Book. A failure to give a Notice of the Commencement Date within the 42 days will constitute a breach of the Contract, which may entitle the Contractor to damages and if the delay is more than 84 days, to termination under Sub-Clause 16.2(e).

The FIDIC 2017 Contracts Guide advises that administration of the Contract may be facilitated by redefining the Commencement Date as a particular calendar date, stated in the Contract Agreement, which has priority over all of the other Contract documents under Sub-Clause 1.5.[10] Fixing the Commencement Date as a defined calendar date may be mandatory in some jurisdictions as an essential condition for a valid contract.[11]

Planning the Commencement, the CA shall:

- obtain the Employer's confirmation that it will be able to fulfil its obligations under Sub-Clause 2.1 related to the access to and possession of the Site;
- get acquainted with the statutory procedures and permits for commencement of the design and the construction;
- check the Employer's compliance with its obligations under Sub-Clause 1.13 (Silver Book: 1.12) related to the commencement (such as zoning plan/environmental procedures, appointment of authorised third parties for building control, etc.);
- check the Contractor's readiness to comply with its obligation under Sub-Clause 1.13 (Silver Book: 1.12) related to the commencement (such as fences, signs, agreements for dumping of construction debris, traffic diversion, etc.)
- prepare the Drawings to be submitted to the Contractor under the Red Book;
- consult with the Parties before issuing the Notice of the Commencement Date for their readiness and understanding of the local mandatory requirements;
- request under Sub-Clause 4.3 that the Contractor submits for consent the name and particulars of the Contractor's Representative, if not named in the Contract;
- request the Contractor to submit the Sub-Clause 4.2 Performance Security, the Sub-Clause 14.2.1 Advance Payment Guarantee, and the Sub-Clause 19.2 insurances;
- ensure that the Sub-Clause 8.1 Notice to Commence is given in writing following the requirements for communications under Sub-Clause 1.3 within the stipulated timeline.

6.3.8 Value Engineering (VE)

VE analyses the functions of systems, equipment, facilities, services, and supplies for the purpose of achieving essential functions at the lowest life-cycle cost while maintaining consistency, quality, reliability, and safety.[12] The ability to influence the project outcome drastically drops with the development of the design.[13] The early involvement of the supply chain is essential for VE. Also, the early involvement of the operation team may bring additional benefits. The planning of early application of VE may influence the project outcomes. Design optimisation may be planned after the elaboration of each design phase with input from the supply chain and operation team, if they are

10 FIDIC, *The FIDIC 2017 Contracts Guide* (FIDIC 2022) 8

11 Ninth Appellative Court Moscow ruled in *Statement No 09АП-11115/2010-ГК for Case No.A40-146012/09-8-1030* that the contract had not been concluded, since the notice to commence to be issued by the contractor within 14 days after the subcontractor receives the Letter of Acceptance is not an inevitable event (under Clause 190 of the *Russian Civil Code*, the time for completion must be either stated in calendar dates or linked to an inevitable event) <https://gkodeksrf.ru/ch-1/rzd-1/podrzd-5/gl-11/st-190-gk-rf> accessed 24 July 2022).

12 PMI Inc., *Construction Extension to the PMBOK® Guide* (PMI 2016) 34

13 See FIDIC, *FIDIC Definition of Services Guidelines. Infrastructure Construction / Civil Works* (FIDIC 2016) figure 3 <https://fidic.org/sites/default/files/_DOS_guide_civil_works_1.pdf> accessed 2 May 2022

appointed. The CA should carefully plan and manage the VE against KPIs, focusing on sustainability and limiting the scope creep, and in accordance with the relevant contract provisions on VE.

6.4 Directing and managing project work

The Contract documents detail the scope of work, performance criteria to be met after completion, and inspection and verification requirements. The CA's role is central in Project Integration Management. It is underpinned by the delegated authority under the following provisions:

- Sub-Clause 3.5 [Engineer's Instructions]/ Silver Book 3.4 [Instructions]: the Contractor shall comply with the instructions on any matter related to the Contract;
- Sub-Clause 3.7 (Silver Book: 3.5) [Agreement or Determination]: the Engineer/ Employer's Representative shall not be deemed to act for the Employer proceeding to agree or determine any matter or Claim. His/her determination is binding on the Parties unless and until it is revised by the dispute resolution forums.

6.4.1 Instructions

When directing the project work, the CA may issue to the Contractor at any time instructions that may be necessary for the execution of the Works, all in accordance with the Contract. It is important for the Contractor to know whether the person giving the instruction is authorised (with a Notice of Delegation). If a Site Manager is instructed by an assistant to demolish some works and execute them in a different manner, he/she has to ask the Contractor's Representative whether the assistant is authorised to issue such instructions. If an instruction is issued by an assistant without the relevant delegation, the cost of demolishing and reinstating the works will be at the Contractor's risk.

If an instruction states that it constitutes a Variation, the change management procedure pursuant to Sub-Clause 13.3.1 [Variation by Instruction] shall apply. If not so stated, and the Contractor considers that:

- the instruction constitutes a Variation; or
- does not comply with applicable Laws; or
- will reduce the safety of the Works; or
- is technically impossible,

the Contractor shall immediately, and before commencing the instructed works, give a Notice to the Engineer (Silver Book: Employer) with reasons. The CA has a week to respond giving a Notice confirming, reversing, or varying the instruction. If he/she does not respond within 7 days, this is deemed revocation.

Even in the case of disagreement with an instruction, the Contractor has to comply with it while following the Clause 20 Claims procedure to seek relief.

6.4.2 Agreement or determination

When directing and managing the project work, the CA has to seek timely resolution of issues and determination of entitlements, following the agreement/determination procedure. The detailed procedure has two stages:

(i) the CA consults with the Parties (jointly or separately), encouraging discussion in an endeavour to reach agreement:
   - if an agreement is reached within 42 days, the CA gives a Notice of the Parties' Agreement, attaching the agreement signed by the Parties;
   - if no agreement is achieved within 42 days, or before that if both Parties advise that it is not possible to be achieved, the CA gives a Notice and proceeds to determine the issue;

(ii) within 42 days, the CA makes a fair determination, taking into account all relevant circumstances, and gives a Notice, attaching the determination.

Other time limits may be proposed by the CA and agreed by both Parties.

An error may be notified by the CA or the Parties in the next 2 weeks; then, the CA has 1 week to correct the agreement or determination (Figure 6.2). An agreement is final and binding. A determination is binding on both Parties and the CA unless and until revised by the DAAB under Sub-Clause 21.4. The failure to comply may be referred directly to arbitration under Sub-Clause 21.6.

If a Party is dissatisfied with the Engineer's determination, it may give a Notice of Dissatisfaction (**NOD**) to the other Party with a copy to the CA within 28 days and after that refer the dispute to the DAAB within 42 days under Sub-Clause 21.4. If no NOD is given within 28 days, the determination becomes final and binding. If a NOD is given and the dispute is not referred to the DAAB within 42 days, the NOD will be no longer valid (Figure 6.3).

6.4.3 Work performance and inspections

The contract documents detail the inspection and verification requirements. The 2017 Editions add new definitions clarifying the scope of checking of a Contractor's submission:

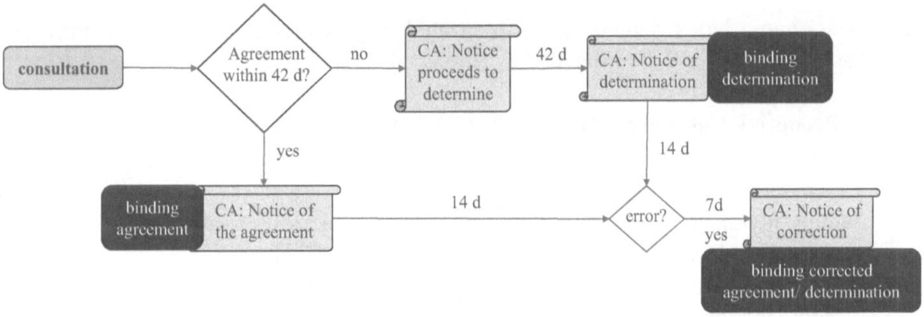

Figure 6.2 Procedure for agreement or determination under Sub-Clause 3.7 (Red/Yellow/ Emerald Books)/ 3.5 (Silver Book)

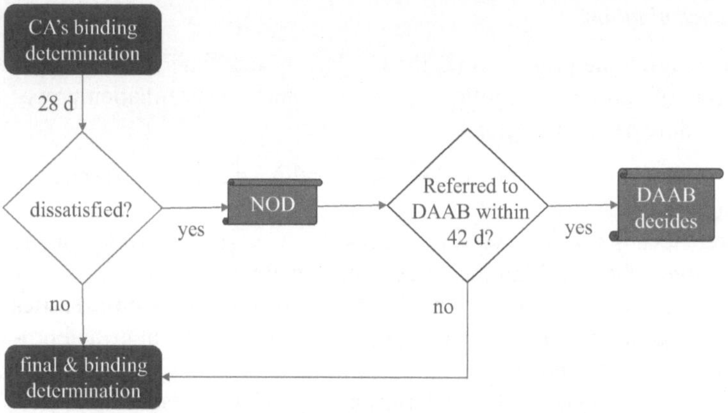

Figure 6.3 Procedure for dissatisfaction with the Engineer's/Employer's Representative's determination

- 'Review': examination and consideration by the CA to assess the compliance with the Contract and/or with the Contractor's obligations under or in connection with the Contract (such as minimum mandatory scope of the design under the local regulations);
- 'No-objection': the CA has no objections to the documents submitted by the Contractor being used for the Works.

Clause 7 stipulates the requirements for the execution of the works and the procedures for inspection and testing, including rejection of non-compliant materials and works. These procedures are the starting point: it is recommended that the CA agree with the Contractor detailed procedures for the requests for inspection and records of the results and keep inspection logs. The results of the inspections are the basis for payments and acceptance of the deliverables. The topic will be further discussed later in relation to quality and communication management.

6.4.4 Contract administration

The CA has many duties related to the integrated project management, which may be grouped as follows (referring to the Yellow Book provisions):

Group 1. Proactive Engineer's duties, where the initiative lies with the Engineer, related to:

- ambiguity/discrepancy in Contract documents under Sub-Clause 1.5;
- error/defect in Contractor's documents under Sub-Clause 1.8;
- Site data and items of reference under Sub-Clause 2.5;
- Engineer's instructions under Sub-Clause 3.5;
- agreement or determination under Sub-Clause 3.7;

- chairing and recording management meetings under Sub-Clause 3.8;
- requesting the Contractor to submit method statements under Sub-Clause 4.1;
- requesting reasonable evidence that the Contractor has paid the nominated Subcontractors under Sub-Clause 4.5;
- co-operation under Sub-Clause 4.6;
- Variation in the case of changes in the technical standards and regulations under Sub-Clause 5.4;
- instruction for additional samples under Sub-Clause 7.2;
- examinations, inspections, measurement, testing, checking the progress of manufacture, under Sub-Clause 7.3;
- rejection of Plant, Materials, design, or workmanship not in accordance with the Contract under Sub-Clause 7.5, and instruction for remedial work under Sub-Clause 7.6 and/or retesting under Sub-Clause 7.4;
- Notice stating the Commencement Date under Sub-Clause 8.1;
- request for update of the Programme under Sub-Clause 8.3;
- advance warning for a future event with probable adverse impact under Sub-Clause 8.4;
- rate of progress under Sub-Clause 8.7;
- suspension instructed by the Engineer under Sub-Clause 8.9;
- Contractor's delay in carrying out the Tests on Completion or the Employer's Personnel's delay in proceeding under Sub-Clause 9.2;
- retesting under Sub-Clause 9.3;
- rejection of the Works or Section under Sub-Clause 9.4/11.4;
- instructing the Contractor to search for the cause of a Defect during the DNP under Sub-Clause 11.8;
- Tests after Completion under Sub-Clause 12.1;
- issuance of the Performance Certificate under Sub-Clause 11.9;
- Variations instructed under Sub-Clause 13.3;
- instructions to use Provisional Sums under Sub-Clause 13.4;
- Variation instructed on a daywork basis under Sub-Clause 13.5;
- Variation instructed for changes in Laws under Sub-Clause 13.6;
- agreement/determination of revised instalments in the Schedule of Payments under Sub-Clause 14.4;
- Notice to Correct under Sub-Clause 15.1;
- valuation after termination under Sub-Clause 15.3;
- inspection of contemporary records under Sub-Clause 20.2.3;

Group 2. Reactive Engineer's duties, where the Engineer responds to requests by the Parties related to:

- clarifications under Sub-Clause 1.5 in the case of ambiguity/discrepancy in Contract documents;
- error/defect in a document under Sub-Clause 1.8;
- error in the Employer's Requirements under Sub-Clause 1.9 or items of reference under Sub-Clause 4.7;

- delayed access to/possession of the Site under Sub-Clause 2.1;
- appointment/replacement of the Contractor's Representative under Sub-Clause 4.3/ Key Personnel under Sub-Clause 6.12 not named in the Tender;
- new Subcontractors under Sub-Clause 4.4;
- co-operation under Sub-Clause 4.6;
- Unforeseeable physical conditions under Sub-Clause 4.12;
- changes to the access routes under Sub-Clause 4.15;
- delivery/removal of Contractor's Equipment under Sub-Clause 4.17;
- the Contractor's measurement apparatus under Sub-Clause 4.19;
- archaeological/geological findings under Sub-Clause 4.23;
- consent to designers under Sub-Clause 5.1;
- Review of the Contractor's Documents under Sub-Clause 5.2.2, including Quality Management (**QM**) System (Sub-Clause 4.9), programme (Sub-Clause 8.3), as-built records (Sub-Clause 5.6), Operation and Maintenance (**O&M**) Manuals (Sub-Clause 5.7), and test reports (Sub-Clause 9.1);
- consent for the Contractor to work outside the normal working hours pursuant to Sub-Clause 6.5;
- tests under Sub-Clause 7.4;
- inspection of Materials, Plant, or work under Sub-Clause 7.3;
- proposal for remedial works under Sub-Clause 7.5;
- advance warning under Sub-Clause 8.4;
- EOT under Sub-Clauses 8.5/8.6;
- payment of Delay Damages under Sub-Clause 8.8;
- Contractor's entitlement under Sub-Clauses 8.10–8.13 as a result of Employer's suspension;
- taking over under Sub-Clauses 10.1/10.2/10.3;
- Employer's interference with the Tests on Completion under Sub-Clause 10.3;
- defects/damages during the DNP under Sub-Clauses 11.2;
- extension of the DNP under Sub-Clause 11.3;
- testing after remedying defects under Sub-Clause 11.6;
- delayed access during the DNP under Sub-Clause 11.7;
- entitlement related to the search of defects/damages during the DNP under Sub-Clause 11.8;
- Tests after Completion under Sub-Clauses 12.2/12.3/12.4;
- VE proposal under Sub-Clause 13.2;
- quotations from suppliers/subcontractors under Sub-Clauses 13.4/13.5;
- changes in Laws under Sub-Clause 13.6;
- Plant and Materials under Sub-Clause 14.5;
- certification of payments under Sub-Clauses 14.6/14.13;
- receipt of any Goods and Contractor's Documents after termination under Sub-Clause 15.2.3;
- valuation after termination under Sub-Clause 15.3/15.6/18.5;
- suspended work/ reduced rate of work under Sub-Clause 16.1;
- obligations after termination under Sub-Clauses 15.2/15.5/16.2/18.5 and release from performance under Sub-Clause 18.6;
- rectification of losses and damages under Sub-Clause 17.2;

- Exceptional Events under Sub-Clause 18.4;
- Claims under Clause 20;
- clarifications and witness statements for dispute resolution under Clause 21.

Group 3. Passive Engineer's duties to observe compliance with contract provisions, rules, and regulations related to the:

- communications pursuant to Sub-Clause 1.3;
- law and language under Sub-Clause 1.4;
- assignment under Sub-Clause 1.7;
- care and supply of Contractor's Documents under Sub-Clause 1.8;
- Employer's use of the Contractor's Documents pursuant to Sub-Clause 1.10;
- Contractor's use of the Employer's documents pursuant to Sub-Clause 1.11;
- Parties' Sub-Clause 1.12 confidentiality obligations;
- Parties' Sub-Clause 1.13 compliance with Laws;
- limitation of liability under Sub-Clause 1.15;
- access to/possession of the Site under Sub-Clause 2.1;
- assistance under Sub-Clause 2.2;
- behaviour of its own and other Employer's Personnel under Sub-Clause 2.3;
- Employer's financial arrangements under Sub-Clause 2.4;
- Employer-supplied Materials/ Employer's Equipment under Sub-Clause 2.6;
- validity of the Sub-Clause 4.2 Performance Security;
- delegation of authority by the Contractor's Representative under Sub-Clause 4.3;
- payments to nominated Subcontractors under Sub-Clause 4.5;
- co-operation under Sub-Clause 4.6;
- Contractor's health and safety obligations under Sub-Clauses 4.8/6.7;
- QM under Sub-Clause 4.9;
- Contractor's obligations related to right of way (Sub-Clause 4.13)/ avoidance of interference (Sub-Clause 4.14)/ access routes (Sub-Clause 4.15)/ transport of Goods (Sub-Clause 4.16)/ Contractor's Equipment (Sub-Clause 4.17)/ security of the Site (Sub-Clause 4.21)/ operations on Site (Sub-Clause 4.22)/ compliance with standards and regulations (Sub-Clause 5.4)/ training (Sub-Clause 5.5)/ as-built records (Sub-Clause 5.6)/ O&M Manuals (Sub-Clause 5.7)/ manner of execution (Sub-Clause 7.1)/ ownership of Plant and Materials (Sub-Clause 7.7)/ payments for natural Materials and disposal of material (Sub-Clause 7.8)/ clearance of Site after the Performance Certificate (Sub-Clause 11.11)/ staff and labour (Clause 6);
- Parties' environmental obligations under Sub-Clause 4.18;
- temporary utilities under Sub-Clause 4.19;
- Contractor's progress reports under Sub-Clause 4.20;
- Employer's Sub-Clause 14.7 payment obligations;
- indemnities under Sub-Clauses 17.3/17.4/17.5/17.6;
- insurance under Clause 19.

Besides these duties under the Contract, the CA manages or monitors interfaces with various stakeholders like Employer's designers, authorised third parties, other contractors, project beneficiaries, local authorities, utility operators, etc. Active problem solving, timely instructions, and risk mitigation are essential for the success of the project.

6.5 Managing project knowledge

Project data are usually recorded in a Project Management Information System (**PMIS**).[14] The PMIS and BIM enhance Project Integration Management. IT tools may help the CA's team to collect data, solve issues, and distribute updated information. Automated reporting on KPI can be part of PMIS. The PMIS should contain as a minimum the following information:

- Contract documents – as listed under Sub-Clause 1.5 with all amendments;
- Correspondence – Notices and other communications under Sub-Clause 1.3;
- Contractor's Documents – the documents prepared by the Contractor with their status – e.g. under Review, with No-objection, approved by building control;
- Engineer's instructions under Sub-Clause 3.5 (Employer's Silver Book: 3.4), including Variations under Sub-Clause 13.3;
- Minutes of meetings – for management meetings under Sub-Clause 3.8, weekly meetings, etc.;
- Progress reports under Sub-Clause 4.20;
- Issue Log;
- Risk Register, Risk Management Plan (included in the Contract documents under Sub-Clause 1.5 of the Emerald Book);
- Quality documents – results of inspections and testing, non-conformity reports, proposals/approval/compliance documents for Plant and Materials, etc. (pursuant to Sub-Clause 4.9, related to Clause 7);
- Records – Contractor's records for resources used under Sub-Clause 6.10, Site Diaries, additional contemporary records in case of Claims under Sub-Clause 20.2.3;
- Payments – measurement sheets (under the Red Book Sub-Clause 12.1), applications for payment under Sub-Clause 14.3, IPCs under Sub-Clause 14.6, invoices, etc.

The Contractor's records under Sub-Clause 6.10 must contain detailed information for the resource usage for each work activity in the Programme, at each work location, and for each day of work. They are appended to the monthly reports, unless otherwise agreed. This is a very burdensome requirement, which will result in a huge volume of information, difficult to check 1 month later. Imagine the volume of the records presenting daily information for each of the classes of Contractor's Personnel, units of Contractor's Equipment/Plant and Materials on a railway project where the Programme has 5,000 activities and the works are being executed on 5 different sections in parallel. It is important to have data for Variations and Claims, but in order to prevent disputes, these data should be agreed within a week. The Internet of Things (IoT), weekly drone footage, daily video recordings, and BIM combined with special data processing software can

14 PMI Inc., *PMBOK Guide* (PMI sixth edition 2017) 26

automate the Contractor's records and warrant the truth of the data. It is recommended that the Specification/Employer's Requirement impose on the Contractor an obligation to register daily in Site Diaries the weather, Unforeseeable events, and the data for the records under Sub-Clause 6.10. If done on paper, the Site Diaries may be filled in by the Contractor's Site Managers at the locations where Site offices are available and countersigned by the Resident Engineers, then scanned and uploaded in the PMIS. If done digitally, the Contractor may submit them to the Engineer by email on the next day (or daily upload them in the PMIS). If the Engineer does not respond within a week, the Engineer should be deemed to have given No-objection to the records. Microsoft® Excel® spreadsheets may be used to fill in the data and then summarise the information for the records.

The CA may allocate the responsibilities for solving problems, planning the Parties' actions, or Engineer's instructions in an **Issue Log.** Issues are events that have happened and need an action plan, while risks have a probability that they will happen and need a mitigation plan. When a risk materialises, the CA should move it in the Issue Log from a risk to an issue. Issues recorded in the Issue Log should be discussed almost every day; risks recorded in the Risk Register should be reviewed at weekly Risk Meetings and at the monthly management meetings. The tracking of risks that have materialised as issues provides important knowledge for risk identification and should be analysed in the lessons learned for future projects.[15]

The CA may keep a confidential **Events Log,** chronologically recording significant events. Such an Events Log will help in writing progress reports and reviewing Claims, but it is more important for the CA to focus on the collaborative risk management and systematic Review/uploading of the Contractor's records in the PMIS, seeking transparency and dispute prevention.

BIM is the best integrator, but it needs collaboration. Section 5.12.4 discusses the information management in FIDIC Contracts with ISO 19650. ISO 19650-2 Clauses 5.4 and 5.5 require during mobilisation the appointing party (e.g. the Employer), the lead appointed party (e.g. the Contractor), and the appointed parties (e.g. the Subcontractors) to collaborate and agree key roles and responsibilities. The information delivery plan (in the form of BIM execution plan, MIDP, and TIDP) outlines coordination and delivery mechanisms.

6.6 Monitoring and control of project work

The performance data is gathered in the PMIS. To become work performance information, the data are compared with the PM plan components: baselines and contractual KPIs.

The passive duties of the CA listed previously secure constant monitoring of the Parties' compliance with the contractual and regulatory requirements.

The Contractor's monthly progress reports under Sub-Clause 4.20 contain the following detailed work performance information:

(a) charts, diagrams, and detailed descriptions of progress, including each stage of design, Contractor's Documents, procurement, manufacture, delivery to Site, construction, erection, testing, commissioning, and trial operation;

15 Duraideivamani Sankararajan and N. K. Shrivastava, Risks vs. Issues (*PM Network* 2012) 26 (6), 28–29 <https://www.pmi.org/learning/library/risks-vs-issues-project-failure-2328> accessed 23 July 2022

(b) photographs and/or video recordings showing the status of manufacture and of progress on and off the Site;
(c) for the manufacture of each main item of Plant and Materials, the name of the manufacturer, manufacture location, percentage progress, and the actual/expected date of commencement of manufacture, Contractor's inspections, tests, shipment, and arrival at the Site;
(d) the Contractor's records for the use of resources under Sub-Clause 6.10;
(e) copies of QM documents, inspection reports, test results, and compliance verification documentation (including certificates of Materials);
(f) a list of Variations, and any Notices of Claim given by either Party under Sub-Clause 20.2.1;
(g) health and safety statistics, including details of any hazardous incidents and activities relating to environmental aspects and public relations;
(h) comparisons of actual and planned progress, with details of any events or circumstances that may adversely affect the completion of the Works in accordance with the Programme and the Time for Completion, and the measures being (or to be) adopted to overcome delays.

The contents of the progress reports may be tailored in the Specification/Employer's Requirements depending on the needs of the project. Lean reporting is recommended, referring to the data regularly uploaded in the PMIS. Instead of presenting tons of raw data, the report should focus on the analysis of the variance in relation to the baselines/KPIs and measures adopted to solve pending issues.

Sub-paragraph (h) of Sub-Clause 4.20 requires the Contractor to include 'details of any events or circumstances which may adversely affect the completion of the Works'. The FIDIC 2017 Contracts Guide explains that an '*event*' is generally understood to be a happening or an occurrence or an incident, whereas a '*circumstance*' is generally understood to be a situation or a set or conditions or a state-of-affairs that relate to, or are associated with, an event or events.[16] This information may be presented by attaching the updated Issue Log and Risk Register.

The CA also reports on a monthly basis. Pursuant to Sub-Clause 4.20, the Contractor has to submit its progress report in the first week of the next month. The CA will need several days after that to check it and prepare its own report. It is recommended that the agreement for services should require the CA to submit a progress report to the Employer by the end of the second week of the next month. The monthly management meeting may be planned in the third week to enable the Employer to review both reports. It is important that the CA gives a Notice to the Contractor of any deficiencies/inconsistency/wrong data found in the Contractor's report before the meeting, and also comments and includes the findings in the minutes of the meeting. The reports and the minutes of management meetings are important contemporary records for the analysis of criticality, causation, and concurrency during the multi-tier dispute resolution process.

The comparison of the performance against the baselines and contractual KPIs will be discussed further later in relation to the other PM Knowledge Areas. The monthly tracking of the schedule and earned value are included in the progress reports.

16 FIDIC, *The FIDIC 2017 Contracts Guide* (FIDIC 2022) 36

6.7 Integrated change control

The key benefit of this process is that it allows documented changes within the project to be considered in an integrated manner while addressing overall project risk, which often arises from changes made without consideration of the overall project objectives or plans.[17] In FIDIC contracts, integrated change control is regulated by the provisions for Variations.

The Variation procedure is further discussed in detail in Chapter 7 in relation to scope management.

6.8 Close project or phase

The key benefits of the process of finalising all activities for the project, phase, or contract are that the information is archived, the planned work is completed, and the resources are released to pursue new endeavours.[18]

The early closing of contracts by termination was discussed in Chapter 5. The closing of the construction phase with the commissioning and taking over and the contract closure after the expiry of all DNPs is discussed here.

6.8.1 Taking over of the Works

Sub-Clause 10.1 sets out the taking-over procedure to be followed when the Works or a Section have been substantially completed by the Contractor in accordance with the Contract. Figure 6.4 presents the sequence of actions:

- the Contractor applies for a Taking-Over Certificate (**TOC**) by giving a Notice to the Engineer (Silver Book: Employer) not more than 14 days before the Works/ Section will, in the Contractor's opinion, be complete and ready for taking over.

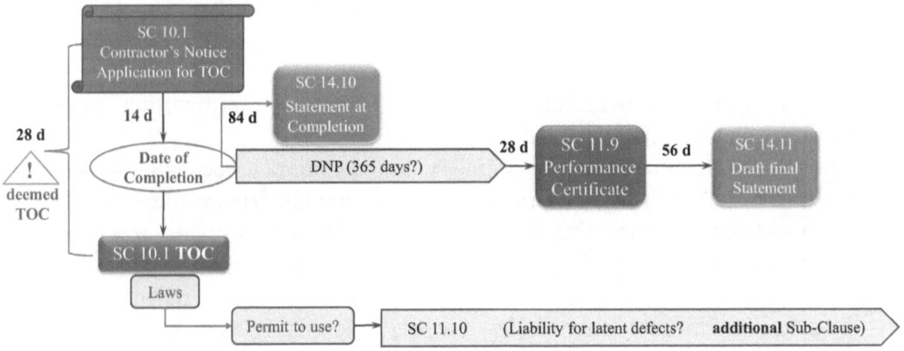

Figure 6.4 Taking over and closing of the FIDIC Contract

17 PMI Inc., *PMBOK Guide* (PMI sixth edition 2017) 113
18 PMI Inc., *PMBOK Guide* (PMI sixth edition 2017) 121

- the Engineer (Silver Book: Employer) issues the TOC or rejects the application within 28 days after receiving the Contractor's Notice.
- if the Engineer (Silver Book: Employer) has not responded within 28 days, and if the conditions for a TOC have been fulfilled, the Works/Section shall be deemed to have been completed on the 14th day after the receipt of the application, and the TOC shall be deemed to have been issued.

The Contractor must demonstrate that the following conditions for a TOC have been fulfilled:

- the Works have been completed in accordance with the Contract, except for any minor outstanding work and defects (as listed in the TOC) that will not substantially affect the **safe** use of the Works or Section for their intended purpose;
- the Works have successfully passed the Tests on Completion;
- the Engineer/Silver Book: Employer has given (or is deemed to have given) a Notice of No-objection to the as-built records and final O&M Manuals for the Works;
- the Contractor has carried out the training (if agreed in the Contract).

The CA checks the as-built documentation, inspects the Works, and prepares a punch list fixing timelines for completion of outstanding work and remedy of defects during the DNP.

During the 4 weeks after the Contractor's application for taking over, the key experts of the Contractor and the Employer have to work in collaboration to agree what has to be done to meet the Contract requirements. Joint inspections, attendance at the Tests on Completion, checking of documentation, and meetings to discuss and update the punch list have to be well organised and documented. Digitalisation, especially BIM, facilitates the process for the as-built documentation and the punch list update.

Usually, the Contractor is optimistic considering when the Works/Section will be completed and ready for taking over. The Contractor is threatened by Delay Damages under Sub-Clause 8.8 and often argues that the conditions are met when they are not, for example:

- The O&M Manuals under Sub-Clause 5.7 are not ready for important Plant, and the Tests on Completion are delayed.
- The as-built documents do not comply with the Contract. Focusing on the completion of the Works, the Contractor often does not allocate resources to prepare them and either declares that this is not the time for office work or presents existing documents not reflecting the as-built changes. The preparation of the as-built documents is an important activity, needing templates agreed during the mobilisation period, timely submission of as-built records for Review, and updating the PMIS. Even with regular preparation and Review, an extra effort is needed during the Tests on Completion, since some adjustments necessitate changes in the as-built records and O&M Manuals.
- According to the Contractor, the Tests on Completion are successful, but the CA gives a Notice stating that some of the results do not comply with the Contract.

However, even if the Works did not pass the Tests on Completion, the Employer may prefer a TOC, pursuant to sub-paragraph (d) of Sub-Clause 9.4.

- Sometimes, it is obvious that the Works are not ready for taking over due to outstanding work that cannot be completed and tested within 2 weeks and/or taking into account the duration of the test programme agreed under Sub-Clause 9.1. In this case, the CA should reject the application for taking over within a few days, giving reasons.
- There might be only minor outstanding work and defects, but this may substantially affect the safe use of the Works or Section for their intended purpose (e.g. the safety signage on a highway is not completed).
- The training under Sub-Clause 5.5 has not been carried out by experienced trainers in accordance with the Employer's Requirements. The attendance of the Employer's Personnel at the Tests on Completion cannot cover the dedicated theoretical training, if required under the Contract.

It is very important that the TOC record the Date of Completion when the conditions for a TOC have been fulfilled. Normally, the TOC will be issued after the Date of Completion, since at least 1 week is necessary for joint inspections, trial operation, and final checking of the updated as-built documentation, agreement on the punch list, and preparation of the TOC (see Figure 6.4).

Commentators explain that the distinction between the Date of Completion and the date of issue of the TOC is clear: for operational matters, such as the clearing of the Site, the issue of the TOC is the trigger, whereas in relation to calculating time periods, such as the DNP or the amount of Delay Damages, the actual Date of Completion is used.[19]

The Date of Completion is an important milestone:

- the starting point of the DNP for the Works or Section;
- the end of the calculation of Delay Damages under Sub-Clause 8.8;
- the end of the Contractor's obligation to submit progress reports under Sub-Clause 4.20 (except for the completion of any outstanding work listed in the TOC);
- within 84 days, the Contractor shall submit a Statement at Completion under Sub-Clause 14.10 showing the value of the work done, any further sums the Contractor considers due, and an estimate of the amounts to become due.

On the issue of the TOC:

- the Contractor shall clear and leave the Site under Sub-Clause 4.22;
- the liability for care of the Works is transferred to the Employer under Sub-Clause 17.2 (see Figure 13.5);
- the Engineer's (Silver Book: Employer's) right to instruct remedial work under Sub-Clause 7.6 and Variations under Clause 13.1 ends;

19 Jacob Sorensen, *FIDIC Silver Book. A companion to the 2017 EPC/Turnkey Contract* (ICE Publishing 2019) 109

- any outstanding advance payments under Sub-Clause 14.2 must be repaid;
- the Contractor's obligation under Sub-Clause 14.4 to submit updated estimates every 3 months when the Contract does not include a Schedule of Payments ends;
- the Contractor shall include the first half of the Retention Money in a Sub-Clause 14.3 Statement;
- the Contractor's obligation to maintain insurance under Sub-Clause 19.2 ceases, except for outstanding works/loss or damage caused by the Contractor.

6.8.2 DNP

The DNP serves to demonstrate how the Works perform during operation, for remedy of defects and completion of minor outstanding works after the TOC. If not stated otherwise in the Contract Data, the DNP is 1 year. During this year, the project is operating in all seasons. Tests after Completion, if any, will prove how the facility is meeting the performance criteria. For energy projects, the DNP is often a year and a half or 2 years in order to measure the availability of the facility and how the Works meet the performance criteria for a longer period. Defects that appear are considered patent, and the Contractor shall remedy them, following the procedures under Clause 11. The DNP may be extended under Sub-Clause 11.3 with the period when the Works/Section/major Plant cannot be used for the intended purposes for defects/damages attributable to the Contractor, subject to Sub-Clause 20.2. The DNP shall not be extended by more than 2 years.

It is important that the facility operator is informed about the Employer's remedies in the case of defects and about the requirement for the Contractor to be notified of a defect under Sub-Clause 11.1 and to be given access pursuant to Sub-Clause 11.7 to the Works and the records of the maintenance, performance, and operation of the Works. The Employer may lose its entitlements where the operator interferes with the Works to remedy the defects without informing the Contractor.

6.8.3 Performance Certificate

The Performance Certificate constitutes final acceptance by the Employer of the Works (see Figure 6.4). Under Sub-Clause 11.9, the Engineer (Silver Book: Employer) has an obligation to issue it within 28 days after the later of:

- the expiry dates of the DNP, or
- the date when the Contractor has supplied all the Contractor's Documents, and completed and tested all the Works (including remedying any defects) in accordance with the Contract.

If the CA fails to issue the Performance Certificate within this period of 28 days, the Performance Certificate shall be deemed to have been issued on the date 28 days after the date on which these conditions are met. The project may have several Sections with different DNPs, some of which may have been extended, so all of them have to expire before the issuance of the Performance Certificate. The CA usually organises a final inspection 1 month before the expiry of the DNPs with representatives of the operator/facility manager. If any defects are found, the Contractor has to remedy them, and the CA has to inspect the Works and issue the Performance Certificate after the remedy of those defects.

On the issue of the Performance Certificate:

- the Contractor's obligation under Sub-Clause 4.2.1 to keep the Performance Security valid ends, and the Employer becomes obliged under Sub-Clause 4.2.3, subject to the Contractor's compliance with Sub-Clause 11.11, to return the Performance Security to the Contractor;
- the Contractor's undertaking under Sub-Clause 5.1 of the Yellow/Silver Books that his/her designers and design Subcontractor will be available for discussions ends;
- the Contractor's obligation under Sub-Clause 5.6 (b) to update the as-built records for the Works ends;
- the Contractor's obligation under Sub-Clause 6.8 to provide superintendence ends;
- the Contractor has to comply with Sub-Clause 11.11 to remove any remaining Contractor's Equipment, surplus material, wreckage, rubbish, and Temporary Works from the Site; reinstate all parts of the Site that were affected by the Contractor's activities and are not occupied by the Permanent Works; and leave the Site and the Works in the condition stated in the Contract (if not stated, in a clean and safe condition).

6.8.4 After the Performance Certificate

The project cannot be closed when the Performance Certificate is issued, since the CA has to manage the following activities:

- checking of the Contractor's Final Statement (see Figure 6.4) and issuance of the Final Payment Certificate or IPC in case of disagreement under Sub-Clause 14.3 (payment under the Silver Book);
- checking of the final clearing of the Site;
- agreement/determination of pending Claims;
- giving clarifications/ attending hearings for dispute resolution, if the DAAB or arbitrators request this;
- preparation of final report under the agreement for services, including lessons learned in accordance with the scope of services;
- permit management, to the extent included in the scope of services.

In some projects, the agreement for services expires during the DNP, since the Time for Completion has been extended without extending the service agreement. In those cases, where the Employer considers that its employees may replace the Engineer, the Employer should give a Notice to the Contractor at least 42 days in advance with the name, address, and relevant experience of the intended replacement Engineer. The Employer has to nominate a professional engineer with suitable qualifications, experience, and competence and to remember that the Contractor may object to the replacement within 14 days under Sub-Clause

3.6 if those requirements are not met. The Silver Book Sub-Clause 3.1 obliges the Employer to give a 14-day Notice in case of replacement of the Employer's Representative and entitles the Contractor to give a reasonable objection without specifying a time limit.

The Contractor's obligations and liabilities in respect of defects depend not only on the FIDIC provisions but also on the applicable Laws. Figure 6.4 presents an example for mandatory process for achieving permit to use, which is the starting date of strict decennial liability in some civil law jurisdictions. The Parties may agree to opt out of the defects liability under the applicable Laws to the extent that they are not mandatory.[20] In the case of strict decennial liability, a warranty period may be added in the definitions of the Special Provisions, stating the commencement and expiry of the mandatory liability period for latent defects.[21] Some employers misuse the FIDIC contracts, making the DNP equal to the decennial liability in the Particular Conditions; this incurs additional costs for extended Performance Security under Sub-Clause 4.2 and Retention Guarantee for the second half of the Retention Money.

6.9 Innovative Project Integration Management

6.9.1 KPIs for integration management

Contractual KPIs are used to measure progress and performance. It is important that KPIs are relevant and proportionate to the size and complexity of the project. With the right KPIs in place, it should follow that contracts are designed to incentivise delivery of the things that matter, minimise perverse or unintended incentives, and promote good relationships. KPIs should drive both a focus on outcomes and continuous improvement.[22]

KPIs and metrics can be displayed in dashboards, scorecards, and reports. Typically, between six and ten KPIs are standard. The most important attribute of a KPI may be that it is actionable.[23] The following KPIs may be used to present the achievements during the implementation of a FIDIC works contract:

1. Cost performance index
2. Schedule performance index
3. Number of accidents
4. Number of Variations
5. Ratio Claim Notices/Claims
6. Greenhouse gas emissions
7. Number of Disputes

---

20 Clause 665 (1) of the *Kazakhstan Civil Code* defines a 10-year warranty period beginning after the acceptance of the works by the employer, which may be changed by legislative act or contract. Additionally, Clause 69 (1) of the *Act on Architectural, Building and Construction Activities* establishes a minimum 2-year warranty period for strict liability after taking over. <https://kodeksy-kz.com/ka/grazhdanskij_kodeks_osobennaya_chast.htm> accessed 25 July 2022 and <https://kodeksy-kz.com/ka/ob_arhitekturnoj_deyatelnosti/69.htm> accessed 7 May 2023

21 *Ordinance 2 from 31 July 2003 for the commissioning of construction works in the Republic of Bulgaria and minimum warranty periods* issued by the Minister of Regional Development and Public Works establishes minimum warranty periods depending on the type of works: 10 years for structural works, 5 years for finishing works, etc. <https://www.lex.bg/laws/ldoc/2135470557> accessed 26 July 2022

22 UK Cabinet Office, *Construction Playbook* (2020) 40 <https://www.gov.uk/government/publications/the-construction-playbook> accessed 20 April 2022

23 Harold Kerzner, *Project Management Metrics, KPIs and Dashboards* (Wiley 2017) 129–130

6.9.2 Tools supporting innovative integration management

Early supply-chain involvement, collaboration, and digitalisation are the most important tools for successful integration management.

The multi-party contract FAC-1 may be used for early Contractor involvement with a FIDIC Red Book, enabling one or two selected Contractors to be paid a fee for improved value solutions during the design process. The optimisation continues after the Contractors select key Subcontractors/suppliers. After the design is agreed, the Contractors price the BQs. The Red Book Contract may commence as a lump-sum contract with the preferred Contractor if the ACA is within the budget.

Contract packages assigned under FIDIC contracts in a complex megaproject may be integrated under the umbrella of FAC-1 for more efficient project integrated management and early supply-chain involvement for innovations.

An advanced BIM allows easy access to updated information and serves as a PM integrator.

The iContract solution would alleviate the CA's onerous tasks currently handled manually, allowing greater speed and accuracy[24] and diverting the effort to Project Integration Management.

6.10 Example Notices

Appendix 1 continues the scenario with examples of Engineer's Notices of the Commencement Date under Sub-Clause 8.1 and delegation of authority under Sub-Clause 3.4.

24 Alan McNamara, 'Automating the Chaos: Intelligent Construction Contracts' Chapter in *Smart Cities and Construction Technologies* (IntechOpen 2020) 10

CHAPTER 7

Project Scope Management with FIDIC Contracts

7.1 Project Scope Management processes

According to the PMBOK® Guide, Project Scope Management includes the processes required to ensure that the project includes all the work required, and only the work required, to complete the project successfully. Managing the project scope is primarily concerned with defining and controlling what is and is not included in the project.[1]

The planning of the scope management begins with the project initiation. The initial studies, followed by feasibility studies discussed in Chapter 5, are the first steps for determining, documenting, and managing stakeholder needs and requirements to meet project objectives. Following the feasibility report recommendations, the procurement strategy is tailored to define the scope. Taking into account different considerations, the project is subdivided into smaller, more manageable components. The work breakdown structure (WBS) is specified in the tender documents. The contracts define the scope of obligations and the scope baseline for the Works, integrating the requirements and the WBS.

1 PMI Inc., *PMBOK® Guide* (PMI sixth edition 2017) 129

DOI: 10.4324/9781003272892-7

Dr. Charrett formulates the following principles of *lex constructionis*[2] on scope, acknowledging the complexity of construction projects leading to scope changes:[3]

1. The contract defines, in writing, the original scope of Works and the known conditions at the Site.
2. The Employer has the right to instruct Variations consistent with the original scope of the Contract, and the Contractor has the obligation to carry out all such Variations.
3. Construction Works must be carried out so as to protect the health and safety of workers.
4. Construction Works must not adversely affect the environment or the interests of third parties.
5. Subject to any contractual requirements, the Contractor selects the methods and timing of the Works.

The FIDIC contracts contain scope validation provisions and change management clauses.

7.2 Defining the scope in FIDIC agreements for services

Defining the scope of professional services can happen in one of two ways:[4]

- a client may require a solution for a problem, the nature of which is unknown or is not defined at the time a consultant is engaged; in this situation, a client and consultant need to define the scope and likely deliverables together before fees are agreed;
- the client is able to define quite precisely the scope, services, and deliverables required from a consultant.

In the first scenario, the consultant submits rates for its experts, and the remuneration is done on a monthly basis following the approval of timesheets for the executed services. In the second scenario, the price for the services may be a lump sum.

Under Sub-Clause 1.1.23 of the White Book, 'Services' means the services defined in Appendix 1 (Scope of Services) to be performed by the Consultant in accordance with the Agreement, which includes any Variations to the Services instructed or arising in accordance with the Agreement. The White Book guidance advises the description in Appendix 1 to be as comprehensive as reasonably practicable and, where beneficial to the understanding of the scope, to define matters to be excluded from the scope. Sub-Clause 3.1 stipulates that where appropriate, the Client shall describe the function and purpose of the Services and state the same explicitly in Appendix 1. In the performance of the Services, the Consultant shall have no other responsibility than to exercise the reasonable skill, care, and diligence to be expected from a consultant experienced in the provision

2 Charles Molineaux, Moving Toward a Lex Mercatoria – A Lex Constructionis (*Int'l J. Arb.* 1997) 14 (1), at 62: 'The *lex constructionis* can perhaps be seen as a force for a valuable consistency in interpretation'

3 Donald Charrett, *FIDIC Contracts in Europe: A Practical Guide to Application* (Informa Law from Routledge, 2023) 13

4 FIDIC, *FIDIC Definition of Services Guidelines. Infrastructure Construction / Civil Works* (2016) <https://fidic.org/sites/default/files/_DOS_guide_civil_works_1.pdf> accessed 2 May 2022

of such services for projects of similar size, nature, and complexity. To the extent achievable using such standard of care, and without extending the obligation of the Consultant, the Consultant shall perform the Services with a view to satisfying any function and purpose that may be described in Appendix 1. Where explicitly described in Appendix 1, the Consultant shall perform the role of the Engineer, Employer's Representative, project manager, or similar as laid down in the Works Contract. The Consultant shall provide construction administration services in accordance with the scope of Services.

The FIDIC Definition of Services Guidelines Infrastructure Construction / Civil Works[5] provides a clear template for the scope of a consultant's work through the different phases of a typical commission: design – procurement – construction. The Guidelines point the user to FIDIC support documents available to assist in evaluating what level of environmental sustainability should be pursued and which design tools, such as Building Information Modelling (BIM), might be applicable. Key requirements for defining the scope of a consultant's work are summarised in Appendix A – The Consultant's Brief, which provides a short guide for clients and their consultants applicable to all types of projects.

7.3 Defining the scope in FIDIC works contracts

A clear definition of the scope of work at the outset – the work that has to be done within the scope of the contract – is essential to certainty of price and schedule.[6]

Signing the Contract Agreement, in consideration of the payments to be made by the Employer, the Contractor takes the responsibility to design to the extent defined by the Contract, execute and complete the Works, and remedy any defects therein, in conformity with the provisions of the Contract.

The General Conditions (GCs) as amended by the Particular Conditions define the scope of the Contractor's obligations. While under the Red Book the Contractor executes and completes the Works in accordance with the Contract and the Engineer's instructions, under the Yellow/Emerald/Gold/Silver Books the Contractor designs and builds the Works and is responsible for their fitness for the intended purposes at completion.[7] The scope of the Gold Book includes the Contractor's further responsibility that the Works remain fit for purpose for 20 years' operation. In the Silver Book, the Contractor accepts full responsibility even for the errors in the Employer's Requirements. The Green Book is drafted for Employer's design, but can easily be adapted to allocate the design risk to the Contractor, leading to a fitness for purpose obligation.[8] The scope of the Red Book

5 FIDIC, *FIDIC Definition of Services Guidelines. Infrastructure Construction / Civil Works* (2016) 6 <https://fidic.org/sites/default/files/_DOS_guide_civil_works_1.pdf> accessed 2 May 2022

6 Philip Loots and Donald Charrett, Reconsidering Procurement Criteria and Procedures for Major Infrastructure Projects; Why the Use of 'Price' as the Determining Factor in Construction Procurement Decisions is Wrong (*ICLR* 2017) 1, 71

7 Sub-Clause 4.1 of the Yellow/Silver/Emerald Books stipulates that when completed, the Works shall be fit for the purpose(s) for which they are intended, as defined and described in the Employer's Requirements (or, where no purpose(s) are so defined and described, fit for their ordinary purpose(s)).

8 Sub-Clause 4.7 [Contractor's Design]of the Green Book 2021 stipulates that the Contractor shall design the Works to the extent (if any) stated in the Contract and the corresponding Works shall be, when completed, fit for the purpose(s) for which they are intended, as defined and described in the Contract (or, where no purpose(s) is (are) so defined and described, fit for their ordinary purpose(s)).

may include design of minor works listed in the Specification, with the linked fitness for purpose obligations.[9]

'Works' means the Permanent Works and the Temporary Works, or either of them as appropriate (and the facility to be operated by the Contractor during the Operation Service Period, Gold Book).[10] The scope of the Permanent Works and their performance when completed is essential for the scope of the Contract and has to be carefully specified.

The Temporary Works are also part of the scope, but their specification is usually the Contractor's responsibility. By definition, the 'Temporary Works' may be of every kind (other than Contractor's Equipment) required on Site for the execution (Gold Book: plus completion and operation) of the Works.[11] Their scope is usually defined in the Contractor's proposal, method statements, and Health and Safety Plan. In some cases, usually under a Red Book, the Contract may specify important Temporary Works, such as construction of temporary roads for a hydro project. The environmental, health and safety regulations and the local authorities impose requirements that impact the scope of the Temporary Works. The Temporary Works may include strengthening and maintenance of access roads and bridges under Sub-Clause 4.15, wheel washers for compliance with Sub-Clause 4.18, temporary utilities under Sub-Clause 4.19, fencing, barriers for the security of the Site under Sub-Clause 4.21, any works necessary for stability of the Works during the construction pursuant to Sub-Clause 4.1, lighting, structures, and railing for safe access and working conditions, as well as roadways, footways, guards, and fences, which may be necessary for the use and protection of the public and of owners and occupiers of adjacent land and property under Sub-Clause 4.8.

Pursuant to Sub-Clause 4.1, the scope includes delivery of the Plant (and spare parts, if any) and Contractor's Documents specified in the Contract and any resources and services, whether of a temporary or a permanent nature, required to fulfil the Contractor's obligations under the Contract.

By definition, the 'Employer's Requirements' describe the purpose(s) for which the Works are intended, and specify the scope, and/or design, and/or other performance, technical, and evaluation criteria, for the Works in the Yellow/Emerald/Silver Books, plus the criteria for the provision of the Operation Service under the Gold Book.[12]

In the Red/Green Books, the Employer sets out the **scope of the Works** (as already designed) **in the Specification and the Drawings**. Under the Red Book, the Specification is generally to be read together with the Drawings prepared by the Employer, though it may define limited design scope to be developed by the Contractor (for example, the foundation design for a Plant delivered by the Contractor). The Green Book Sub-Clause 1.1.40 defines 'Specification' as the document specifying the Works and including the Employer's design and/or the Employer's requirements if any design is to be carried out by the Contractor.

9 Unlike the Green Book, where the Guidance Notes advise how to use it for a full Contractor's design scenario, the GP3 will be infringed if the Red Book scope includes Contractor's design for the majority of the Works (see *The FIDIC Golden Principles* (2019) 9)

10 Sub-Clause 1.1.87 (Red)/1.1.89 (Yellow)/1.1.79 (Silver)/1.1.103 (Emerald)/1.1.82 (Gold Book)

11 Sub-Clause 1.1.80 (Red)/1.1.82 (Yellow)/1.1.72 (Silver)/1.1.95 (Emerald)/1.1.79 (Gold Book)

12 Sub-Clause 1.1.33 (Yellow)/1.1.31 (Silver)/1.1.39 (Emerald)/1.1.36 (Gold Book)

In the Yellow/Emerald/Silver Book Schedule of Performance Guarantees,[13] the Employer should clearly specify the performance criteria that the Works or any Section will be required to achieve when completed, and also may define any Performance Damages for compensation when permissible lower values are measured.

Sub-Clause 4.11 stipulates that unless otherwise stated in the Contract, the ACA (Silver Book: Contract Price stated in the Contract Agreement) 'shall be deemed to cover all the Contractor's obligations under the Contract and all things necessary for the proper execution of the Works in accordance with the Contract'. This means that the scope of the Contractor's expressed and implied obligations covers all that is needed for the completion of the Works within their defined scope, unless the Contract provisions allow reduction of the scope of the Contractor's obligations (for example, the Employer may supply some of the materials under Sub-Clause 2.6 or provide temporary utilities under Sub-Clause 4.19). The scope of the Works also depends on the:

- responsibilities for connections to utilities; and
- conditions for taking over under Clause 10, listed in Chapter 6.

The Red/Yellow/Emerald Book scope is based on the use of Site Data (including data about the surface, underground, climatic, and environmental conditions) as stipulated under Sub-Clause 4.10. The Contractor cannot blindly rely on the data submitted by the Employer under Sub-Clause 2.5, but is responsible for interpreting them, inspecting and examining the Site, and finding other available information (related to access, Site surroundings, necessary work and Goods, Laws of the Country, etc.) to the extent that was practicable taking into account cost and time. Under the Silver Book, the Contractor is responsible not only for interpreting the Site Data submitted by the Employer but also for verifying it, and has no relief in the case of Unforeseeable physical/ exceptionally adverse climatic conditions (see Figures 13.7 and 13.9).

7.4 Decomposition of the scope in FIDIC works contracts

The scope can be decomposed into lower levels of detail using a WBS. The WBS is a hierarchical decomposition of the total scope of work to be carried out by the project team to accomplish the project objectives and create the required deliverables.[14]

The planned work is contained within the lowest level of WBS components, which are called work packages. A work package can be used to group the activities where work is scheduled and estimated, monitored, and controlled. In the context of the WBS, work refers to work products or deliverables that are the result of activity and not to the activity itself.[15] For example, the work package 'Signalling' in a railway project under a Yellow Book Contract includes as a deliverable the design to be submitted for Review, then the delivery of the Plant, and the completed works as result of the assembly and testing.

13 Sub-Clause 1.1.74 (Yellow)/1.1.64 (Silver)/1.1.87 (Emerald Book) defines it as 'the document(s) entitled schedule of performance guarantees (if any) in the Schedules showing the guarantees required by the Employer for performance of the Works and/or the Plant or any part of the Works (as the case may be), and stating the applicable Performance Damages payable in the event of failure to attain any of the guaranteed performance(s).'

14 PMI Inc., *PMBOK® Guide* (PMI sixth edition 2017) Glossary

15 PMI Inc., *PMBOK® Guide* (PMI sixth edition 2017) 157

The level of detail for work packages will vary with the size and complexity of the project. As the work is decomposed to greater levels of detail, the ability to plan, manage, and control the work is enhanced.[16] This is the approach of the Red Book, where the WBS is presented in the BQs after the elaboration of the detailed design. The BQs follow the WBS, but the scope baseline depends on the requirements contained in the Specification and Drawings, which have higher priority than the BQs under Sub-Clause 1.5.

However, excessive decomposition can lead to non-productive management effort, inefficient use of resources, decreased efficiency in performing the work, and difficulty aggregating data over different levels of the WBS.[17] Under a Yellow Book Contract, the work packages usually do not contain such details as the Red Book BQs, since the design is only conceptual when the tender documents are prepared. Further decomposition will be possible when the Contractor's design has successfully undergone the Engineer's Review. This technique is sometimes referred to as rolling wave planning.

Achieving the desired visualisation, analysis, monitoring, evaluating, and reporting of projects' performance using the BIM model requires the adoption of a BIM-friendly WBS. The BIM-friendly WBS will become the basis for associating the different project records with the BIM objects. The benefits of having a BIM-friendly WBS can also be associated with time-lapse cameras installed on construction sites.[18]

7.5 Scope validation in FIDIC works contracts

The scope validation is the formal acceptance of the completed project deliverables. The key benefit of this process is that it brings objectivity to the acceptance process and increases the probability of final acceptance by validating each deliverable. This process is performed periodically throughout the project as needed.[19]

The Engineer (Silver Book: Employer) validates important Contractor's submissions, such as design, as-built documents, O&M Manuals, and programme, through the formal Review process. 'Review' is defined in the 2017 Books as examination and consideration by the Engineer/Employer of a Contractor's submission in order to assess whether (and to what extent) it complies with the Contract and/or with the Contractor's obligations under or in connection with the Contract.[20] The validation is formalised with a Notice of No-objection or a Notice stating the extent of non-compliance.

The construction cannot commence before the validation of the relevant design. Pursuant to Sub-Clause 5.2.2 (Yellow/Silver Book)/4.4.1 (Red Book), if the Engineer (Silver Book: Employer) gives no Notice within the Review Period, then he/she shall be deemed to have given a Notice of No-objection to the Contractor's Document, and the Contractor can proceed with the execution of the design for the sake of efficiency. However, even if a Notice of No-objection was previously (deemed to have been) given to the design, which is later found to contain an error, omission, etc., the Contractor is required to correct and revise the Contractor's Document, and to resubmit it for Review to the Engineer/Employer at

16 PMI Inc., *PMBOK® Guide* (PMI sixth edition 2017) 160

17 PMI Inc., *PMBOK® Guide* (PMI sixth edition 2017) 160

18 PMWeb, 'The Great Value of a BIM-Friendly WBS to Better Evaluate Construction Project Performance During the COVID-19 Era and Beyond' < https://pmweb.com/better-evaluate-construction-projects-performance-with-bim-friendly-wbs/> accessed 31 July 2022

19 PMI Inc., *PMBOK® Guide* (PMI sixth edition 2017) 163

20 Sub-Clause 1.70 *Red*/1.71 *Yellow*/1.1.82 *Emerald*/1.61 *Silver Book*

the Contractor's risk and cost under Sub-Clause 5.8 of the Yellow/Silver Books. If there is mandatory third-party validation of the design under the Laws of the Country, it is recommended that the Contract specify how this procedure is integrated in the Review process.[21] Where the third party may be a private company, it may be part of the Engineer (if the Engineer cannot be authorised, a local authorised company may be appointed by the Engineer as a Sub-Consultant or a partner in the joint venture [JV]). Where the building control is privatised, it is not recommended that the Employer appoint separately an Engineer and an authorised third-party validator, since this may create discrepancies in the requirements and delays during the design validation process.[22] For a Silver Book contract, the third party may be appointed by the Employer or by the Contractor, if the applicable law does not require the third party to be independent from the Contractor.

The FIDIC 2017 Contracts Guide comments that besides the verification for compliance with the Contract, the Review of the Contractor's Documents serves to give the earliest possible opportunity for the CA to consider whether the proposed Works are what the Employer actually requires and if not, to initiate a Variation changing the scope baseline.[23]

Constant quality control is performed by inspection and testing. The Employer's Personnel carry out compliance assessment of the deliveries/ completed works/ test results with the requirements of the Contract. The inspections under Sub-Clause 7.3 serve for verification of any Materials, Plant, or work before it is to be covered up, put out of sight, or packaged for storage or transport. The testing under Sub-Clause 7.4 applies to all tests before taking over. When the specified tests have been passed, the Engineer should endorse the Contractor's test certificate, or issue a test certificate to the Contractor, to that effect. If, as a result of an examination, inspection, measurement, or testing, any Plant, Materials, Contractor's design, or workmanship is found to be defective or otherwise not in accordance with the Contract, the CA shall give a Notice to the Contractor under Sub-Clause 7.5 rejecting the item of Plant, Materials, design, or workmanship that has been found to be defective.

The Engineer measures the quantities under the Red Book or when required for any payment under the other Books.[24]

The taking-over procedure under Sub-Clause 10.1 enables validation at substantial completion of the Works or Section. The Engineer (Silver Book: Employer) verifies the completed Works, the results of the Tests on Completion, the as-built records, the O&M Manuals, and the training for compliance with the requirements of the Contract.

Third-party validation may be mandatory under the Laws of the Country as a condition for operation of the Works. Regulators perform additional validation for high-risk projects to issue a licence to operate.

---

21 For example, Article 23 of the *Urban Planning Code of the Republic of Uzbekistan* (May 2021) requires design examination (to be provided by the Employer) by specialised structural units of a specially authorised state body, authorised ministries, state committees and departments, expert commissions or expert groups, as well as accredited legal entities (expert organisations), as a condition for coordination, approval, and further implementation. <https://lex.uz/docs/5437769> accessed 13 February 2023

22 #24 of the Final Provisions of the *Bulgarian Spatial Development Act* (SG 12/2001 amended 2007) stipulates that FIDIC contracts may be used for projects funded entirely or partially by international financial institutions (IFIs)/EU funds; for the design conformity assessment and construction supervision, the FIDIC Engineer must be licensed under the terms and conditions of Article 167 or must assign the implementation of these activities to a subcontractor, registered pursuant to this Act. <https://cpcp.mrrb.government.bg/cms/assets/Laws/SPATIAL%20DEVELOPMENT%20ACT.pdf> accessed 2 August 2022

23 FIDIC, *The FIDIC 2017 Contracts Guide* (FIDIC 2022) 224

24 Under Sub-Clause 12.3 (*Red Book*)/13.8 (*Emerald Book*)

The results of the trial operation stage of the Tests on Completion under Clause 9 and the Tests after Completion under Clause 12 serve for validation of the performance of the Works to be verified against the performance criteria, included in the Employer's Requirements and the Schedule of Performance Guarantees of the Yellow/Emerald/Silver Books. If the Works, or a Section, fail to meet the performance criteria, and applicable Performance Damages are set out in the Schedule of Performance Guarantees, Sub-Clauses 9.4, 11.4, and 12.4 entitle the Employer to claim payment of these Performance Damages by the Contractor in full satisfaction of this failure. If the Contractor pays the Performance Damages to the Employer during the DNP, then the Works or Section shall be deemed to have passed the Tests after Completion. The Performance Damages are applicable for lower performance parameters in a specified range acceptable for the Employer and in compliance with the regulations in force. Where the failure is outside the specified range, the Works or Section shall not be accepted.

The final inspection after the expiry of the DNPs before the issuance of the Performance Certificate validates the compliance of the Works with the requirements of the Contract. Pursuant to Sub-Clause 11.9, only the Performance Certificate constitutes acceptance of the Works.

If the validation demonstrates nonconformity with the requirements of the Contract, the Employer is protected by the:

- Contract provisions for remedy of defects and retesting providing different reliefs (Sub-Clauses 7.5/7.6, Clauses 9/11/12);
- Sub-Clause 14.6.2 provision enabling temporary withholding of amounts from future payments for non-compliant works until the remedy of the default;
- Performance Security provided by the Contractor under Sub-Clause 4.2 to be valid until the Performance Certificate is issued;
- Performance Damages specified in the Schedule of Performance Guarantees under Sub-Clause (Yellow Book) 1.1.74/(Silver Book) 1.1.64/(Emerald Book) 1.1.87;
- Retention Money retained from each payment, 50 per cent paid after the TOC and the rest after the expiry of the DNPs.

7.6 Scope control

7.6.1 Scope creep management

Scope control includes monitoring the status against the scope baseline and managing changes. The Contract documents specify the scope baseline. Controlling the project scope ensures that all requested changes and recommended corrective or preventive actions undergo the integrated change control process addressed in Chapter 6. Scope control is critical for the project success. Scope creep can derail the project.[25]

Often, scope changes are approved without evaluating the downstream impact on work packages that have not started yet.[26] The CA must understand the causes for scope creep and how to use the FIDIC tools to manage it within acceptable limits:

25 PMI Inc., *Construction Extension to the PMBOK® Guide* (PMI 2016) 47

26 Harold Kerzner, *Project Management Metrics, KPIs and Dashboards* (Wiley 2017) 60

- **poor understanding of requirements** – the CA may issue clarification or instruction under Sub-Clause 1.5;
- **poorly defined requirements** – identified issues to be discussed and solved in collaboration, if necessary, with Variations under Clause 13.3 and DAAB assistance under Sub-Clause 21.3;
- **unclear scope limits** – connection points for infrastructure and interfaces with other contractors should be clearly specified in the Contract; management meetings under Sub-Clause 3.8 (Silver Book: 3.6) with the participation of the Employer's other contractors, the personnel of legally constituted public authorities, and/or private utility companies may help solve the issues;
- **complexity** – advance warning under Sub-Clause 8.4 and regular Contract Risk Management meetings (Sub-Clause 1.16, Emerald Book) may help solve issues with minimum scope creep;
- **lack of resources** – Variations under Clause 13.2/13.3 must be used with caution;
- **poor communications** – regular meetings under Sub-Clause 3.8 (Silver Book: 3.6), timely Notices under Sub-Clause 1.3, Project Management Information Systems (PMIS) are the key;
- **misunderstanding expectations** – visualisation helps the stakeholders to understand the project; the Sub-Clause 4.4.1 (Red Book)/5.2 (Yellow/Emerald/Silver) Review of Contractor's Documents enables the Employer to scrutinise the design before construction;
- **perfectionism** – overperformance leads to higher cost and delay and may be admissible only in the case of whole-life improved value;
- **changes in Laws** – the procedure under Sub-Clause 13.6 should be followed;
- **errors/omissions** – the Employer is responsible for the design under the Red Book, and the Engineer shall instruct the Contractor under Sub-Clause 3.5 in case of errors; Sub-Clause 1.9 of the Yellow/Emerald Books gives the procedure in case of errors in the Employer's Requirements;
- **deception:** tenderers see gaps in the requirements and do not comment on the errors in order to make bigger profits – the foreseeability test may prove that the error in the Employer's Requirements under Sub-Clauses 1.9 of the Yellow/Emerald Books or the adverse physical conditions under Sub-Clause 4.12 of the Red/Yellow/Emerald Books were reasonably foreseeable for an experienced contractor before submitting the tender, then no EOT/compensation should be due;
- **attempt to escape Delay Damages creating unnecessary additional work** – the CA should reject changes that are not necessary if the Value Engineering (VE) benefits are not related to the whole life of the Works;
- **additional needs of the Employer** – under Sub-Clause 13.1, the Contractor may object to Unforeseeable changes having regard to the scope and nature of the Works, and the CA may cancel or vary the instruction;
- **unrecognised accumulation of changes** – Sub-Clause 3.5 (Silver Book: 3.4) entitles the Contractor, before commencing any work related to an instruction, to give a Notice that it constitutes a Variation; the CA shall respond within 7 days, confirming, reversing, or varying the instruction, and otherwise, shall be deemed to have revoked it;

- **poor change control** – the Variation procedure under Sub-Clause 13.3 should be strictly followed and the list of Variations included in the Sub-Clause 4.20 monthly progress reports.

Construction projects get more complex and vulnerable to scope creep. The larger the project, the greater the likelihood that scope creep will occur. The scope boundary can drift during the implementation of the project because, as we get further into the project and gain more knowledge, we identify unplanned additions to the scope. Each stakeholder's vision of the project must be aligned. When scope creep occurs and scope changes are necessary, the scope boundary can move.[27] However, the scope boundary may not be able to move if the financing institutions object.

For major construction projects, scope creep is inevitable due to changing market demands, new sustainability targets, Exceptional Events causing disruption of the supply chain, changed environment, and volatility of prices. Robust change management is essential to prevent uncontrolled scope creep. FIDIC works contracts include a detailed procedure for Variations under Sub-Clause 13.3. When the Contractor disagrees with the instructed Variation, a Claim may follow. The Claim may escalate into a Dispute where an agreement cannot be reached following the Sub-Clause 20.2 Claims procedure.

For robust change control, it is recommended that the Issue Log discussed in Chapter 6 should be cross-referenced with the Variation Log, and the Variation Log should track the proposal/agreement process and have cross references with the Claims Register where applicable.

7.6.2 White Book change control

Sub-Clause 1.1.25 defines 'Variation' as any change to the Services instructed or approved as a Variation under Sub-Clause 5.1. A Variation **shall not substantially change the extent or nature of the Services.**

Sub-Clause 1.1.26 defines 'Variation Notice' as a written communication identified as a Variation Notice and issued in accordance with the provisions of Sub-Clause 1.3. Sub-Clause 5.1.1 empowers the Client to issue a Variation Notice at any time prior to completion of the Services, with the restriction that the Variation must not change substantially the extent or nature of the Services. A Variation may amend the scope of Services, amending Appendices 1 or 2, or omitting part of the Services no longer required. It may change the specified sequence/timing/method.

The Consultant shall not make changes to the Services unless one of the routes presented in Figure 7.1 is followed:

1. The Client issues a Sub-Clause 5.1.1 Variation Notice:
   - the Consultant may promptly object, giving a Sub-Clause 5.1.4 Notice that it does not possess the relevant skills/resources or considers that the Variation will substantially change the extent or nature of the Services;

27 Harold Kerzner, *Project Management Metrics, KPIs and Dashboards* (Wiley 2017) 59

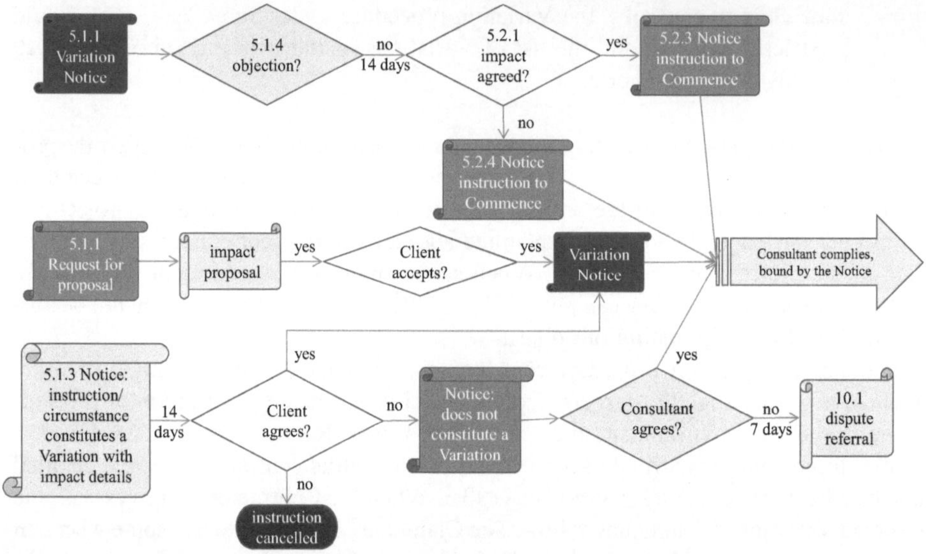

Figure 7.1 Variation procedure under the White Book Clause 5

- if it does not promptly give such a Notice, the Parties shall agree the impact of the Variation on the price, Programme, Time for Completion, and other parts of the Services;
- the Client shall issue an instruction to the Consultant to commence work on the Variation under Sub-Clause 5.2.3 in the case of agreement, or under 5.2.4 where no agreement is reached within 14 days or where it is not practicable to establish the impact in advance.

2. The Client requests a Variation proposal:
   - the Consultant submits the proposal estimating the impact;
   - the Client may issue a Variation Notice accepting the proposal.
3. The Consultant gives a Notice under Sub-Clause 5.1.3 as soon as reasonably practicable where it considers that any instruction/direction from the Client or other circumstance constitutes a Variation, including details of the estimated impact:
   - within 14 days, the Client shall agree and issue a Variation Notice, or disagree, cancelling the instruction/direction or giving a Notice explaining why the instruction/direction/circumstance does not constitute a Variation;
   - the Consultant shall comply, unless it refers the matter as a dispute under Clause 10 within 7 days.

7.6.3 Change control in works contracts

7.6.3.1 Admissibility of scope changes

'Variation' is defined as any change to the Works, and under the Yellow/Emerald/Silver Books also to the Employer's Requirements, **instructed as a variation** under Clause 13.[28]

28 Sub-Clause 1.1.86 (*Red Book*)/1.1.88 (*Yellow Book*)/1.1.56 (*Silver Book*)/1.1.102 (*Emerald Book*)

Figure 7.2 presents the integrated change control algorithm:

1. In order to prevent scope creep, the proposed change must be **necessary** for the completion of the Works or to improve the value of the Works;
2. The change shall be **admissible** under the local Laws (for example, complying with the zoning plan, the environmental law, etc.);
3. The Variation procedure may be used only where the change is **within the scope of the Contract.** Changes to the Contract scope shall be proceeded as Contract Amendments or new contracts, depending on the procurement rules of the financing institution and/or applicable laws. The Silver Book Employer's Representative may have the full authority of the Employer under Sub-Clause 3.1, while the Engineer is not authorised to amend the Contract (Red/Yellow/Emerald Book Sub-Clause 3.2).

The third question leads to debates: what does it mean to be within the scope of the Contract? The Red Book Sub-Clause 13.1 has an open list for the admissible changes, including changes to the quantities, characteristics, levels, positions, or dimensions of any item of work, or to the sequence or timing of the execution of the Works; omission of any work (unless it is to be carried out by others without the agreement of the Parties); any additional work, Plant, Materials, or services **necessary** for the Permanent Works, including any associated Tests on Completion, boreholes, and other testing and exploratory work. The other books where the Contractor is responsible for the design do not include a similar list. In these, the Employer's Requirements specify the scope baseline for the Works. It may be argued that a Variation is admissible if the changed Works are not so different as to be considered Unforeseeable, having regard to the scope and nature of the Works described in the Employer's Requirements.

If the change is **not** within the scope of the Contract, and the Works may be put into operation without it, a separate restricted or open tender procedure may be used, depending on the applicable procurement rules. Pursuant to Article 72 of *Directive 2014/24/EU*,[29] contracts may be modified without a new procurement procedure where the mod-

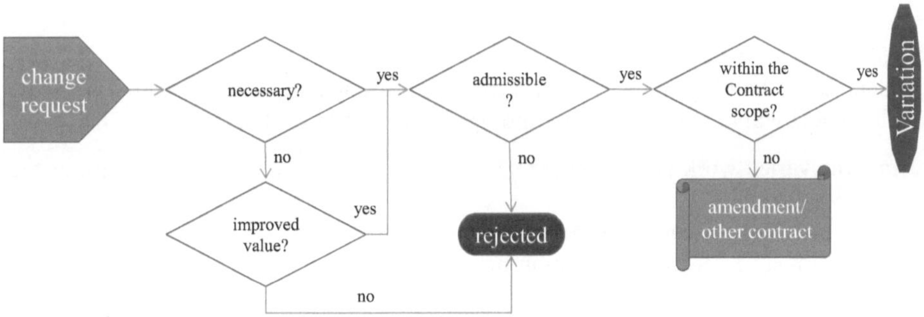

Figure 7.2 Change control under FIDIC works contracts

29 *Directive 2014/24/EU of the European Parliament and of the Council of 26 February 2014 on public procurement* <https://eur-lex.europa.eu/legal-content/EN/TXT/?uri=CELEX%3A02014L0024-20220101> accessed 6 July 2022

ifications, irrespective of their value, are not substantial. A modification of a contract shall be considered to be substantial where it renders the contract materially different in character from the one initially concluded, including considerable extension of the scope of the contract. The Variation procedure under FIDIC contracts must comply with the legislation transposing the EU Directive where the employers are public contracting authorities in EU member states.

7.6.3.2 Case study – scope management in a mega innovative public project

Fusion for Energy (**F4E**) is the European domestic agency responsible for the EU's contribution to the International Thermonuclear Experimental Reactor (**ITER**), the world's largest scientific partnership. F4E applies EU public procurement rules but is not subject to the EU Public Procurement Directives. F4E procures the ITER site infrastructure using more than 15 FIDIC-based contracts with tailored change management procedures. In order to provide flexibility in contract implementation, F4E endeavours to add all changes foreseeable at the tender stage to the initial scope in the tender documents, and additionally provides contingency funds, as part of the initial Contract, for unforeseen changes to be instructed as Variations. A Variation may be issued following a multi-tier change management procedure. Before issuing a Variation, the Engineer submits to the Employer an impact analysis. The Engineer issues the Variation only after the F4E Authorising Officer's approval, and in certain cases, this includes agreement of the Contractor on the impacts of the Variation. During contract implementation, the contingency funds may be increased, when necessary, in compliance with the public procurement rules on contract modification, with amendments to the Contract before the implementation of the change. If advertised in the initial tender documents with the conditions described therein, additional works of similar nature may be ordered subject to a negotiated procedure. However, alteration of the overall nature of the initial contract is not permitted. As a last resort to ensure compliance with public procurement rules on contract modification, termination of the Contract may take place to allow the additional and remaining scope to be retendered in open competition.[30]

7.6.3.3 Variation procedure

The Contractor shall be bound by each instructed Variation and shall execute it with due expedition and without delay, unless it promptly gives a Notice that:

(a) the varied work is Unforeseeable having regard to the scope and nature of the Works;
(b) the Contractor cannot readily obtain the Goods required for the Variation;
(c) it will adversely affect its Environmental, Health, and Safety (EHS) obligations;

and for the Yellow/Emerald/Silver Books:

(d) it will have an adverse impact on the achievement of the Performance Guarantees; or

30 Karoly Tamas Olajos, JD/MBA/LLM/PMP, ITER Site & Buildings Legal Affairs Group Leader, F4E, presentation at the FIDIC International CUC London 2015

(e) it may adversely affect the Contractor's obligation to complete the Works so that they shall be fit for the purposes for which they are intended.

The Engineer (Silver Book: Employer) shall respond promptly by giving a Notice cancelling, confirming or varying the instruction. The Contractor shall comply with the instructed Variation (see Figure 7.3).

If the change goes through the three gateways shown in Figure 7.2, the Variation procedure shall be followed with a:

- Variation by instruction under Sub-Clause 13.3.1 in the case of urgency and necessity or
- Variation by request for proposal before instruction under Sub-Clause 13.3.2.

Under Sub-Clause 13.3.1, the Contractor shall proceed with execution of the Variation and in parallel submit within 28 days its proposal for agreement/determination of EOT and adjustment to the Contract Price.

If the Contractor has submitted a VE proposal under Sub-Clause 13.2, where it is consented to, the procedure continues as a Variation by instruction. The consent shall be at the sole discretion of the Employer, with focus on the whole-life value. The benefits may be acceleration; reduction of the cost to the Employer of executing, maintaining, or operating the Works; or improvement of the efficiency or value to the Employer of the completed Works. The initial VE proposal usually includes only concept/basic design developed at the Contractor's risk and cost. The Contractor develops its proposal further after the consent. The agreement/determination shall include consideration by the CA of the sharing (if any) of the benefit, costs, and/or delay between the Parties.

Under Sub-Clause 13.3.2, the Contractor submits the proposal and does not execute the Variation until the agreement/determination process is over. An accepted proposal is, itself, not enough to constitute an instruction, so it will still be necessary to refer back to the instruction provisions.[31] In the case of rejection, the Contractor shall be entitled

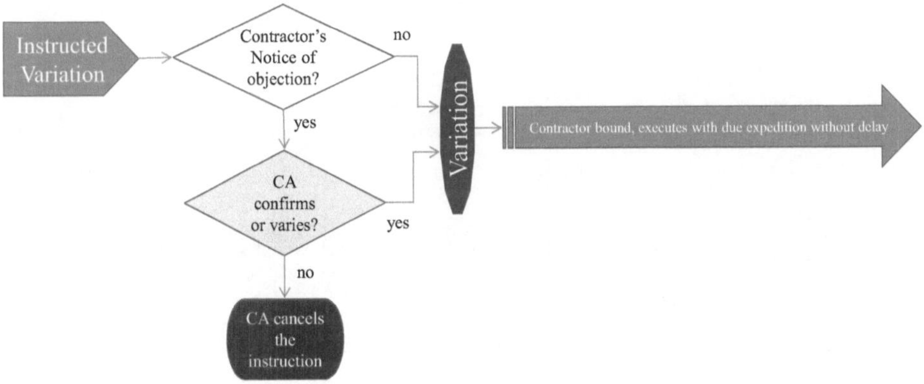

Figure 7.3 Objection procedure under Sub-Clause 13.1

31 Corbett & Co., *FIDIC 2017 – A Practical Legal Guide* (Corbett & Co. 2020) 329

to claim under Sub-Clause 20.2 payment of Cost incurred as a result of submitting the proposal requested by the CA, unlike the VE proposal initiated by the Contractor and prepared at its risk.

Figure 7.4 shows that often, the Contractor must execute the instruction before the cost is agreed or determined. However, payment will be due under Sub-Clause 14.3 (i) for the completed work instructed under Sub-Clause 13.3. Until the adjustments for the Variation are agreed/determined, the Engineer shall assess a provisional rate/price for the purposes of IPC under the Red/Yellow/Emerald Book. Similarly, the Employer shall assess a provisional rate or price for the purposes of interim payment under the Silver Book.

7.6.3.4 Scope control

It is important to know the power and authority of the Engineer and the Employer's Representative, the requirements of the financing institutions, and the applicable laws related to Variations. In many projects, the need for changes leads to instructions for Variations, and the Contractor is obliged to execute the varied works, but their certification and payment is delayed due to the Employer's fear of sanctions for infringement of the public procurement rules or lack of contingency funds. Sometimes, the Engineer as Employer's Personnel, forgetting its neutral role in certification, instead of pricing the Variation with a provisional rate until the price is agreed/determined, requests the Contractor to delete the amounts for varied works from the Sub-Clause 14.3 Statement. This is a breach of contract, since payment for executed varied works is due under Clause 14. Such misuse of FIDIC contracts often leads to the Contractor's suspension under Sub-Clause 16.1. After accumulation of significant unpaid amounts for executed varied works, Contractors consider termination under Sub-Clause 16.2. Some Employers wrongly believe that those provisions are not applicable when they comply with the terms for payment under Sub-Clause 14.7 for the works included in the scope baseline.

Scope control is important to minimise underperformance and overperformance exceeding the requirements. After the Contract is entered into:

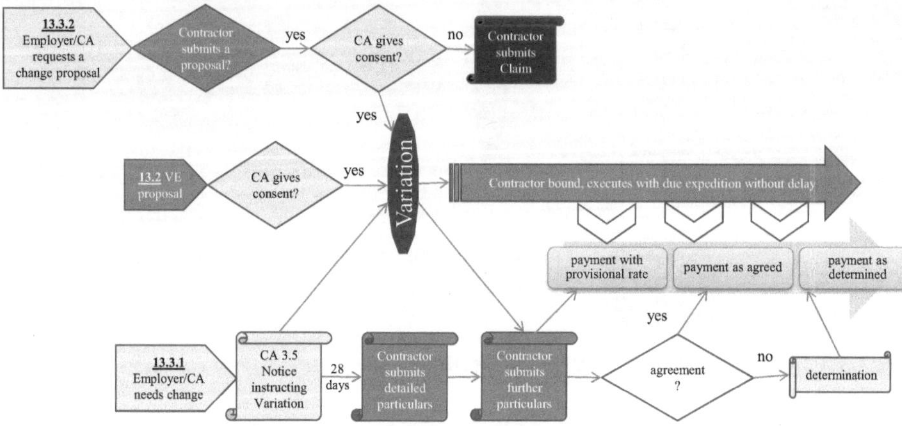

Figure 7.4 Variation procedure in FIDIC works contracts

- the Employer does not have to accept an inferior quality, but the Contractor's proposal to use inferior quality may be treated as a VE proposal under Sub-Clause 13.2, and if accepted, the Parties may share the saving in cost, leading to a reduction in the Contract Price; and
- the Contractor does not have to provide a superior quality, unless and until instructed to do so as a Variation under Sub-Clause 13.3.1. In that case, the Contractor may be entitled to an EOT and/or an adjustment to the Contact Price.[32]

7.6.3.5 Application of the Variation procedure

The following provisions (examples from the Yellow Book) expressly refer to the Variation procedure in the case of:

- Sub-Clause 1.9 – measures that the Contractor is instructed to take in case of errors in the Employer's Requirements;
- Sub-Clause 3.5 – any instruction constituting a Variation;
- Sub-Clause 4.2.1 – incurred Cost as a result of increase of the Performance Security;
- Sub-Clause 4.7.3 – measures that the Contractor is required to take to rectify errors in the items of reference;
- Sub-Clause 4.12.3 – instruction for dealing with Unforeseeable physical conditions constituting a Variation under Sub-Clause 13.3.1;
- Sub-Clause 5.4 – changes in applicable standards/regulations after the Base date;
- Sub-Clause 7.2 – additional samples;
- Sub-Clause 7.4 – varied test arrangements or instructed additional tests;
- Sub-Clause 8.4 – advanced risk mitigation;
- Sub-Clause 8.7 – revised methods/ instructed acceleration;
- Sub-Clause 8.12(b) – omission of part of the Works suspended for more than 84 days;
- Sub-Clause 11.2 – remedy of defects attributable to a cause for which the Contractor is not responsible;
- Sub-Clause 11.4 – omission of part of the Works, which cannot be used for its intended purposes;
- Sub-Clause 13.6 – adjustments for changes in Laws;
- Sub-Clause 17.2 – instructed rectification of loss/damage to the Works, Goods, and Contractor's Documents caused by an Employer's risk event.

The Contractor may give a Notice if it considers that any instruction of the Engineer under Sub-Clause 3.5 (Silver Book: Employer under Sub-Clause 3.4) constitutes a Variation: for example, if the Contractor's Documents comply with the Contract, but the Engineer instructs alternative design or requests additional documents that are not in the scope of the Contractor's Documents.

The CA as representative of the Employer controls whether or not a change is to be implemented as a Variation. The Contractor cannot execute any changes before the CA's instruction.

32 FIDIC, *The FIDIC 2017 Contracts Guide* (FIDIC 2022) 61

The procedure in the case of Unforeseeable physical conditions under Sub-Clause 4.12 of the Red/Yellow/Emerald Books may seem an exception to the rule that the Contractor is to wait for an instruction before dealing with changes: the Contractor has an obligation to notify the Engineer and continue the execution of the Works, using appropriate proper and reasonable measures enabling the Engineer to inspect and investigate them. In some cases, this will mean emergency measures for strengthening/dewatering an excavation, in other cases, changing the Contractor's Equipment/method of work. When an experienced contractor encounters adverse physical conditions, it is expected to use its expertise to overcome such conditions and to execute and complete the Works in accordance with the Contract. However, the Engineer may consider giving instructions for better risk mitigation. If any such Engineer's instruction constitutes a Variation, then Sub-Clause 13.3.1 applies, and the Contractor may be entitled to EOT/compensation where it has suffered delay/ incurred costs complying with the instruction. This entitlement to EOT and/or an adjustment to the Contract Price for a Variation is in addition to the Contractor's entitlement to make a Claim for EOT/Cost under Sub-Clause 4.12.4.[33] This provision may be amended by the governing law, which may impose an obligation on the Contractor to suspend the Works and wait for an instruction.[34] When substantial changes to the structure of the foundations are necessary due to the Unforeseeable conditions, the design revision/ Review/approval procedure may lead to suspension of the impacted part of the works. The construction of this part may commence after the design is approved (submitted by the Engineer with the instruction for Variation under the Red Book or under the Yellow/ Emerald Book when the CA gives a Notice of No-objection to the Contractor's design, subject to approval procedure under the local Law).

Sub-paragraph (f) of Sub-Clause 4.20 requires the Contractor to include a list of Variations in the progress report. Usually, the CA keeps a Variation Log, registering and following up instructions and proposals for Variations under Sub-Clause 13.3. The Variation Log is a spreadsheet including columns for the proposed/provisional/agreed cost/EOT and source of financing (contingencies, Provisional Sums, savings, etc.).

7.7 Innovative Project Scope Management

7.7.1 Outcome-based approach

The UK Construction Playbook advises shared focus on outcomes, rather than scope, to unlock innovation and drive continuous improvement. The outcome-based approach should assess the whole-life value, performance, and cost. The requirements should not be overly prescriptive to allow innovative solutions.[35]

Clear and measurable outcomes may be specified in the Yellow/Emerald/Silver Book Employer's Requirements and Schedule of Performance Guarantees. Carbon footprint reduction targets for construction activities may be set up in the Specification/Employer's

33 FIDIC, *The FIDIC 2017 Contracts Guide* (FIDIC 2022) 192

34 For example, Article 670 of the *Civil Code of the Republic of Uzbekistan* (1997) obliges the Contractor to report to the Employer in the case of circumstances necessitating additional work/costs and to suspend the Works after 10 days if the Employer does not reply (another time period may be provided under a Law or the Contract) with entitlement to prolongation costs

35 UK Cabinet Office, *Construction Playbook* (2020) 24 <https://www.gov.uk/government/publications/the-construction-playbook> accessed 20 April 2022

Requirements and measured on a quarterly basis. Chapter 15 recommends drafting of Particular Conditions introducing sustainability targets, and carbon emissions incentives and damages.

7.7.2 Tools supporting innovative scope management

The improved quality of information in projects using BIM can result in a significant reduction in Variations.[36] BIM helps in the definition of the requirements with the early visualisation of the project. BIM supports efficient change management with early clash detection and shared updated information.

Using BIM, IT experts may demonstrate in minutes during a 'hackathon' the impact of changes requested by the Employer[37] to facilitate the decision for instruction of Variations under Sub-Clause 13.3.1.

Lean construction methods help to focus on the outcomes reducing waste.

PMIS and AI, combined with drone monitoring and video recording, support scope control and may automate the scope validation:

- AI uses field data captured by drone for automatic volume calculations in infrastructure projects, material count, progress analyses, showing deviations from the plan/scope, and detecting quality issues. Automatic generation of as-built digital twins enables progress tracking and comparison with design, delivering reports within hours.[38]
- Wearing a helmet-mounted 360° camera is all that a platform needs to create a digital twin of the Site and produce detailed reports, outlining deviations from the design/schedule and flagging uncompleted tasks.[39]

Optimising change management through the speed and accuracy of a digital solution would be beneficial to the decision-making process.[40]

7.8 Example Notices and logs

Appendix 1 continues with the following examples:

- Engineer's Notice of Variation under Sub-Clause 13.3.1;
- Contractor's Notice of objection to the Engineer's Notice under Sub-Clause 13.1 (c);
- Engineer's Notice of confirmation of the instructed Variation;

36 FIDIC, *Yellow Book* GPPC (FIDIC 2022) 57

37 Krzysztof Wojslaw, FIDIC webinar on 4 April 2023 'BIM and Digital Technologies – Optimise Traditional Ways of Working or Make a Paradigm Shift?' <https://fidic.org/node/40429> accessed 8 April 2023

38 DatuBIM presented at Construction 4.0 by ConTech (2021) <https://www.youtube.com/watch?v=iq-QRHbnkXQ> accessed 12 August 2022

39 Buildots UK presented at Construction 4.0 by ConTech (2021) <https://www.youtube.com/watch?v=iq-QRHbnkXQ> accessed 12 August 2022

40 Alan McNamara, 'Automating the Chaos: Intelligent Construction Contracts' Chapter in *Smart Cities and Construction Technologies* (IntechOpen 2020) 11

- Engineer's Notice of consent to the Contractor's VE proposal under Sub-Clause 13.2;
- Issue Log;
- Variation Log;
- Engineer's Notice of adjustment to the execution of the Works as a result of a change in the Laws under Sub-Clause 13.6 and request for proposal under Sub-Clause 13.3.2;
- Contractor's Notice of objection to the Engineer's request for proposal;
- Engineer's Notice of confirmation of the request for proposal and instruction under Sub-Clause 13.4 (a).

CHAPTER 8

Project Schedule Management with FIDIC Contracts

8.1 Project Schedule Management processes

According to the PMBOK® Guide, Project Schedule Management includes the processes required to manage the timely completion of the project.[1] These overlapping and interacting processes include planning of procedures, defining and sequencing of activities, estimating activity durations, developing the schedule (taking into account the resource requirements and schedule constraints), and controlling the schedule, managing the changes to the baseline.

The term 'schedule' is used in the PMBOK in relation to the project schedule, defined as 'an output of a schedule model that presents linked activities with planned dates,

1 PMI Inc., *PMBOK® Guide* (PMI sixth edition 2017) 173

DOI: 10.4324/9781003272892-8

durations, milestones, and resources'. The 'schedule model' is defined as 'representation of the plan for executing the projects' activities including durations, dependencies, and other planning information, used to produce a project schedule along with other scheduling artifacts'.

In FIDIC Works Contracts, the capitalised term 'Schedule' is used for all documents entitled schedules prepared by the Employer and completed by the Contractor, and included in the Contract. The schedules may include those listed in Table 5.1 related to payments, damages, and guarantees, and may include a Schedule for Completion in the meaning of the PMBOK definition. The 'Completion Schedule' is a defined term in the Emerald Book for the document included in the Contract stating the Time for Completion for each of the Works/Sections/Milestones, setting out logical sequential links between each Time for Completion[2] and activities comprising the required items of work linked to the Sections and Milestones.[3] The 'Programme' is a defined term in the FIDIC 2017 Red/Yellow/Silver Books and Emerald Book, meaning a detailed time programme prepared and submitted by the Contractor to which the Engineer (Silver Book: Employer) has given (or deemed to have given if silent during the time limit for Review) a Notice of No-objection under Sub-Clause 8.3 (Programme). The FIDIC initial programme is submitted within 28 days after receiving the Notice to commence and shall be consistent with any Schedule of Completion that is part of the Contract (like the Completion Schedule under the Emerald Book).

The non-capitalised term 'schedule' used further in this chapter shall be read in the PMBOK meaning, which is related to the project time management, governed by:

- any contractual Schedule of Completion and the FIDIC Sub-Clause 8.3 [Programme] for the Works Contracts;
- the requirements in Appendix 4 [Programme] part of the Agreement for Services and Sub-Clause 4.3 [Programme] including the minimum scope of the Programme as a defined term for the document submitted by the Consultant within 14 days of the Commencement Date under the White Book.

The FIDIC Subcontracts include Subcontractor's back-to-back obligations to those of the Contractor under the relevant Red/Yellow Book to provide a Subcontract Programme under Sub-Clause 8.3. The Sub-Consultancy Agreement includes back-to-back obligations with the White Book related to the Sub-Consultant's Programme and compliance with the Consultant's Programme under Sub-Clause 4.3. FIDIC users must remember that the supply change management has a key role for the proper schedule management.

The schedule management considerations include:

- completion milestones imposed by the Employer (e.g. related to credit agreements or interfaces with other contractors);
- resources needed for each activity;
- local regulations in force related to the built environment (e.g. prohibition of building during the summer in the sea resorts or in the rivers during the fish spawning season);

2 FIDIC, *Emerald Book* (FIDIC 2019) 1.1.8
3 FIDIC, *Emerald Book* (FIDIC 2019) Guidance 4.6.1

- influences of weather conditions (e.g. rainy season or severe winter);
- constructability or critical sequence constraints (e.g. partial use of roads by the public);
- allowance for operating of facilities during construction (e.g. reconstruction of a railway station);
- considerations of external stakeholders (e.g. quiet hours in a residential neighbourhood);
- restrictions arising from third parties (permits, design approvals, right of way acquisitions);
- availability and procurement of special equipment and materials (such as a turbine);
- mandatory requirements for labour visas, import/export of materials and equipment, etc.;
- other local, state, federal, or international regulations (e.g. mandatory procedure for commencement of construction, time needed for building control procedures, COVID restrictions).

8.2 Project Schedule Management in FIDIC agreements for services

The Client should include in the White Book Appendix 4 [Programme] the following:

- any key dates and interface obligations;
- requirements on the order or sequence of activities;
- review and approval periods for the Services;
- any particular software to be used for the Programme;
- information to be included in the monthly reports for tracking of the progress against the Programme.

Sub-Clause 4.2 sets out the timeframe for the performance of the Services. The Consultant shall:

- commence the performance of the Services as soon as is reasonably practicable after the Commencement Date identified in the Particular Conditions; where no date is identified, 14 days after the Effective Date (the date of the latest signature necessary to complete the formal Agreement);
- complete the whole of the Services within the Time for Completion stated in the Particular Conditions, or as may be amended in accordance with the Agreement, calculated from the Commencement Date.

Pursuant to Sub-Clause 4.3, the Programme to be submitted by the Consultant within 14 days of the Commencement Date shall include as a minimum:

- the order and timing in which the Consultant intends to carry out the Services in order to complete them within the Time for Completion;
- any key dates stipulated in the Agreement;

- the key dates when the Client or third parties shall give to the Consultant decisions/consents/approvals/information;
- any other requirements stated in Appendix 4.

The Consultant shall proceed in accordance with the Programme, unless the Client, within 14 days of receiving it, gives Notice stating the extent to which it does not comply with the Agreement.

The schedule control provisions include the following obligations:

- the Consultant shall keep the Programme under review and shall update it to comply with the Agreement (Sub-Clause 4.3.1);
- the Parties shall promptly give Notice to each other of any events or circumstances that may adversely affect or delay the Services or lead to an increase in the cost of the Services (Sub-Clause 4.3.3).

The Consultant shall be entitled to extension of time (EOT) under Sub-Clause 4.4 if and to the extent that completion of the Services is or will be delayed by:

a) a Variation to the Services;
b) any delay, impediment, or prevention caused by or attributable to the Client, or the Client's other consultants, contractors, or other third parties;
c) an Exceptional Event; or
d) any other event or circumstance giving an entitlement to EOT under the Agreement.

Sub-Clause 4.5 outlines the schedule control procedure in case of the Consultant's delay for any reason that does not entitle the Consultant to an EOT:

- the Client may give Notice that the rate of progress is insufficient to ensure compliance with the Time for Completion; then
- the Consultant shall:
  - revise the Programme; and
  - issue a Notice describing the intended measures to achieve completion of the Services within the Time for Completion.

The 2017 Sub-Consultancy Agreement Clause 4 includes similar GCs related to schedule management. The Sub-Consultant has the obligation to produce a Sub-Consultant's Programme for monitoring and recording purposes. However, the Sub-Consultant shall be bound by and shall comply at all times with the Consultant's Programme submitted to the Client under the White Book Agreement. Any EOT under the Sub-Consultancy Agreement shall have due regard to the Consultant's Programme and any constraints therein.

Suspension of all or part of the Services is possible under the White Book:

- under Sub-Clause 6.1.1 by the Client at its sole discretion and for any reason by giving 28 days' Notice;
- under Sub-Clause 6.1.2 by the Consultant:

- for delayed payment giving 7 days' Notice;
- where an Exceptional Event arises, giving a Notice as soon as reasonably practicable;
- where the Client fails to comply with Sub-Clause 2.4 (Client's Financial Arrangements).

EOT under Sub-Clause 4.4 shall reflect the effect of the suspension on the Programme. The Consultant has an obligation to take reasonable measures to mitigate the effects of the suspension of the Services.

Suspension of the Services is possible under the Sub-Consultancy Agreement Sub-Clause 6.1:

- by the Consultant:
  - at its sole discretion and for any reason by giving 14 days' Notice;
  - immediately if the Services under the (main) Agreement are suspended;
- by the Sub-Consultant where an Exceptional Event arises giving Notice immediately, if the Consultant agrees that an Exceptional Event has arisen.

The Time for Completion shall be amended in accordance with Sub-Clause 4.4 to reflect the effect of the suspension on the Sub-Consultant's Programme.

8.3 Planning Schedule Management in FIDIC Works Contracts

The schedule planning depends on the contractual requirements for the components of the schedule, activity definition, resources, cost loading, and performance evaluation.[4] The most important components of the schedule in FIDIC contracts are the Time for Completion and the Commencement Date, when the countdown begins. The procedure for notifying the Commencement Date is discussed in Chapter 6. The Time for Completion is the time commitment agreed by the Parties under the Contract for completion of the Works or Section for the purposes of taking over under Sub-Clause 10.1. Planning the schedule requires including all activities agreed under the Contract for achieving completion, like passing the Tests on Completion, receiving Notices of No-objection for as-built records and O&M Manuals, training the Employer's Personnel, and other applicable mandatory procedures under the local Laws (e.g. authorised laboratory testing, third-party certification).

If the Works are to be completed by the Contractor and taken over by the Employer in stages, it is important that the Employer defines in the Contract Data each stage as a Section with a separate Time for Completion. The Employer may use the completed Section, or invite another contractor to commence finishing works, or sell the Section. Whether used or not, if the relevant Section is taken over, a Taking-Over Certificate (TOC) should be issued, certifying the Date of Completion for the relevant Section in accordance with Sub-Clause 10.1.

Sub-Clause 10.2 enables taking over parts of the Works when the Employer wishes to use them. The Employer may need some parts of the Works completed due to interfaces

4 PMI Inc., *Construction Extension to the PMBOK® Guide* (PMI 2016) 52

with other contractors/utility operators. If the Employer wishes to have such parts ready on agreed dates without using them, then each such part of the Works should be clearly described in the Specification/Employer's Requirements as a 'Milestone'. To incentivise the Contractor, the GPPC proposes a new Sub-Clause 4.24 to be added to the Special Provisions, linking the Contractor's failure to achieve the Milestones to delay damages.

When the Employer wishes to have the Works or a Section completed before the Time for Completion, then an additional Sub-Clause may be added in the Special Provisions giving the Contractor incentives for early completion. The GPPC suggests a provision entitling the Contractor to a bonus payment.

The critical path method (**CPM**) is preferred for FIDIC schedules, created under Sub-Clause 8.3 [Programme]. The critical path determines the Time for Completion, since it is the longest sequence of logically linked activities. There may be several critical paths with zero **float**. The float is the amount of time that the activity may be delayed beyond its early start/finish dates without delaying the Time for Completion. A delay of any activity on the critical path without acceleration or re-sequencing measures will cause delay to the Time for Completion. It is important for the planning to be realistic, taking into account all known constraints, Site data, and procedural timelines, and then to optimise the resources. The Contractor should include risk contingencies in the Programme, allowing time reserve for Contractor's risk events (e.g. for severe climate conditions, which often happen in winter, so cannot be categorised as exceptionally adverse climatic conditions entitling the Contractor to EOT under the Red/Yellow/Emerald Book Sub-Clause 8.5). The risk contingency shall not be treated as a float owned by the Employer but as an activity on the critical path that cannot be quantified with limited maximum duration. The CA shall carefully Review a programme in which all activities are on the critical path. The CA may reject a programme where the links are not logical as required under Sub-Clause 8.3(g) and the critical paths are obviously manipulated to provide maximum opportunity for the Contractor to claim EOT. The early appointment of key Subcontractors/suppliers giving inputs for the preparation of a realistic Programme is of utmost importance.

The Programme to be submitted under Sub-Clause 8.3 is not part of the Contract but a schedule management tool. It has the following primary purposes:

- as a PM device to inform the CA of the Contractor's intended sequence and timing of the activities so that the CA can plan the activities that involve the input or participation of the Employer's Personnel or others (e.g. submission of drawings under the Red Book, Review periods, access to the Site, etc.);
- to allow the CA to monitor the Contractor's progress;[5]
- to be used as a baseline for assessment of EOT when Employer's risks eventuate.

The FIDIC 2017 Contracts Guide strongly recommends the preferable programming software to be clearly identified in the Specification/Employer's Requirements.[6] Primavera, developed by Oracle Corporation, and Microsoft Project are widely applied by FIDIC users. They enable schedule and resource management complying with the requirements

5 Ellis Baker and others, *FIDIC Contracts: Law and Practice* (Informa Law from Routledge 2009) para 5.47

6 FIDIC, *The FIDIC 2017 Contracts Guide* (FIDIC 2022) 275

of Sub-Clause 8.3 to show the earliest and latest start and finish dates for each activity, calculating the float and the critical paths, taking into account the work days of rest and holidays inserted in the work calendar.

The initial Programme must include all the information and details as listed under Sub-Clause 8.3. They include the intended order of carrying out of the Works and the dates for access to the Site, the Commencement Date, the Time for Completion of the Works and any Section, Review periods, the sequence and timing of testing, and key delivery dates for Plant and Materials. The Programme must be accompanied by a supporting report, describing the major stages and methods for execution of the Works, with estimation of the resources for each major stage.

The author recommends that the Particular Conditions amend Sub-Clause 8.3, adding a requirement for a resource-loaded schedule. The GCs require lists of estimated number of each class of Contractor's Personnel/type of Contractor's Equipment in the supporting report, with no connection to activities and parts of the Site. Showing the resource planning for each activity enables:

- better schedule control;
- timely resource management;
- better cost management, including easier calculation of delay and disruption costs;
- prevention of disputes.

The Construction Extension advises the WBS to be integrated to meet time, cost, and responsibility commitments,[7] which can best be done by linking the activities with the time constraints, resources, and costs.

The FIDIC 2017 Contracts Guide suggests that for simple projects, the Employer may consider simplifying the requirements for the Contractor's Programme.[8] Later this chapter discusses how the schedule management may be innovated using the Last Planner® System (LPS). The integration of LPS may be done by amending Sub-Clause 8.3 in the Particular Conditions and adding detailed requirements in the Specification/Employer's Requirements.

8.4 Developing project schedule in the FIDIC Programme

8.4.1 Sub-Clause 8.3 requirements

Using the agreed software, the Contractor shall submit an initial programme for the execution of the Works to the CA within 28 days after receiving the Notice to commence under Sub-Clause 8.1. The Contractor should take into account:

- the scope of the Works defined under the Contract;
- the minimum contents of the Programme listed in Sub-Clause 8.3;

7 PMI Inc., *Construction Extension to the PMBOK® Guide* (PMI 2016) 53
8 FIDIC, *The FIDIC 2017 Contracts Guide* (FIDIC 2022) 277

- the constraints related to the Site, as stipulated under Sub-Clause 4.10, including climatic conditions and local seasonal restrictions;
- any other constraints and milestones set out under the Contract or instructed by the CA;
- the normal working hours complying with Sub-Clause 6.5;
- the local regulations, especially the minimum duration of the agreement and permitting procedures, complying with Sub-Clause 1.13 (Red/Yellow/Emerald)/1.12 (Silver Book);
- the availability of temporary utilities complying with Sub-Clause 4.19;
- the available resources and any limitations related to their allocation and procurement, including Subcontractors appointed under Sub-Clause 4.4 (Red Book 5.1), design teams under Sub-Clause 5.1 (Yellow/Silver/Emerald Books), requirements for staff and labour under Clause 6, access routes under Sub-Clause 4.15, and supply of Materials under Sub-Clauses 7.1 and 7.2.

The attributes of the activities include durations, costs, labour hours, and quantities.[9] It is important that the Contract specify which attributes have to be presented in the schedule developed under Sub-Clause 8.3. The attributes determine the weight of each deliverable or activity and enable tracking progress.

In order to estimate the duration of the activities, the Contractor has to estimate the amount of work effort required and assign resources to each activity. Changing the key resources allocated to an activity will usually have an effect on the duration, but this is not always the case. Resource optimisation methods are applied, adjusting the start and finish dates of activities to level/smooth the resource use.

Activities should be sequenced in a logical manner, determining predecessors and successors. The type of relationship, such as 'finish-to-start', 'start-to-start', or 'finish-to-finish', should also be determined with necessary leads or lags. A **lead** is the amount of time a successor activity can be advanced with respect to a predecessor activity. A **lag** is the amount of time a successor activity has to be delayed with respect to a predecessor activity. Mandatory dependencies are those that are legally or contractually required or inherent in the nature of the work (e.g. the foundations can only be executed after the excavation; pursuant to Sub-Clause 5.2 of the Yellow/Silver/Emerald Books, the Works can start only after the Notice of No-objection is given or deemed to have been given by the CA for the Contractor's design, following a 21-days' Review Period; the commencement of the relevant works may depend also on a mandatory third-party compliance assessment of the design).

The schedule baseline is the approved version of the schedule model. Sub-Clause 8.3 sets out the procedure:

- the CA shall Review the initial programme within 21 days;
- the initial programme shall become the Programme if the CA gives a Notice of No-objection or is silent, and it will serve as schedule baseline;
- the Contractor shall proceed in accordance with the Programme.

9 PMI Inc., *Construction Extension to the PMBOK® Guide* (PMI 2016) 54

8.4.2 Case study – project schedule management

Figure 8.1 presents a case study inspired by a real project using a Red Book Contract for a Regional Waste Management Centre. The scope includes remediation of existing landfills in two towns: town K is the regional city, and town M is a smaller one. The Works include reshaping of the old waste, then construction of a top capping system with leachate and gas collection, surface water drainage system, and finally, hydro-seeding. The town K part of the Site is on the critical path, since the landfill body surface is more than twice as large as the one near town M. The Time for Completion is agreed under the Contract to be 360 days, which is the duration of the activities on the critical path for town K. The duration of the activities for town M is 200 days. They start 100 days after the activities in town K, since the Contractor has planned to finish the shaping of the waste in K with a special machine for 100 days and then transfer it to reshape the waste in town M. The relation between part K and part M is 'SS+100 d' (start-to-start plus 100 days' lag). According to the initial planning, part M will be ready 60 days earlier than the Time for Completion, which is the float for M.

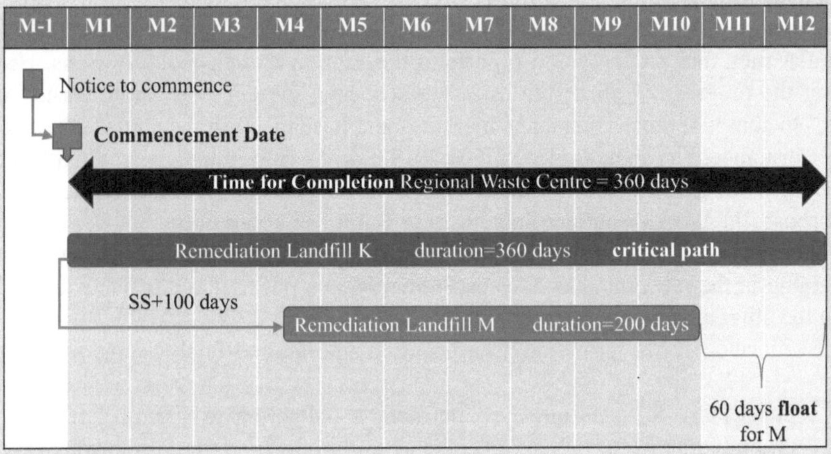

Figure 8.1 Case study: Master Schedule in a FIDIC Contract

8.5 Controlling the Schedule in FIDIC Works Contracts

8.5.1 Contractor's obligations for compliance with the schedule baseline and the CA's control

Sub-Clauses 8.1 and 8.2 set out three separate obligations of the Contractor related to the schedule baseline:

i. to commence the execution of the Works on, or as soon as is reasonably practicable after, the Commencement Date and then
ii. to proceed with the Works with due expedition and without delay and
iii. to complete the whole of the Works, and each Section (if any), within the Time for Completion, including completion of all work that is stated in the Contract as

being required for the Works or Section to be considered to be completed for the purposes of taking over under Sub-Clause 10.1.

In the Sub-Clause 4.20 progress reports, the Contractor has an obligation to compare the actual and planned progress, and give details of any events or circumstance which may adversely affect the completion of the Works in accordance with the Programme and the Time for Completion, including measures adopted to overcome delays. In each progress report, the Contractor shall include Sub-Clause 6.10 records of the resources used for each work activity shown in the Programme, at each work location, and for each day of work.

The Programme is the basis for monitoring the Contractor's progress and planning the CA's activities and obligations. It is also important for the CA regularly to check the Contractor's progress reports and resource records and notify the Contractor in the case of missing or wrong details. In addition, during the monthly meeting under Sub-Clause 3.8 (Silver Book: 3.6), the Contractor reports progress, and the CA comments and records any reasons for delays. The use of Site Diaries and Issue Logs, discussed in Chapter 6, facilitates schedule control. The reports, minutes of meetings, Site Diaries, and related correspondence constitute contemporary records as defined under Sub-Clause 20.2.3 of particular importance for delay and disruption Claims.

The CA monitors progress and reports to the Employer on a monthly basis. Besides tracking the progress of all activities in the schedule, the CA may apply earned value analysis to compare the performance measurement baseline with the actual schedule and cost performance. The standard ISO 21508:2018 'Earned value management in project and programme management' may be applied for review of past and forecast performance. For this purpose, the Works Contract Specification/Employer's Requirements, besides stating the programming software, must include a requirement for the Contractor to submit the Programme in the relevant native format (not only as a pdf file) and to include the costs among the other attributes of the activities.

Clause 8 includes further provisions for schedule monitoring and timely change management:

- Sub-Clause 8.3 sets out the Contractor's obligation to submit a revised programme that accurately reflects the actual progress of the Works whenever any programme ceases to reflect actual progress or is otherwise inconsistent with the Contractor's obligations; it also empowers the CA when the Programme ceases to reflect actual progress:
  - to give a Notice to the Contractor, then:
  - the Contractor has 14 days to submit a revised programme, then;
  - the CA has 14 days for the Review;
  - after a Notice of No-objection is given (or deemed to be given if the CA is silent), the revised programme becomes the Programme;
- under Sub-Clause 8.4, each Party and the CA may give advance warning of any known or probable future events or circumstances that may delay the execution of the Works or a Section; then, the CA may request the Contractor to submit a proposal under Sub-Clause 13.3.2 to avoid or minimise the delay;
- under Sub-Clauses 8.5 and 8.6, the Time for Completion should be extended if and to the extent that completion for the purposes of taking over is or will be delayed by listed causes that are the Employer's responsibility (related to any impacted activities on the critical path(s), including those labelled 'Contractor's contingency');

- Sub-Clause 8.7 stipulates that when progress has fallen behind the schedule baseline, the CA may instruct the Contractor to submit a revised programme describing the revised methods that the Contractor proposes to adopt in order to expedite progress to complete within the Time for Completion.

8.5.2 Entitlement to EOT

The Time for Completion stated in the Contract Data may be extended under Sub-Clause 8.5 to include any EOT due for excusable delays. **Entitlement to an EOT arises only if the critical path is or will be affected.** Table 8.1 demonstrates that the grounds for EOT under the Silver Book are considerably less than those under the Red/Yellow/Emerald Books.

When determining each EOT under Sub-Clause 20.2 [Claims For Payment and/or EOT], the CA shall review previous determinations under Sub-Clause 3.7 (Silver Book: 3.5), and may increase, but shall not decrease, the total EOT. There is no requirement to comply with Sub-Clause 20.2 in the case of Variations, where the Sub-Clause 13.3 procedure is followed. If a delay caused by a matter that is the Employer's responsibility is concurrent with a delay caused by a matter that is the Contractor's responsibility, the Contractor's entitlement to EOT shall be assessed in accordance with the rules and procedures stated

Table 8.1 Provisions entitling the Contractor to EOT

Sub-Clause	*Grounds for EOT*	*Silver Book*
Red 1.9	Delayed Drawings/instructions	No
Yellow/Emerald 1.9	Errors in the Employer's Requirements	No
1.13 (Silver 1.12)	Employer's delay to obtain permits/approvals	Yes
2.1	Late access/possession of the Site	Yes
4.6	Delay due to instruction for Unforeseeable co-operation	Yes
4.7	Error in the positions, levels, dimensions, or alignment	No
4.12	Unforeseeable physical conditions	No
4.15	Changes to the access route by the Employer or a third party	Yes
4.23	Archaeological and geological findings	Yes
7.4	Delayed tests	Yes
Red 8.5a	Additional quantities	No
8.5/13.3	Delay due to Variation	Yes
8.5	Exceptionally adverse climatic conditions (Unforeseeable having regard to available climatic data)	No
8.5	Unforeseeable shortages in resources caused by epidemic or governmental actions	No
8.5	Delays by Employer, the Employer's Personnel, or the Employer's other contractors on the Site	Yes
8.6	Delays caused by Authorities	Yes
8.10/8.12	Suspension instructed by the Employer	Yes
10.3	Interference with Tests on Completion	Yes
13.6	Adjustments for changes in Laws	Yes
16.1	Suspension by the Contractor	Yes
16.2	Termination by the Contractor, 14 days between the Notices	Yes
17.2	Rectification of loss/damage	Yes
18.4	Consequences of Exceptional Events	Yes

in the Special Provisions (if not stated, as appropriate, taking due regard of all relevant circumstances). The UK SCL Delay and Disruption Protocol[10] may be included as a reference where the governing law does not impose mandatory rules.

8.5.3 Case study – Master Schedule with Employer's delay

Figure 8.2 shows a development of the Red Book case study where the Employer has delayed the access to the part M of the Site by 240 days. Since part M is not on the critical path, at the beginning, the delayed access does not have any impact. After Day 100, the delay causes disruption, since the special machine completing the reshaping of the waste on part K cannot start this activity on part M, as planned. Whether the special machine is owned by the Contractor or rented, there will be disruption costs while it is idle. After the float is over, the activities on part M become on the critical path, and the Contractor will be entitled to an EOT to the extent that completion will be delayed. Under Sub-Clause 2.1, the Contractor may be entitled to Cost Plus Profit for this period as well, based on the incurred prolongation costs. The Contractor is entitled to 80 days' EOT under Sub-Clause 2.1, subject to Sub-Clause 20.2, since this is the delay on the critical path caused by the Employer's risk event:

100 days (to planned start on M) + 200 days (part M duration) + 60 days (float) = 360 days (critical path = original Time for Completion)

240 days (before actual start of M) + 200 days (part M duration) = 440 days (actual duration)

440 days (actual duration) – 360 days (initial Time for Completion) = 80 days EOT

So, the Time for Completion becomes 440 days, including 80 days for delayed access to part M. The Contractor's entitlement to prolongation and disruption costs incurred due to the Employer's delay is discussed in Chapter 9.

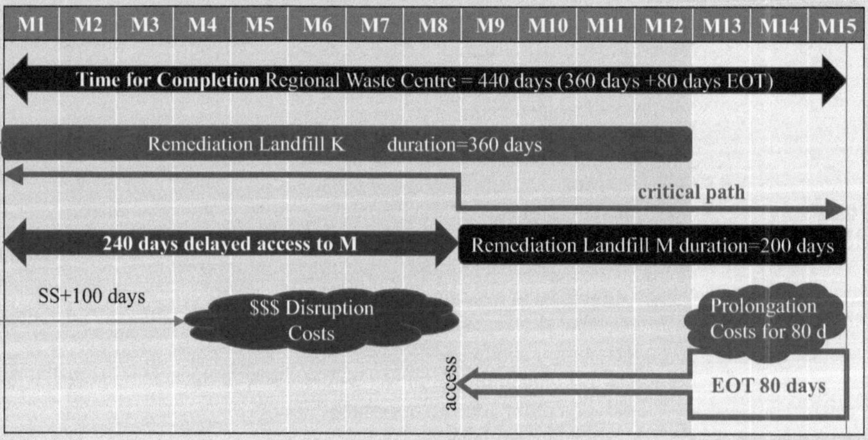

Figure 8.2 Case study: Master Schedule with Employer's delay

10 <https://www.scl.org.uk/sites/default/files/documents/SCL_Delay_Protocol_2nd_Edition_Final.pdf> accessed 8 April 2023

8.5.4 Non-excusable delays

Clause 8 provides remedies for the Employer in case of the Contractor's non-excusable delays, incentivising the Contractor to comply with the time-related obligations under Sub-Clauses 8.1/8.2:

- Sub-Clause 8.8 entitles the Employer to payment of Delay Damages by the Contractor in the amount stated in the Contract Data for every day which elapses between the relevant Time for Completion and the relevant Date of Completion of the Works or Section;
- Sub-Clause 8.7 entitles the Employer to payment of incurred additional costs due to adopted revised methods by the Contractor in order to expedite progress, in addition to Delay Damages (e.g. for additional time spent for supervision outside normal working hours).

The Delay Damages shall not relieve the Contractor from the obligation to complete the Works. The Employer has additional remedies for the Contractor's failure to complete within the Time for Completion:

- a prolonged delay in completion of the Works when the Delay Damages exceed the maximum amount may expressly entitle the Employer to terminate the Contract under Sub-Clause 15.2.1 (c);
- a failure to complete the Works (but not Sections) within the Time for Completion will also impact on the Contractor's entitlement to price escalation amounts under Sub-Clause 13.7 in favour of the Employer.

8.5.5 Case study – Master Schedule with Employer's and Contractor's delays

Figure 8.3 shows a further development of the Red Book case study where the Employer has delayed the access to part M of the Site by 240 days. The Time for Completion becomes 440 days,

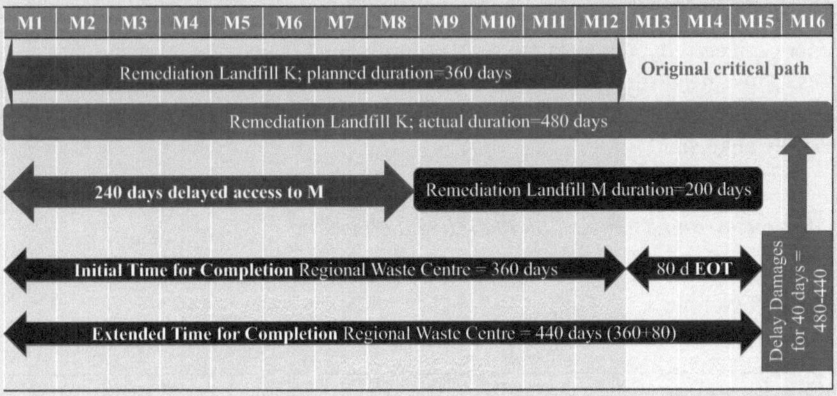

Figure 8.3 Case study: Master Schedule with Employer's and Contractor's delays

including 80 days' delay on the critical path. At the same time, the Contractor is in delay for the part K, which is completed within 480 days. The Delay Damages due will be calculated for 40 days:

480 days (actual duration part K) – 440 days (Time for Completion) = 40 days (non-excusable delay)

8.5.6 Delay analysis

The SCL Protocol recommends a common-sense approach to delay analysis, especially establishing concurrency.[11] When agreeing/determining the entitlement to EOT under Sub-Clause 8.5, the CA must establish **separately for each Employer's risk event** whether and to what extent completion for the purposes of taking over is or will be delayed. EOT can be claimed either when it is clear that there will be delay (prospective delay) or when the delay has at least started to be incurred (a retrospective delay).[12] The SCL Protocol describes different methods for delay analysis. The method may be agreed under the contract, taking into account the specific requirements for the schedule management. Different methods may be more efficient in different situations, but for the sake of dispute prevention, it is recommended that the method should be agreed in advance if not stated in the Contract.

Critical path analysis is not limited to the use of specialist programming software. While such software can provide a powerful analytical tool, the critical path may on occasion be more reliably established through a practical analysis of the relevant facts or by analysis of production and/or resource data.[13] The case law confirms that 'the reliability of any sophisticated impact analysis is only as good as the data put in'.[14]

Delay due to Employer's risk event does not automatically lead to EOT. As demonstrated with the example presented in Figure 8.2, the CA shall agree/determine the extent to which the critical path is impacted. The CA has to monitor the schedule baseline on a monthly basis. When the Programme ceases to reflect actual progress, the CA shall request a revised Programme that accurately reflects the critical path(s). Sub-Clause 8.3 sets out additional requirements for a revised Programme to present the actual progress to date for each activity, any delay to such progress, and the effects of such delay on other activities. After the 14-day Review period, if the CA gives (or is deemed to have given) a Notice of No-objection to the revised programme, it becomes the Programme to be used for the delay analysis.

8.5.7 Suspension and its impact on the schedule baseline

The FIDIC GCs stipulate provisions for suspension by the Parties and related changes to the schedule baseline:

11 SCL Delay and Disruption Protocol (2017) 10.11

12 *Obrascon (OHL) v Her Majesty's Attorney General for Gibraltar* [2014] EWHC 1028 (TCC) 312

13 SCL Delay and Disruption Protocol (2017) 11.4 (c)

14 *Skanska Construction UK Ltd vs Egger Ltd* [2004] EWHC 1748 (TCC)

- Sub-Clause 8.9 authorises the CA to instruct the Contractor to suspend the progress of part or all of the Works, stating the cause of the suspension:
  - the Contractor shall not be entitled to an EOT when in default (e.g. in case of: unsafe working; working with non-compliant/missing Contractor's Documents; covering defects; breach of Laws related to the performance);
  - if the Contractor suffers delay from complying with a CA's instruction for suspension under Sub-Clause 8.9 and/or resuming the work under Sub-Clause 8.13, it shall be entitled to an EOT, subject to Sub-Clause 20.2.
- Sub-Clause 16.1 entitles the Contractor to suspend work (or reduce the rate of work) 21 days after giving a Notice to the Engineer (Silver Book: Employer) unless and until the Employer has remedied certain defaults (delayed certification/payment; failure to provide reasonable evidence of financial arrangements; non-compliance with a CA's agreement/determination/ DAAB's decision, constituting a material breach of the Employer's obligations under the Contract). If the Contractor suffers delay as a result of suspending work (or reducing the rate of work), it shall be entitled to an EOT, subject to Sub-Clause 20.2.

Sub-Clause 8.12 sets out a procedure in the case of suspension continuing for more than 84 days:

- the Contractor may give a Notice to the Engineer (Silver Book: Employer) requesting permission to proceed;
- if the Engineer (Employer) does not give a Notice within 28 days under Sub-Clause 8.13 [Resumption of Work], the Contractor may:
  (a) agree to a further suspension, in which case the Parties may agree the EOT and/or Cost Plus Profit, and/or payment for suspended Plant and/or Materials; or
  (b) after giving a (second) Notice, treat the suspension as an omission of the affected part of the Works under Sub-Clause 13.3.1 with immediate effect (including release from any further obligation to protect, store, and secure those Works). If the suspension affects the whole of the Works, the Contractor may give a Notice of termination under Sub-Clause 16.2.1.

8.6 Schedule Management under the Emerald Book

The Emerald Book Contractor's Tender shall include production rates for each excavation drive or zone, based upon the conditions described in the Geotechnical Baseline Report (GBR). The estimated quantities of the excavation, ground support, ancillary measures, and lining, as well as the corresponding production rates, shall be included in the Schedule of Baselines (Sub-Clause 8.2.2), and the corresponding sequential logical links between Milestones shall be included in the Completion Schedule (Sub-Clause 8.2.1).

All subsurface physical conditions not addressed in the GBR shall be considered Unforeseeable. In the case of encountering more onerous conditions that impact on the critical path, the Time for Completion should be extended, while in the case of less onerous conditions, the Time for Completion should be reduced.

8.7 Innovative Schedule Management

8.7.1 Scheduling software

Scheduling software has the capability to help plan, organise, and adjust the sequence of the activities; insert the logical relationships, lead and lag values; and differentiate the different types of dependencies. It enables tracking of progress and reporting the resource allocation and task assignment. Software can link the schedule with the three-dimensional (3D) model and automatically generate notices to the supply chain.

8.7.2 Last Planner® System (LPS)

The CPM logic focuses on managing dates, but the longer and the more detailed the forecast, the more wrong it is likely to be. LPS helps to move from a push to a pull approach. LPS supports logistics planning involving much more than just the due dates and the sequence of tasks.[15] LPS is based on a 'Should-Can-Will-Did' system of project planning instead of the CPM 'Shall Do' approach. In order to ensure the most effective and efficient use of capacity, the work that SHOULD be performed by a certain date must be available to be performed (CAN) without any blockage or interruption, i.e. constraint.[16] LPS focuses on making a realistic 6–8-week Look-Ahead Schedule with detailed weekly plans in discussion with the last planners (trade foremen in charge of the execution of the works in the next look-ahead window). Eliminating final constraints, the foremen prepare the assignments for the workers. In this way, the workers are neither overloaded nor idle; they only do what they have promised, and this helps to keep a track of the productivity. LPS utilises the 'just in time' tool, since all the project participants sit together to form the Look-Ahead Schedule, wherein continuous improvement is built into the process.[17]

With LPS, the schedule control becomes proactive rather than reactive. It is compared to a fire prevention system, while the traditional approach is compared to fire-fighting.[18] LPS can be integrated in FIDIC contracts, replacing the second paragraph of Sub-Clause 8.3 with requirements for a Programme in the form of a milestone Master Schedule. The Contractor shall then submit resource-loaded detailed Look-Ahead Schedules several weeks before the relevant look-ahead period commences. Defining the look-ahead period in the range of 6 to 8 weeks, long-lead items need to be identified and presented in the initial milestone Master Schedule. Detailed requirements referring to LPS should be drafted in the Specification/Employer's Requirements.

---

15 Katarzyna Ćwik and Jerzy Rosłon, 'Last Planner System in Construction' (DOI: 10.1051/matecconf/201711700032) <https://www.matec-conferences.org/articles/matecconf/pdf/2017/31/matecconf_rsp2017_00032.pdf> accessed 1 September 2022

16 Glenn Ballard and Iris Tommelein, 'Current Process Benchmark for the Last Planner System' (2016) http://p2sl.berkeley.edu/wp-content/uploads/2016/10/Ballard_Tommelein-2016-Current-Process-Benchmark-for-the-Last-Planner-System.pdf> accessed 12 September 2022

17 Ankit Bhatla and Fernanda Leite, *Integration framework of BIM with the Last Planner System*TM (Proceedings for the 20th Annual Conference of the International Group for Lean Construction) <https://www.academia.edu/2851478/INTEGRATION_FRAMEWORK_OF_BIM_WITH_THE_LAST_PLANNER_SYSTEM_TM> accessed 12 September 2022

18 Alan Mossman, 'Last Planner®: 5 + 1 Crucial & Collaborative Conversations for Predictable Design & Construction Delivery' (2015) <https://www.researchgate.net/publication/235791767_Last_PlannerR_5_1_crucial_collaborative_conversations_for_predictable_design_construction_delivery_Dec_2015> accessed 12 September 2022

8.7.3 4D Building Information Modelling (BIM)

Four-dimensional (4D) BIM links the 3D digital model with the schedule. It incorporates start and finish dates for the supply and installation of construction components. It enables collaboration between the design and construction teams and the development of a realistic schedule. 4D BIM may be used with Lean planning tools, especially to form the look-ahead plans. Integrating as-built information helps schedule control.

8.7.4 4D visualisations

4D simulations facilitate schedule development:[19]

- planning the schedule and Site arrangement;
- presenting construction sequencing;
- identifying potential conflicts and finding potential optimisations;
- facilitating interdisciplinary coordination and review;
- communicating the Programme to the project stakeholders;
- providing an essential overview of the construction process to eliminate congestion and improve logistics, health, and safety.

4D simulation may be used for schedule control:

- to show delayed, on time, or accelerated activities;
- to visualise schedule changes by updating the planned schedule up to the as-built conditions;
- to show the cause of an event, the responsible party, and the percentage of that responsibility with visual indicators;[20]
- for delay analysis to identify criticality of components with colour coding based on the float of each activity.[21]

8.7.5 AI combined with digital tools

AI can optimise changing workflow through multitask learning to prioritise the most relevant tasks in schedule planning and control.[22] AI saves time and other resources, improving the value by:

---

19 NIRAS A/S, '4D Visualisation of the Building Process' <https://www.niras.com/sectors/building/bim-vdc-and-ict-management/4d-visualisation-of-the-building-process/> accessed 12 September 2022

20 Michel Guévremont and Amin Hammad, 'Review and Survey of 4D Simulation Applications in Forensic Investigation of Delay Claims in Construction Projects' (ASCE 2020) <https://ascelibrary.org/doi/10.1061/%28ASCE%29LA.1943-4170.0000391> accessed 11 September 2022

21 Michel Guévremont and Amin Hammad, 'Criticality Visualization Using 4D Simulation for Major Capital Projects' (2017) <http://simulation.su/uploads/files/default/2017-guevremont-hammad.pdf> accessed 11 September 2022

22 McKinsey & Company, 'McKinsey Technology Trends Outlook 2022. Applied AI' (2022) <https://www.mckinsey.com/~/media/mckinsey/business%20functions/mckinsey%20digital/our%20insights/the%20top%20trends%20in%20tech%202022/McKinsey-Tech-Trends-Outlook-2022-Applied-AI.pdf> accessed 3 September 2022

- helping designers to access the database;
- detection of clashes and allowing fast changes of plans;
- allowing more parameters to be considered in less time;
- tracking instant progress of construction against the plan stored in the BIM cloud using frequent image recording and 360-degree laser scans done by robots and drones;
- detecting variance automatically by comparing the digital twin and the as-built;
- performing automatic changes in the construction schedule in the BIM cloud;
- helping workers find tools faster with the use of the image recognition software;
- informing stakeholders instantly and automatically about progress;
- performing troubleshooting at an early stage;
- optimising materials delivery.[23]

Project teams can leverage generative-scheduling tools to test millions of schedule configurations and find the optimal work sequence and resources for a construction project in a matter of minutes.[24]

8.7.6 Distributed Ledger Technology (DLT), Internet of Things (IoT), and iContracts

The UK Weather Ledger project is a practical demonstration of DLT and IoT applied to automate the entitlement to an EOT.[25] The Weather Ledger facilitates schedule management:

- saving time for collecting weather reports and compiling records;
- preventing disputes, since the Parties use the same validated data, and the entitlement to an EOT is calculated transparently as agreed under the Contract;
- advance weather reports distributed from the Weather Ledger improve the daily planning and the mitigation of adverse weather impact.

A similar approach may be used for FIDIC Red/Yellow/Emerald Book contracts in the case of exceptionally adverse climatic conditions entitling the Contractor to an EOT under Sub-Clause 8.5. IoT weather sensors may be used to capture Site weather parameters. Sub-Clause 8.5 (c) may be tailored to fit into a smart contract (law-to-code binding).

IContracts can analyse in real time the entitlements resulting from codified risk events where the IoT sensors provide data. Also, iContracts may compare the actual contract

23 viAct, 'Artificial Intelligence: How Will It Affect Construction Industry' (2021) <https://www.viact.ai/post/artificial-intelligence-how-will-it-affect-construction-industry> accessed 17 August 2022

24 McKinsey & Company, 'Generative Scheduling: Saving Time and Money in capital Projects' (2022) <https://www.mckinsey.com/capabilities/operations/our-insights/generative-scheduling-saving-time-and-money-in-capital-projects> accessed 9 October 2022

25 Digital Catapult, 'The Weather Ledger' (2021) <https://www.digicatapult.org.uk/news-and-insights/publications/post/the-weather-ledger/> accessed 11 September 2022

performance from the data received with the contract requirements and allow the CA/Parties to address the situation.[26]

8.8 Example Notices

Appendix 1 includes examples of Engineer's Notice of non-compliance of the initial programme pursuant to Sub-Clause 8.3 and Notice for slow progress under Sub-Clause 8.7.

26 Alan McNamara, 'Automating the Chaos: Intelligent Construction Contracts' Chapter in *Smart Cities and Construction Technologies* (IntechOpen 2020) 12

CHAPTER 9

Project Financial and Cost Management with FIDIC Contracts

 DOI: 10.4324/9781003272892-9

9.1 Project Financial Management with FIDIC Conditions of Contract

9.1.1 Project Financial Management processes

The PMBOK® Guide does not include a section dedicated to Project Financial Management, but its Construction Extension has one. It presents project financial management planning, monitoring and control. This Knowledge Area covers the processes to acquire and manage the financial resources for the project. It is more concerned with revenue sources and monitoring net cash flows for the construction project than with managing the day-to-day costs covered in Section 9.2.

9.1.2 Project Financial Management planning

Project Financial Management planning identifies the sources of financing and provides all financial requirements, assigning roles, responsibilities, and reporting obligations.

9.1.2.1 Financing arrangements

The FIDIC Procurement Procedures Guide recommends a financing time schedule to be prepared in conjunction with the project time programme, securing the flow of the funds to meet the payment milestones.[1] Based on the cost plan, the financing plan is drawn up to cover the total amount of finance required, including financing costs, taxes like VAT, and any contingency reserve.

The financing model should take into account requirements for costs in one currency to be financed in that currency.[2] Sub-Clause 7.3 of the White Book stipulates that applicable currencies shall be stated in Appendix 3. It provides relief in the case of issues related to the currency inhibiting the Consultant in the performance of the Services or resulting in financial disadvantage to a Consultant from another country. Sub-Clause 14.15 of the Red/Yellow/Emerald/Silver Books includes payment arrangements to be used where the Contract Price shall be paid in more than one currency named in the Contract Data.

1 *FIDIC Procurement Procedures Guide* (2011) 4.2
2 Yescombe, *Principles of Project Finance* (Elsevier 2014) 353

If loans or grants are necessary, the applicant must meet the eligibility criteria and strictly follow the requirements of the financing institution. Eight MDBs recently signed licence agreements with FIDIC for the use of FIDIC Contracts. Each of them has its own sets of standard procurement documents, including Particular Conditions, amending the FIDIC GCs, to be used by the beneficiaries.[3]

In corporate finance, loans are primarily lent against a company's balance sheet and financial projections extrapolated from its past cash flow and profit record.[4]

Project finance is an alternative method of funding in which the lender looks primarily to the revenues generated by a single project, both as the source of repayment and as security for the exposure. The borrower is usually a Special Purpose Vehicle (**SPV**) that is not permitted to perform any function other than develop, own, and operate the project. Project financing is used for large, complex, and expensive installations that might include, for example, power plants, chemical processing plants, mines, transportation infrastructure, environment, and telecommunications infrastructure, often executed with FIDIC Contracts. This will be discussed further under public–private partnerships (PPPs).

9.1.2.2 Case study – SPVs

For example, a global energy corporation enters a new country with the goal of developing a portfolio of photovoltaic parks (**PVPs**). The corporation finds feasible projects and registers for each of them a SPV with the owner of the land. Each SPV signs a grid connection agreement with a power transmission or distribution company, a power purchase and power balancing group agreement with a licensed power trader, and a loan agreement with a commercial bank. Before signing the loan agreement for a PVP, the bank requires due diligence to be performed. Each SPV is the Employer under a FIDIC Silver Book Contract for the design, construction, and commissioning of the relevant PVP.

9.1.2.3 Financing provided by the Employer

In traditional projects, the Employer pays the Contractor. The FIDIC Employer may be a private or a public entity (local authority, ministry, agency, municipal or state water company, etc.).

The private funds may be a combination of loans, retained profits, financial reserves, or new capital. For example, the USA Micron Corporation Inc. decided to use FIDIC Conditions of Contract for their projects all over the world. In 2021, they intended to finance several projects with their own funds.

Public employers use funds from taxation or similar levies, grants, or loans. If financing is to be procured from aid agencies, development banks, export credit agencies, or other international financial institutions (IFIs), the Special Provisions may need to incorporate the IFI's special requirements. FIDIC lists what those requirements may cover and advises

3 For example, the World Bank Standard Procurement Documents, *Request for Bid Works (Without Prequalification)* (2021). New definitions are added in the Special Provisions: '"**Bank**" means the financing institution (if any) named in the Contract Data' (Sub-Clause 1.1 89); '"**Borrower**" means the person (if any) named as the borrower in the Contract Data' (Sub-Clause 1.1.90)

4 Yescombe, *Principles of Project Finance* (Elsevier 2014) 6

making reference to the institution to ascertain them, and seeking approval of the draft tender documents.[5]

Where the financing is provided by commercial banks, those banks may need to ensure that the Contractor's rights to additional costs are very restricted, preferring the Contract to be based upon the Silver Book, albeit this may lead to higher tender prices for the Works.[6]

Sub-Clause 2.4 of the White Book obliges the Client to submit to the Consultant, within 28 days of receiving a request, reasonable evidence that financial arrangements have been made to enable timely payments as stipulated in the Agreement. The failure to comply entitles the Consultant to:

- suspend the Services under Sub-Clause 6.1.2(c);
- EOT and adjustment of the agreed remuneration under Sub-Clause 6.3.3 to reflect the impact of suspension and resumption of the Services;
- terminate in case of suspension longer than 42 days, giving 14 days' Notice under Sub-Clause 6.4.2(b);
- adjustment of the agreed remuneration under Sub-Clause 6.5.3 to reflect the impact of termination.

Sub-Clause 2.4 sets out the Employer's obligation to include details for the financial arrangements in the Contract Data of works contracts. The FIDIC 2017 Contracts Guide advises:

> If funding is being provided as a bank loan, or a loan or grant from an international financing institution (such as one of the bilateral or multilateral development banks) it may be possible to provide tenderers with copies of the written loan or grant agreement from the Employer's bank or funding agency. In this way tenderers would be provided with details of the financing arrangements, including the timing and amounts of moneys to be released, particularly if such release is in instalments. Or, if funding is coming from a national or governmental source, an officially authorised formal letter confirming the arrangements and the availability of funds should be provided so that tenderers know that adequate funding has been secured.[7]

The Employer's failure to provide reasonable evidence in accordance with Sub-Clause 2.4 entitles the Contractor to:

- suspension/reduction of the rate of work after 21 days' Notice under Sub-Clause 16.1(b) until reasonable evidence is received;
- EOT and/or Cost Plus Profit for the period of suspension, subject to Sub-Clause 20.2;
- 14 days' Notice of termination under Sub-Clause 16.2.1(a) (42 days after giving a Notice under Sub-Clause 16.1(b));
- termination under Sub-Clause 16.2.2 by giving a second Notice if no payment is received after the 14 days;
- EOT and/or Cost Plus Profit for the 14 days' period under Sub-Clause 16.2.2, subject to Sub-Clause 20.2.

5 FIDIC, *Yellow Book* GPPC 46
6 FIDIC, *Yellow Book* GPPC 46
7 FIDIC, *The FIDIC 2017 Contracts Guide* (2022) 113

A recent case[8] serves as a warning to Employers who take a relaxed view towards their obligation under Sub-Clause 2.4. The Privy Council found that the arbitrator's conclusion that insufficient evidence had been provided was one of fact, not law, and therefore, it was not open to a court to interfere with, or set aside.[9] According to the arbitrator, the Contractor was entitled to terminate the Contract under Sub-Clause 16.2 due to the Employer's breach of Sub-Clause 2.4:

> The mere fact than an Employer is wealthy is inadequate for the purpose of Sub-Clause 2.4. Similarly, the mere fact than an Employer has good reasons for wanting a project completed does not itself mean he has made and maintained the necessary financial arrangements.

9.1.2.4 Case studies – public infrastructure

The New Europe Bridge connecting Bulgaria and Romania is an example of public infrastructure financed from different sources using the FIDIC Yellow Book for the bridge and the Red Book for the adjacent infrastructure. The 1,971 m-long bridge was financed from EU ISPA (Instrument for Structural Policies for Pre-accession) grant, EIB credit, German KfW grant and loan, French AFD grant, and the Bulgarian budget (€60.8m out of €282m).[10]

Another example for financing of public infrastructure from different sources is the construction of the motorway on Corridor Vc in the Federation of Bosnia and Herzegovina (**FBiH**) using the FIDIC Red and Yellow Books for different sections. In accordance with the *Law on roads of FBiH*, financial resources are provided from public revenues (excise duties, tolls, use of road land), loans, and donations (grants). In the period from 2008 to 2019, agreements have been concluded with the EIB, the EBRD, the Organization of the Petroleum Exporting Countries (**OPEC**) Fund, and the Kuwait Fund for Arab Economic Development. Since the construction of Corridor Vc is financed from funds by IFIs that contribute to the Western Balkans Investment Framework (**WBIF**), the FIDIC Employer JP Autoceste is able to apply for grants from the EU. After the adoption of the Economic and Investment Plan for the Western Balkans, the EC adjusted the WBIF application criteria, allowing 40 per cent co-financing for road infrastructure (instead of the previous 20 per cent), so JP Autoceste could apply for more grants.[11]

9.1.2.5 Financing provided by the Contractor

The Contractor may be prepared to finance the project with a bank loan. The financing bank may require the Employer to make interim payments, although a large proportion of the Contract Price might be withheld until the Works are completed. Since the Contractor would then have to arrange financing to cover the shortfall between the payments and the Contractor's outgoings, the Contractor (and the financing bank) would probably require some form of security from the Employer, guaranteeing payment when due. FIDIC Books

8 *NH International (Caribbean) v National Insurance Property Development Company (Trinidad & Tobago)*[2015] UKPC 37

9 Victoria Tyson, 'Employers Beware' (Corbett & Co. 2015) <https://www.corbett.co.uk/employers-beware/> accessed 2 January 2023

10 Road Traffic Technology, 'New Europe Bridge (formerly Danube Bridge 2)' <https://www.roadtraffic-technology.com/projects/danubebridge/> accessed 1 January 2023

11 JP Autoceste FBiH d.o.o. Mostar, Project Progress Report Construction of Motorway on Corridor Vc in the Federation of B&H as of June 30th 2021 <https://www.jpautoceste.ba/wp-content/uploads/2021/09/EBRD-Q2-2021.pdf> accessed 1 January 2023

include a sample form of payment guarantee by Employer and an example Sub-Clause for Contractor's finance.[12]

9.1.2.6 Long time for payment and late payments lead to self-financing

FIDIC Sub-Clause 14.2 includes provisions for an interest-free loan for mobilisation (and design, if any), if the amount is stated in the Contract Data. This amount is often 15 per cent of the ACA (Silver Book: Contract Price), which cannot secure the cash flow needed in the first months if significant payments have to be made to suppliers. The default time for payment under Sub-Clause 14.7 is 56 days for the Contractor and 70 days for the Subcontractor (Subcontracts for Red/Yellow Book 1999) after the receipt of the statement for the work executed in the previous period. So, the Contractor is paid 2–3 months after the Materials are paid and delivered and the work done, and the Subcontractor even later. If the Time for Completion is 20 months and the average monthly payment is 5 per cent, about 10 per cent has to be financed by the Contractor/Subcontractor. Sub-Clause 1.7 enables the Contractor to leverage its cash flow with loans, using future payments as security in favour of a bank or financial institution. Contractors often use the payment from one project to finance other projects, which can be prevented if payments are transferred to a project bank account (**PBA**).

Contractors prefer 'pay when paid' provisions in the Subcontracts, which is the case in FIDIC Subcontracts, where payments may be withheld/deferred under Sub-Clause 14.7 for amounts not certified/paid under the Main Contract.[13] Delayed payments to Subcontractors lead to numerous disputes and insolvency.[14]

Cash flow issues escalate when the Employer cannot secure additional funds to finance Variations/Changes in Laws or Cost under Clause 13. The Employer's failure to pay within the Sub-Clause 14.7 time limits often results in low quality of Materials and workmanship, delay and disruption due to delayed supplies, and additional costs. It may lead to insolvency of the Contractor and/or key Subcontractors, and termination.

The Red/Yellow/Emerald/Silver Books stipulate the following remedies for Contractors suffering from payments delayed beyond the Sub-Clause 14.7 periods:

- entitlement to financing charges for delayed payment under Sub-Clause 14.8, without submitting a Claim under Sub-Clause 20.2, in parallel with the other remedies;
- suspension/reduction of the rate of work, giving 21 days' Notice under Sub-Clause 16.1(c), until the payment is received;
- EOT and/or Cost Plus Profit for the period of suspension, subject to Sub-Clause 20.2;
- 14 days' Notice of termination under Sub-Clause 16.2.1(c) (42 days after the expiry of the time for payment);

12 FIDIC, *Yellow Book* GPPC (2022) 46, 68

13 The GPPC gives example provisions for jurisdictions where pay-when-paid terms are unenforceable, such as the UK, Canada, and Australia; see Subcontract for Plant and Design-Build (FIDIC 2019) GPPC 22

14 Sam Bolton and others, 'Late Payment to Subcontractors in the Construction Industry' (*Journal of Legal Affairs and Dispute Resolution in Engineering and Construction* 2022) 14 (4): Analysis of payments in 30 UK projects demonstrates late payment in 77 per cent of the subcontracts

- termination under Sub-Clause 16.2.2 by giving a second Notice if no payment is received after the 14 days' Notice;
- EOT and/or Cost Plus Profit for the 14 days' period, subject to Sub-Clause 20.2.

Sub-Clause 14.8 stipulates rules for calculation of the financing charges for delayed payments in breach of Sub-Clause 14.7. The GPPC suggests, as an alternative to the Sub-Clause 14.8 compensation, to draft a special provision for payment of the Contractor's actual financing Costs, taking account of local financing arrangements.[15]

The following may be done to minimise the need for Contractor's/Subcontractor's self-financing (and reduce tender prices):

- the default time for certification under Sub-Clause 14.6 (Red/Yellow/Emerald Books) may be reduced to 21 days and the time for payment be reduced to 28 days under Sub-Clause 14.7 of the Red/Yellow/Emerald/Silver Books and to 35 days under the relevant FIDIC Subcontracts;
- Sub-Clause 14.7 in the Subcontracts may be substituted with the example provision from the GPPC,[16] abolishing the pay-when-paid terms;
- Sub-Clause 14.7 in the Main Contracts and Subcontracts may be amended in the Particular Conditions, introducing a PBA to secure simultaneous payments to the Contractor and key Subcontractors/suppliers;
- Sub-Clause 14.7 in the Main Contracts and Subcontracts may be amended for the use of blockchain for instant invoicing and payments (see Chapter 9).

Directive 2014/24/EU on Public Procurement[17] enables direct payments to subcontractors, applicable for the PBA and blockchain. According to Article 71, paragraph 3,

> Member States may provide that at the request of the subcontractor and where the nature of the contract so allows, the contracting authority shall transfer due payments directly to the subcontractor … Such measures may include appropriate mechanisms permitting the main contractor to object to undue payments. The arrangements concerning that mode of payment shall be set out in the procurement documents.

This option may be integrated in the Contract with a special provision amending Sub-Clause 14.7.

In EU states, the FIDIC Sub-Clause 14.7 payment terms need to comply with the *Directive 2011/7/EU on Late Payment*, which stipulates maximum payment duration of 60 days for transactions between enterprises, and 30 days for contracts between public authorities and companies, with a possible extension to 60 days in specific cases (unless both parties expressively agreed to a different timeline that was not grossly unfair).[18] According to *Directive 2011/7/EU*, the payment terms start as soon as the verification period is success-

15 FIDIC, *Yellow Book* (FIDIC 2022) GPPC 45

16 FIDIC, *Conditions of Subcontract for Plant and Design-Build* (2019) Guidance 22

17 *Directive 2014/24/EU of the European Parliament and of the Council of 26 February 2014 on public procurement and repealing Directive 2004/18/EC Text with EEA relevance* (2014 OJEU L94/136) <https://eur-lex.europa.eu/legal-content/EN/TXT/PDF/?uri=CELEX:32014L0024> accessed 2 January 2023

18 *Directive 2011/7/EU of the European Parliament and of the Council of 16 February 2011 on combating late payment in commercial transactions* (2011 OJ EU L48/1) <https://eur-lex.europa.eu/LexUriServ/LexUriServ.do?uri=OJ:L:2011:048:0001:0010:en:PDF> accessed 2 January 2023

fully concluded. However, Employers appear to increasingly use the verification period as a tool to purposefully delay their payment obligations.[19] For example, while the time for payment under the Red/Yellow Books Sub-Clause 14.7 for EU co-financed water projects in Bulgaria is reduced to 30 days, Sub-Clause 14.6 of the Particular Conditions allows the Engineer, in the case of a significant error in the Statement or supporting documents, to return them to the Contractor, and the time for certification begins after the resubmission.[20]

9.1.2.7 Public–private partnerships (PPPs)

PPPs provide a mechanism to access private capital for the construction of infrastructure and for transferring the management and the risk of the provision of public services to the private sector.[21] The private-sector SPV finances, operates, and maintains public infrastructure, and is paid for its use; the asset concerned usually reverts to public-sector control/ownership at the end of the contract term. There are two main PPP models:

- 'Concessions': construction or refurbishment of public infrastructure with revenue derived from tolls, fares, or similar payments by users ('User Charges');
- 'Private Finance Initiative (**PFI**) Model': construction or refurbishment of a public building or other public infrastructure with revenue derived from payments by a Contracting Authority ('Service Fee').[22]

9.1.2.8 Case study – PPPs

Fraport Twin Star Airport Management AD is the German-Bulgarian SPV awarded the 35-year concession for the seaside airports in Bulgaria. During the first 14 years of the concession, the SPV invested over €200 million in infrastructure and technical improvements at Varna and Burgas airports.[23] The Concessionaire acts as the Client in the FIDIC White Book agreements for design and supervision services, and as the Employer in the FIDIC Silver Book contracts for the new terminals and Red Book contracts for renovations and reconstructions. The Concessionaire used loans from commercial banks.[24]

19 European Builders Confederation, 'Position Paper – EBC position on the Late Payment Directive' (2019) <https://www.ebc-construction.eu/wp-content/uploads/191025-Position-Paper-Late-Payment_EN.pdf> accessed 2 January 2023

20 Agency for Public Procurement, Standard Documents <https://www2.aop.bg/obrazci-i-spisyci/standartizirani-dokumenti/> accessed 28 December 2022

21 Sara Valaguzza and Eduardo Parisi, *Public Private Partnerships* (Edward Elgar Publishing Limited 2020)

22 Yescombe, *Principles of Project Finance* (Elsevier 2014) 16

23 <https://www.fraport-bulgaria.com/en/> accessed 3 January 2023

24 <https://www.boyanov.com/boyanov-co-advised-fraport-twin-star-airport-management-ad-on-a-eur-70-9-million-deal/> accessed 3 January 2023

9.1.3 Project Financial Management monitoring and control

Effective project financial control and monitoring is better achieved when a project produces regular progress reports.[25] The Consultant supplies in the monthly progress reports the information specified under Appendix 4 of the White Book. The Red/Emerald/Yellow/Silver Books Sub-Clause 4.20 obliges the Contractor to produce monthly progress reports, including comparisons of actual and planned progress and lists of Variations and Notices of Claim, which may lead to an increase of the Contract Price.

Throughout the term of the financing, the beneficiary will be required to give reports to the sponsors, lenders, and their advisors. The lenders regularly check the latest information on the project before advancing further tranches of debt. The FIDIC Employers may align the reporting requirements in the CA's Agreement for Services (White Book Appendix 4) to the reporting requirements of the financing institutions. The Engineer/Employer's Representative usually prepares special reports and justifications where changes to the financial arrangements are necessary.

The financial institutions may appoint a Lenders' Engineer as part of their due diligence procedure. Once the project construction is under way, the Lenders' Engineer receives regular information on progress, and provides regular reports to the lenders, highlighting any particular problems. The Lenders' Engineer monitors the construction but otherwise does not in any way supervise or control the construction process, which remains the responsibility of the beneficiary.[26]

Financial and cost control ensure that all items are within budget and the financial cash forecast.[27] A Schedule of Payments can be very useful not only for the Contractor but also for the Employer. Sub-Clause 14.4 of the Red/Emerald/Yellow/Silver Books obliges the Contractor to submit non-binding estimates of the payments expected to become due for each quarterly period if the Contract does not include a Schedule of Payments. The first estimate shall be submitted within 42 days after the Commencement Date. Revised estimates shall be submitted at intervals of 3 months until the issue of the TOC for the Works. The CA usually reviews the Contractor's estimates and gives an independent forecast in the progress reports.

The CA verifies the physical status and the contract value of the completed works and regularly reports the certified/paid amounts against the planned cash flow. Figure 9.1 shows an **S-curve** for rehabilitation of critical infrastructure project with Time for Completion 5 months, using the FIDIC Red Book. The Engineer reports weekly the contract value of the executed Works compared with the planned Works. Since the cost estimates making up the ACA are tied to the activities in the Sub-Clause 8.3 Programme (79 in this case), this enables a time distribution of the cumulative costs, which is typically displayed as an S-curve. For projects with a Time for Completion of several years, the planning and reporting are done on a monthly basis, and often the actual costs are higher than planned during the execution of the underground works, due to adjustments for Unforeseeable physical conditions. The planned and the actual contract value may be used for financial and cost control, and also to calculate the Schedule Performance Index (**SPI**), using the Earned value management (**EVM**) method discussed in Subsection 9.2.7.1.

25 PMI Inc., *Construction Extension to the PMBOK® Guide* (PMI 2016) 165
26 Yescombe, *Principles of Project Finance* (Elsevier 2014) 97
27 PMI Inc., *Construction Extension to the PMBOK® Guide* (PMI 2016) 165

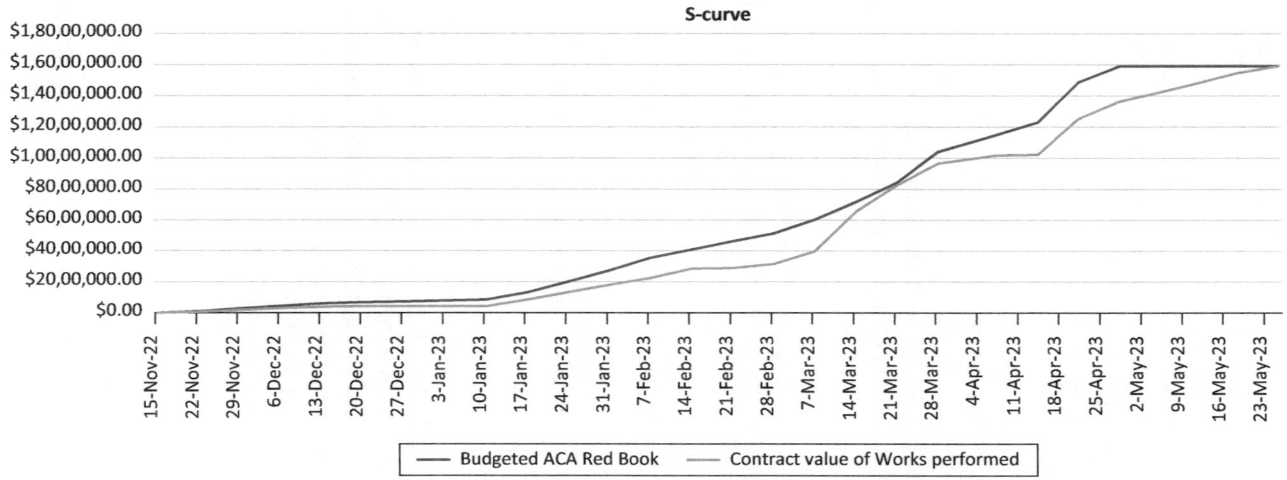

Figure 9.1 S-curve weekly reporting Red Book ACA budget versus Works performed

Credit to Evgeni Tsvetanov

Sub-Clause 2.4.2 of the White Book obliges the Employer to give a Notice to the Consultant with detailed particulars before making any material change to its financial arrangements. The Consultant may be entitled to suspend the Services under Sub-Clause 6.1.2(c) if not satisfied with the proposed change, acting reasonably.

Sub-Clause 2.4 in the works contracts also includes provisions for changes in the Employer's financial arrangements. The Employer is obliged to maintain financial arrangements enabling payment of the Contract Price, which may increase due to Variations, Changes in Laws/Cost under Clause 13, and Claims under the Contract or in connection with the Contract, subject to Sub-Clause 20.2. In the case of any material change to the financial arrangements or in the Employer's financial situation, the Employer shall immediately give a Notice to the Contractor with detailed supporting particulars. The Contractor may request and the Employer shall provide within 28 days reasonable evidence that financial arrangements have been made and are being maintained if:

- the price of an instructed Variation exceeds 10 per cent or the accumulated value of Variations exceeds 30 per cent of the Red/Emerald/Yellow Book ACA (Silver Book: Contract Price stated in the Contract Agreement);
- the Contractor does not receive payment in accordance with Sub-Clause 14.7; or
- becomes aware of a material change in the financial arrangements not notified by the Employer.

Internal and external audits ensure that correct accounting methods and financial practices are being maintained. External audits are often a statutory requirement of the local government or are mandatory under the financing conditions arranged with lenders.[28] For example, the Particular Conditions amending the Red Book GCs in the Standard Documents of the World Bank add under Sub-Clause 1.17 an obligation for the Contractor to

> permit, the Bank and/or persons appointed by the Bank to inspect the site and/or the accounts, records and other documents relating to the procurement process, selection and/or contract execution, and to have such accounts, records and other documents audited by auditors appointed by the Bank.[29]

BIM, PBA, and blockchain secure transparency and protection against misuse of funds. Recent research by Algan Tezel et al.[30] demonstrates that blockchain can revolutionise how the supply chain is evaluated, valued, and compensated, increasing the transparency and traceability of payments in streamlined processes. Project financing and transaction costs will be reduced substantially, which will help with the inclusion of small organisations in project delivery. Similarly, new asset funding opportunities will emerge with blockchain (e.g. crowdfunding of assets through tokenisation). On the commercial front, enhanced financing and allocation of payments in projects can be realised through blockchain, which will help organisations record more manageable cash flows.

---

28 PMI Inc., *Construction Extension to the PMBOK® Guide* (PMI 2016) 165

29 World Bank, *Standard Procurement Documents, Request for Bid Works (Without Prequalification)* (2021)

30 Algan Tezel and others, 'Preparing Construction Supply Chains for Blockchain Technology: An Investigation of Its Potential and Future Directions' (2020) <https://d-nb.info/1214055044/34> accessed 4 January 2022

9.2 Project Cost Management with FIDIC Contracts

9.2.1 Project Cost Management processes

According to the Construction Extension to the PMBOK® Guide, Project Cost Management in construction includes cost estimating, budgeting, cost monitoring, and control, and further entails managing the day-to-day project costs. This is considerably different from the financial management discussed in Section 9.1, dealing with revenue sources for financing the construction project, its return on investment, its cash flow, and its investment payback analysis.

The ability to influence the cost is greatest during the project preparation phase, which makes early scope definition critical.[31] Cost estimating and budgeting set the framework for the cost management. Cost control provides a mechanism to monitor and control project costs in order to deliver the project within the budget.[32] The goal of staying within the budget should not restrict innovations, since they can bring long-term benefits, including cost savings. Spending more for comparison of alternatives/ design review/ higher quality results in lower costs for operation, maintenance, and decommissioning. The whole-life costs and sustainability impact are important considerations when analysing a Value Engineering (VE) proposal during the project preparation or later under Sub-Clause 13.2 of the works contracts.

9.2.2 Cost estimating

A cost estimate is a quantitative assessment of the likely costs for resources required to complete the activity. It is a prediction that is based on the information known at a given point in time. Cost estimates include the identification and consideration of costing alternatives to initiate and complete the project. Cost trade-offs and risks should be considered, such as make versus buy, buy versus lease, and the sharing of resources in order to achieve optimal costs for the project.[33]

According to the FIDIC Procurement Guide, the accuracy of the preliminary cost estimate in the Feasibility Study Report aims to be 20–30 per cent.[34] Financing geotechnical studies and survey of the selected Site, the Employer, supported by the consultant, prepares more detailed estimates and plans the risk allocation. Risks cost money, so the budgeting starts when the works packages are developed and the contract forms selected.

Published information about productivity, cost indices for wages/materials, and resource cost rates is often available from commercial sources that track and provide average costs for material, labour, and equipment. Project-specific considerations for the adjustment of standard costs include:

- Site conditions;
- labour market;
- Site access restrictions;

31 FIDIC, *FIDIC Definition of Services Guidelines. Infrastructure Construction / Civil Works* (2016) figure 3 <https://fidic.org/sites/default/files/_DOS_guide_civil_works_1.pdf> accessed 2 May 2022

32 PMI Inc., *Construction Extension to the PMBOK® Guide* (PMI 2016) 63

33 PMI Inc, *PMBOK® Guide* (PMI sixth edition 2017) 241

34 FIDIC, *FIDIC Procurement Procedures Guide* (FIDIC 2011) 13

- restricted working hours;
- proximity to facilities;
- equipment and materials logistics;
- climate;
- local community and social group impact;
- Health, Safety, Security, and Environmental (HSSE) regulations and commitments; and
- geotechnical data.

The estimating techniques mostly used in construction are analogous, parametric, and bottom-up. Techniques that produce more accurate estimates require more information about the project and take more time and resources.[35] As the project develops, and additional or more accurate information becomes available, it is inevitable that the estimate becomes more accurate, known as rolling wave estimating. These revised costs should be used for the next estimating stage, but once the actual final budget stage has been reached, and the price has been accepted by the client, any further cost refinements will only be useful for updating the monthly cost estimate, which may affect the profit or loss without changing the cost baseline.[36]

9.2.2.1 Analogous estimating

Analogous cost estimating (referred to as preliminary, conceptual, top-down) uses values, or attributes, of a previous project that are similar to the current project. These values/attributes may include scope, cost, budget duration, and measures of scale (e.g. size). Comparison of these project values/attributes becomes the basis for estimating the same parameter or measurement for the current project.[37] Industry publications provide estimating data. Industry sector historical indices and consumer price indices should be consulted for the most up-to-date cost and pricing data.[38]

For example, the recently published cost for wind parks is $1.3 million per megawatt (**MW**) of electricity-producing capacity.[39] Indicative construction costs of various types of buildings for a typical specification per square metre of gross internal floor area are usually available in each country/region, but they have to be adjusted for the specific conditions, size, value, and expected price escalation.[40]

9.2.2.2 Parametric estimating

Parametric estimating uses a statistical relationship between relevant historical data and other variables (e.g. square footage in construction) to calculate a cost estimate for project work. This technique can produce higher levels of accuracy depending on the

35 PMI Inc., *Construction Extension to the PMBOK® Guide* (PMI 2016) 66

36 Albert Lester, *Project Management, Planning, and Control* (Elsevier Ltd. 2014) 62 <http://dx.doi.org/10.1016/B978-0-08-098324-0.00013-5> accessed 19 August 2022

37 PMI Inc,, *PMBOK® Guide* (PMI sixth edition 2017) 244

38 PMI Inc., *Construction Extension to the PMBOK® Guide* (PMI 2016) 66

39 Dan Blewett, 'Wind Turbine Cost: How Much? Are They Worth It In 2022?' (Weather Guard Lightning Tech 2021) <https://weatherguardwind.com/how-much-does-wind-turbine-cost-worth-it/> accessed 18 August 2022

40 Costmodelling Limited, 'Typical UK Construction Costs of Buildings' (2022) <https://costmodelling.com/building-costs> accessed 18 August 2022

sophistication and underlying data built into the model. Parametric cost estimates can be applied to a total project or to segments of a project, in conjunction with other estimating methods.[41] The construction industry uses software applications providing local data and cost publications provided by specialised publishers and professional associations. Commonly used parametric indicators are ratio of steel in concrete, percentage of reinforcement steel for different components of structure, etc.[42] Lester affirms the degree of accuracy for parametric estimating to be ±10–20 per cent.[43]

9.2.2.3 Bottom-up estimating

Bottom-up techniques may be used when the detailed design and specification are ready. The cost of individual work packages or activities is estimated to the greatest level of specified detail. The detailed cost is then summarised or 'rolled up' to higher levels for subsequent reporting and tracking purposes.[44] Higher accuracy is achieved in the range of ±5 per cent.[45]

9.2.3 Budgeting

A project budget can be established once an estimate is approved. This involves aggregating the estimated costs of activities or construction work packages.[46] Work packages most often include all labour, material, equipment, and subcontractor costs. The budget for a Red Book contract with Employer's design includes detailed BQs for all activities, while the budget for lump sum contracts includes estimates for works packages.

An adequate **contingency** allowance must be added to the final estimate to cover for possible, probable, and unknown risks, which could be technical, political, environmental, administrative, etc., depending on the results of a more formal risk analysis.[47]

A separate Schedule may include Provisional Sums priced by the Employer for foreseeable risk events such as:

- crossing of unknown infrastructure;
- connections to utilities where the agreements with the utility operators are preliminary and new requirements may be enforced after the Base Date.

Besides amounts for probable risks, a Schedule of Provisional Sums priced by the Employer may include:

- estimates for payments to nominated Subcontractors;
- an estimated amount for 50 per cent of the DAAB's fees;

41 PMI Inc., *PMBOK® Guide* (PMI sixth edition 2017) 244

42 PMI Inc., *Construction Extension to the PMBOK® Guide* (PMI 2016) 66

43 Albert Lester, *Project Management, Planning, and Control* (Elsevier Ltd. 2014) 59 <http://dx.doi.org/10.1016/B978-0-08-098324-0.00013-5> accessed 19 August 2022

44 PMI Inc., *PMBOK® Guide* (PMI sixth edition 2017) 244

45 Albert Lester, *Project Management, Planning, and Control* (Elsevier Ltd. 2014) 59 <http://dx.doi.org/10.1016/B978-0-08-098324-0.00013-5> accessed 19 August 2022

46 PMI Inc., *Construction Extension to the PMBOK® Guide* (PMI 2016) 70

47 Albert Lester, *Project Management, Planning, and Control* (Elsevier Ltd. 2014) 63 <http://dx.doi.org/10.1016/B978-0-08-098324-0.00013-5> accessed 19 August 2022

- optional items such as services/supplies/works which:
  - are necessary, but the Employer would like to select them after entering the Contract;
  - may be necessary, but the Employer is not certain yet (e.g. due to developing technologies).

While Sub-Clause 1.1 defines Provisional Sum as 'a sum specified in the Contract', the FIDIC 2017 Contracts Guide goes further, stating that 'each Provisional Sum included in the Contract is typically not priced by the tenderers when submitting their tenders, it is a sum stated by the Employer (in one of the Schedules)'.[48] 'Typically' means, that often the Provisional Sums are used with pricing at the Employer's risk under Sub-Clause 13.4 (b), adding the Contractor's overhead and profit on top of the actual costs. However, under Sub-Clause 13.4 (a) the Engineer (Silver Book: Employer) may instruct a Provisional Sum to be used for work executed and paid adjusting the Contract Price under Sub-Clause 13.3.1 [Variation by Instruction]. This means that the rates and unit prices priced at the Contractor's risk from a Red Book BQs or Yellow/Emerald/Silver Book Schedule of Rates and Prices may be used. A Schedule of Rates and Prices will not be priced competitively, unless the Employer has inserted estimated quantities for probable risk events, and the tenderers multiply them by their unit prices/rates to add these amounts to their financial proposals (to be part of the ACA).

So, the tender documents may include as Provisional Sums:

- a Schedule of Rates and Prices for Provisional Sums priced by the Contractor (for probable risks and optional Variations);
- a Schedule of Provisional Sums priced by the Employer (for probable risks and for options which cannot be priced by the tenderers).

The Schedule of Provisional Sums may be a combination of both. 'Schedule' by definition is a document entitled schedule prepared by the Employer and completed by the Contractor. The best approach will be the Employer to price those items for which there is not enough information and to leave the others which may be competitively priced to the tenderers to complete.

Usually, 10 per cent is added on the top of all costs as a contingency to be spent for unforeseen works only if instructed under Sub-Clause 13.4 [Provisional Sums]. Some may argue that Provisional Sums shall be always specified and never used for unknown risks. However, the Sub-Clause 1.1 definition is broad, allowing use for 'execution of any part of the Works or for the supply of Plant, Materials or services under Sub-Clause 13.4'. The use of the Sub-Clause 13.4 procedure for unknown risks is not in conflict with its requirements:

- to use the Provisional Sum in whole or in part in accordance with the CA's instructions;
- the total sum paid to the Contractor to include only such amounts to which the Provisional Sum relates, as instructed.

48 FIDIC, *The FIDIC 2017 Contracts Guide* (2022) 57

Sub-Clause 13.5 [Daywork] gives an additional mechanism for adding a contingency for work of a minor or incidental nature. For this purpose, the Daywork Schedule should include estimated quantities to be multiplied by the rates priced by the Contractor and thus the amounts will be added to the financial proposal (to be included in the ACA). The procedure is user-friendly:

- the Engineer (Silver Book: Employer) may instruct that a Variation shall be executed on a daywork basis;
- the work shall then be valued in accordance with the Daywork Schedule;
- for Goods not priced in the Daywork Schedule, quotations from suppliers/subcontractors may be required, and the CA has 7 days to respond, similarly to the Sub-Clause 13.4 procedure;
- the Contractor delivers each day to the CA statements including records of the resources used in executing the previous day's work to be countersigned by the CA;
- the agreement/determination procedure will follow in the case of disagreement with the statements;
- the Contractor submits for payment priced statements of the agreed/determined resources with all applicable invoices/vouchers/accounts/receipts for any Goods used that are not priced in the Daywork Schedule.

The cost baseline is the approved version of the project budget that includes contingency reserves but excludes management reserves. Management reserves are an amount of the project budget withheld for management control purposes and are reserved for unforeseen work that is within the scope of the project. When an amount of management reserves is used to fund unforeseen work, this amount is added to the cost baseline, thus requiring an approved change to the cost baseline.[49] The Contractor does not know the amount of the management reserve, but Sub-Clause 2.4 entitles it to request reasonable evidence that financial arrangements have been made in case of instructed Variation which may exhaust the available contingency reserve.

To summarise, depending on the project risks, the consultant preparing the budget for the Employer may add the following contingency reserves:

- Schedule of Provisional Sums for Nominated Subcontractors (priced by the Employer);
- Schedule of Provisional Sums for optional works and 50 per cent of the DAAB's fees for reimbursement to the Contractor (priced by the Employer);
- Schedule of Provisional Sums (Red Book BQs) for Probable Risks (priced by the Contractor where clear definitions/details are available), for example, for infrastructure:
  - crossing/relocation of unknown infrastructure (illegal or new built after the Site survey);
  - dewatering of the Site;

49 PMI Inc., *PMBOK® Guide* (PMI sixth edition 2017) 252

    - transportation and backfilling of material complying with the specified requirements (if part of the excavated material is not suitable as initially expected);
    - dealing with adverse soil conditions;
    - additional external quality control by a certified body;
- Daywork Schedules in the amount of 4–5 per cent:
    - Daywork Schedule for Labour;
    - Daywork Schedule for Equipment;
    - Daywork Schedule for Materials;
- 10 per cent contingency above the total amount for mitigation of unknown risks.

Figure 9.2 presents recommended contingencies included in the budget and the FIDIC Red/Yellow/Emerald Book ACA (Silver Book: Contract Price). The cost baseline prepared with the tender documents plus management reserves, if any, forms the budget for financing. The ACA (Silver Book: Contract Price) is usually less than the estimated cost baseline as a result of the competition. However, if the estimated budget is not realistic or becomes inadequate in case of drastic price escalation, the tender procedure may be cancelled due to lack of funds.

The Contract Price is the amount that is due and payable by the Employer to the Contractor for providing the Works in accordance with the Contract. The FIDIC Contracts are not fixed price contracts, because the amount offered by the tenderer is subject to adjustment in specified circumstances, for example, for Variations instructed by or on behalf of the Employer. In the Silver Book, the Contract Price is fixed to a greater extent

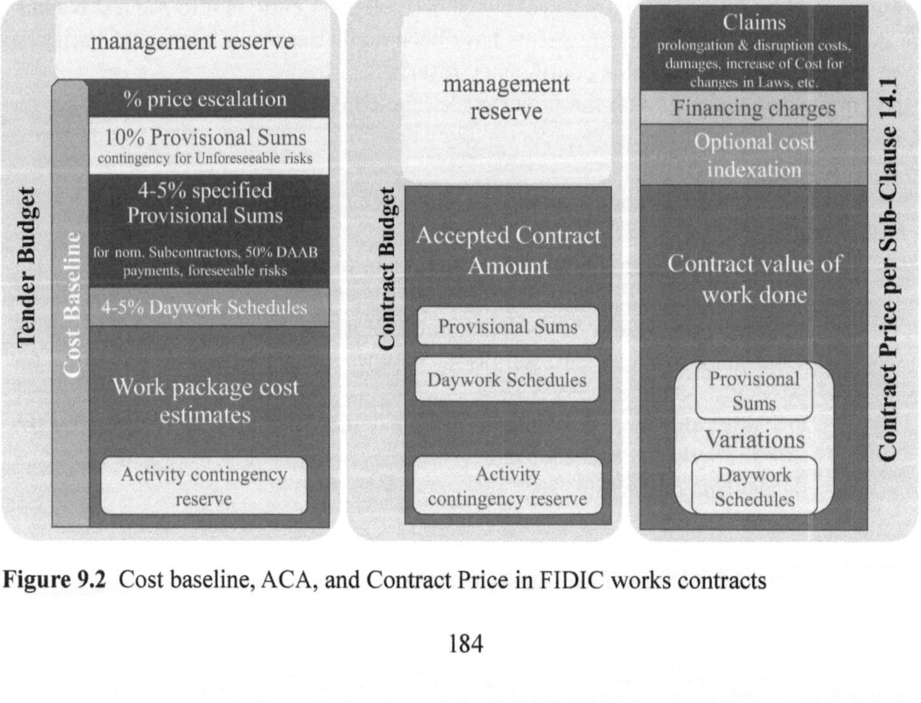

Figure 9.2 Cost baseline, ACA, and Contract Price in FIDIC works contracts

than under the other books, because the circumstances in the GCs that allow for adjustment are considerably more limited.[50]

The quantity surveyor preparing the BQs for a Red Book contract may increase the main quantities by 5 per cent as an activity contingency reserve in the budget.

Depending on the risk allocation, experienced contractors usually increase the lump sum prices in their proposal with a risk contingency from 5 per cent for a Yellow Book contract to 20–30 per cent in a Silver Book contract. Due to the fair risk allocation under the Emerald Book, a minimum risk contingency is expected to be included in the Contractor's Proposal.

9.2.4 Cost management in FIDIC agreements for services

The estimation of the fees may be done in two ways:

1. A work programme is prepared, allocating tasks to the team and estimating the resource input in man hours; the fee is estimated by multiplying the man hours necessary to complete the services by the rates of the relevant experts.
2. The fee is calculated as a percentage of the construction value.

The White Book Sub-Clause 7.1.1 stipulates that the Client shall pay the Consultant for the Services (including Variations) in accordance with the details stated in Appendix 3 [Remuneration and Payment], which should cover:

- agreed remuneration, whether lump sum or schedule of rates or any combination;
- terms of payment (percentage fees, timescale, lump sum);
- rates and prices for Variations/Exceptional Costs;
- times for payment if not 28 days;
- process for submission of invoices and methods of payment;
- price changes, inflation, etc., if applicable;
- currencies of payment;
- financing charges for delayed payment;
- taxation additional to payments, if any (e.g. VAT);
- allowable expenses.

The initially agreed fees may be increased in the case of:

- Variations instructed/approved under Sub-Clause 5.1;
- Exceptional Costs under Sub-Clause 7.1.2 for:
  - the extra time spent by the Consultant's personnel; and
  - the cost of all other expenses **reasonably** incurred by the Consultant;
- other amounts due under the Agreement pursuant to Sub-Clause 7.1.3 (e.g. as decided by the Adjudicator).

Sub-Clauses 5.2 and 7.1.2 stipulate that the value of Variations and Exceptional Costs shall be determined:

50 FIDIC, *FIDIC 2017 Contracts Guide* (FIDIC 2022) 12

- at the rates and prices stated in Appendix 3;
- where the rates and prices are not applicable, then new rates and prices shall be agreed by the Parties;
- if agreement is not reached within 14 days of the issue of the relevant Notice, then reasonable rates and prices shall be applied.

9.2.5 Cost control in FIDIC works contracts

9.2.5.1 Adjustment of the Contract Price

Cost control is the process of monitoring the status of the project to update the project costs managing changes to the cost baseline.[51] The key benefit is to recognise in a timely manner the variance from the plan in order to take corrective and preventive actions to minimise project cost risk. Many complex and mega projects run the risk of cost overruns due to ineffective cost control.[52]

Section 13.8.4 explains the different methods of determining the Contract Price from the perspective of price risk allocation. Sub-Clause 14.1 stipulates that the Contract Price shall be subject to adjustments, additions (including Cost or Cost Plus Profit to which the Contractor is entitled), and/or deductions in accordance with the Contract. Any quantities set out in a Schedule are estimated quantities, not actual/correct quantities, to be used only for the purpose stated in the Schedule.

Figure 9.2 shows that usually, the Contract Price is increased as a result of Variations, Claims, and cost indexation. After the contingency is spent, a formal approval is needed to amend the Contract, adding an amount to the cost baseline from the management reserve.

The Sub-Clause 13.3 Variation procedure is outlined in Chapter 7, the Sub-Clause 20.2 Claims procedure in Chapter 16, while the valuation of Variations and Claims is discussed in the following Sections.

The optional Sub-Clause 13.7 enables the amounts payable to the Contractor to be adjusted for rises or falls in the cost of labour, Goods, and other inputs to the Works, when schedules of cost indexation are included in the Contract for the calculation of the additions/deductions. The governing law may provide additional relief in the case of cost escalation.

9.2.5.2 Measurement of quantities

In a 'remeasurement contract', the payment due to the Contractor is calculated by measuring the actual physical quantity of each item of work that the Contractor has executed and then applying the appropriate rate per unit quantity. This is the payment method under the Red Book GCs. Clause 12 [Measurement and Valuation] stipulates the procedure presented in Figure 9.3:

Option 1:

- the Engineer gives a Notice to the Contractor of not less than 7 days of the part to be measured and specifies the time and place on Site;
- if the Contractor fails to attend, the measurement shall be deemed to have been made in its presence and accepted as accurate.

51 PMI Inc., *PMBOK® Guide* (PMI sixth edition 2017) 257
52 PMI Inc., *Construction Extension to the PMBOK® Guide* (PMI 2016) 71

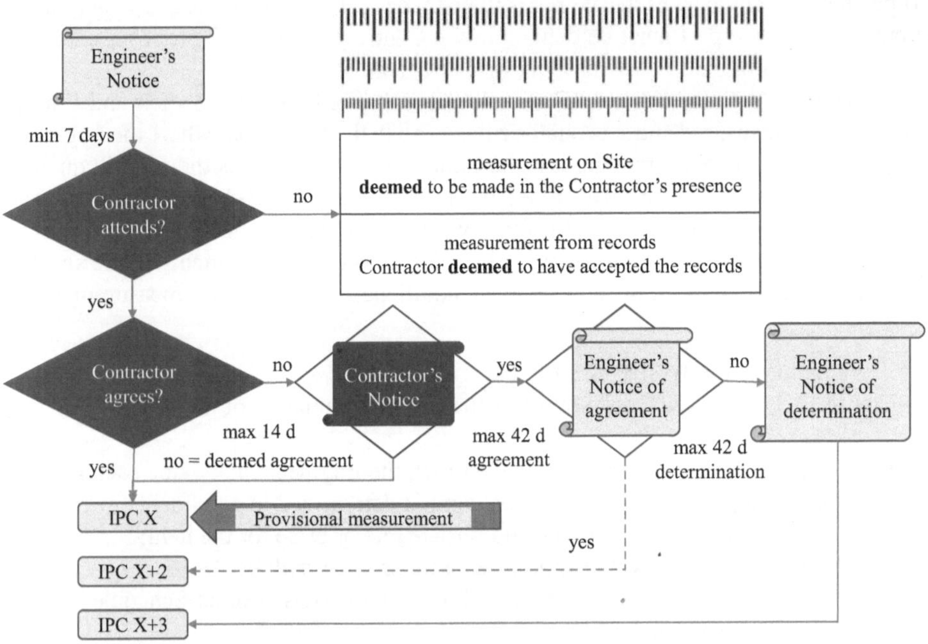

Figure 9.3 Procedure for measurement under the Red Book Sub-Clause 12.1

Option 2:

- the Engineer shall prepare the records for any part identified in the Specification to be measured from records;
- the Engineer gives a Notice to the Contractor of not less than 7 days, stating when and where the Contractor's Representative shall attend to examine and agree the records with the Engineer;
- if the Contractor fails to attend at the notified/agreed time and place, the Contractor shall be deemed to have accepted the records as accurate.

In case the Contractor attends but disagrees:

- the Contractor shall give a Notice to the Engineer setting out the reasons of disagreement (if the Contractor does not give such a Notice within 14 days, the Contractor shall be deemed to have accepted the measurement as accurate);
- after receiving a Contractor's Notice of disagreement, unless at that time such measurement is already subject to the last paragraph of Sub-Clause 13.3.1 [Variation by Instruction], the Engineer shall agree or determine the measurement under Sub-Clause 3.7;
- **until such time as the measurement is agreed or determined, the Engineer shall assess a provisional measurement for the purposes of IPCs.**

Sub-Clause 12.2 of the Red Book stipulates that the method of measurement shall be as stated in the Contract, and measurement shall be made of the net actual quantity of each item of the Permanent Works.

If payment for any part of the Works is to be made on the basis of measurement under the other books, that part must be defined in the Contract. The GPPC proposes example wording for an additional Sub-Clause for measurement.

Sub-Clause 13.8 of the Emerald Book stipulates that only the Excavation and Lining Works are to be measured and sets out the rules. Unlike the Red Book, where the Engineer is responsible for the measurement, it is the Contractor that measures the as-built quantities, which are then agreed or determined by the Engineer. The GPPC recommends that users do not amend this regime. All the other Works are to be paid on a lump sum basis. If other works are specified to be measured, the rules for their measurement will be similar to the Red Book and may be introduced by an additional Sub-Clause for measurement.

9.2.5.3 Valuation of the Works – Red Book

Sub-Clause 12.3 of the Red Book stipulates the rules for valuation of the Works:

- except as otherwise stated in the Contract, the Engineer shall value each item of work by applying the measurement agreed/determined in accordance with Sub-Clauses 12.1 and 12.2, and the appropriate rate or price for the item;
- for each item of work, the appropriate rate or price shall be:
    - the rate or price specified for such item in the BQs or other Schedule, or
    - if there is no such item, specified for similar work;
- any BQ/Schedule item of work for which no rate or price is specified shall be deemed to be included in other rates and prices;
- new rates and prices shall be derived from:
    - any relevant rates/prices specified in the BQs/Schedule, with reasonable adjustments, or
    - if no specified rates/prices are relevant: the reasonable Cost of executing the work, plus the applicable percentage for profit stated in the Contract Data (if not stated, 5 per cent), taking account of any other relevant matters.

A new rate or price shall be appropriate for an item of work if:

(a) the item
- is not identified in any Schedule, and
- no rate or price for this item is specified in any Schedule, and
- no specified rate or price is appropriate because the item of work is not of similar character, or is not executed under similar conditions, as any item in the Contract;

(b) all four criteria are met:
  (i) the measured quantity is changed by more than 10 per cent from the quantity of this item in the BQs or other Schedule,
  (ii) this change in quantity multiplied by the specified rate or price for this item exceeds 0.01 per cent of the ACA,
  (iii) this change in quantity directly changes the Cost per unit quantity by more than 1 per cent, and
  (iv) this item is not specified as a 'fixed rate item', 'fixed charge', or similar; and/or

(c) the work is instructed under Clause 13 and sub-paragraph (a) or (b) above applies.

The difficulties come with (b) (iii), since some Employers wrongly believe that in case of significant increase of the quantity, the rate may be reduced. This is not the case for works like excavation, since the Cost per unit quantity does not change. An example where the actual Cost may be reduced is a ten-span steel roof structure assembled on the Site in blocks with installed services upon specially designed Temporary Works. If a Variation is instructed to increase the structure by 2 spans, the Cost per tonne may be reduced, since the cost of the Temporary Works will be distributed over 12 instead of 10 spans.

If, for any item of work, the Engineer and the Contractor are unable to agree the appropriate rate or price:

- the Contractor shall give a Notice to the Engineer setting out the reasons why the Contractor disagrees;
- after receiving the Notice, unless at that time such rate or price is already subject to Variation by instruction, the Engineer shall agree or determine the appropriate rate or price; and
- until such time as an appropriate rate or price is agreed or determined, the Engineer shall assess a provisional rate or price for the purposes of the IPC.

The Engineer should not forget that cash flow is the lifeblood of the Contractor. The robust cost control procedures take several months in case of disagreement, so provisional measurement and provisional rates and prices must be applied for payment of any executed item of work without delay. This is also applicable for executed varied works instructed under Clause 13.

Sub-Clause 14.1 (d) stipulates that the Contractor shall submit to the Engineer, within 28 days after the Commencement Date, a proposed breakdown of each lump sum price (if any) in the Schedules. Tenderers may be required to include in their Tenders detailed price breakdowns of each rate. This information can be useful for pricing of Variations and Claims.

9.2.5.4 Valuation of the Works – Yellow/Silver Book

Normally, a Yellow/Silver Book contract is based on a lump sum price with little or no measurement. The Contractor thus takes the risk of changes in cost arising from the Contractor's design. The lump sum price may consist of two or more amounts, quoted in the currencies of payment (which may, but need not, include the Local Currency). Usually, the breakdown of the lump sum price follows the work breakdown structure (WBS).

Under the Yellow Book, the Contract Price shall be the lump sum ACA and be subject to adjustments, additions (including Cost or Cost Plus Profit to which the Contractor is entitled under these Conditions), and/or deductions in accordance with the Contract. The Contract Price shall be stated in the Contract Agreement under the Silver Book, where the Contractor has significant risk exposure.

If any part of the Works is to be paid according to quantity supplied or work done, the provisions for measurement and valuation shall be as stated in the Particular Conditions. The Contract Price shall be valued accordingly, subject to adjustments in accordance with the Contract.

9.2.5.5 Valuation of the Works – Emerald Book

The Conditions of Contract for Underground Works provide that only the Excavation and Lining Works are to be measured and paid for using rates and prices set out in the BQs. Their cost will largely depend on the actually encountered subsurface physical conditions and/or ground reaction to such works.

Due to the typically long duration of underground works and the high cost of investment in equipment, an important portion of the cost to the Contractor, such as, e.g., depreciation of and interest for equipment, overhead charges, etc., is time-related, while the remainder of the cost depends on the volume or quantity of the performed works. Hence, the adjustment of Time for Completion (if any) due to the difference between the contractually agreed and the encountered subsurface physical conditions should also lead to an adjustment of the Contractor's remuneration for time-related charges. The BQ for Excavation and Lining Works should therefore distinguish between fixed rate items, time-related rate items, and quantity-related rate items for the performance of the Works (Sub-Clauses 13.8.2 and 13.8.3):[53]

- Fixed rate items are not subject to adjustments for changes in quantity, but are linked to the relevant periods according to the Completion Schedule. They include:
    - the provision of Contractor's Equipment, facilities, and supervision;
    - facilities required by the Employer's Personnel;
    - fees for relevant Contractor's overhead charges.
- Time-related rate items depend on the length of time necessary for the execution of the works and include the extended or reduced availability and maintenance of the fixed rate items expressed in units for extension or reduction for a calendar day.
- Quantity-related rate items are valued by multiplying the measured quantity by the rate or unit price.

Value-related charges for compliance with the contractual provisions such as insurance, securities, permits, levies, etc. shall be included in these rates.

9.2.5.6 Valuation of the Works – Green Book

Clause 8 stipulates that the Works shall be valued as stated in the Contract Data, subject to adjustments in accordance with the Contract. The Contract Data provide for five main different valuation options, and the Employer is to select one option or a combination of several options:

- Option A. Lump sum price with single payment: recommended for very minor works where Variations are not anticipated, and the Works will be completed in a short period, requiring only one payment to the Contractor;

53 FIDIC, *Emerald Book* (FIDIC 2019) <https://fidic.org/books/conditions-contract-underground-works-2019-emerald-book> accessed 22 August 2022

- Option B. Lump sum price with stage payments: suitable for a larger contract, where Variations and stage payments would be required, and the Employer does not have the resources to prepare its own BQs;
- Option C. Lump sum price with BQ: a Lump sum offer based on quantities prepared by the Employer; the Contractor bears the risk of work quantities; the BQs are primarily used for pricing and payment purposes, as well as managing Variations;
- Option D. Remeasurement with BQ: suitable for infrastructure works designed by the Employer; uses the default mechanism under the Red Book;
- Option E. Cost plus: used where the extent of work cannot be ascertained before the Contract is placed; an estimate prepared by the tenderer will be replaced by the actual Cost of the Works; the value of the Works executed shall be calculated and paid for on a monthly basis, on the basis of the Cost Plus Profit of executing the Works, taking account of any other relevant matters.

9.2.5.7 Valuations of Variations

The Sub-Clause 13.3 Variation procedure shall be used for adjustment of the Contract Price in case of instructed varied works (including VE). Chapter 7, Section 7.6.3.5 lists the provisions expressly referring to the Variation procedure. Figure 7.4 shows that the Contractor may commence the execution of an instructed Variation before the costs are agreed or determined. Provisional rates shall be used for payment of varied works until the valuation is done.

The Employer may instruct the Yellow/Silver/Emerald Book tenderers to present detailed price breakdowns, including quantities, unit rates, and other pricing information. This data can be used to value Variations, and for the assessment of interim payments, but may not have been priced competitively. When the tender documents are being prepared, the Employer must therefore decide whether to be bound by the tenderer's price breakdowns.[54]

Sub-Clause 13.3.1 sets out the rules for valuation of Variations under the Yellow/Silver/Emerald Books:

1. If a 'Schedule of Rates and Prices' is included in the Contract for each item of work, the appropriate rate/price shall be:
   i. the rate/price specified for such item in the Schedule of Rates and Prices, or
   ii. if there is no such item, the rate or price specified for similar work; however,
   iii. a new rate/price shall be appropriate if the work is not of similar character, or is not executed under similar conditions, as any item in the Contract.
2. Each new rate or price shall be derived from any relevant rates or prices in the Schedule of Rates and Prices, with reasonable adjustments taking account of all relevant circumstances. If no rates or prices are relevant for the derivation of a new rate or price, it shall be derived from the Cost Plus Profit of executing the work.
3. If no 'Schedule of Rates and Prices' is included in the Contract, the valuation of Variations shall be derived from the Cost Plus Profit of executing the work.

54 FIDIC, *Yellow Book* (FIDIC 2022) GPPC 40

Until such time as the adjustments to the Contract Price and the Schedule of Payments are agreed or determined, the CA shall assess a provisional rate or price for the purposes of payment.

Under Sub-Clause 13.3.1(c), whenever the omission of any work forms part (or all) of a Variation, and if:

- the Contractor has incurred or will incur cost that, if the work had not been omitted, would have been deemed to be covered by a sum forming part of the ACA (Silver Book Contract Price); and
- the omission of the work has resulted or will result in this sum not forming part of the Contract Price,

this cost may be included in the Contractor's proposal. If the Parties have agreed to the omission of any work that is to be carried out by others, the Contractor's proposal may also include the amount of any loss of profit and other losses and damages suffered (or to be suffered) by the Contractor as a result of the omission.

The Red Book Sub-Clause 13.3.1 refers to Clause 12 for the valuation of Variations using measured quantities. Sub-Clause 12.4 [Omissions] provides for compensation for cost incurred for omitted works. If the Parties have agreed for the omitted works to be performed by others, Sub-Clause 13.3.1(c) provides for additional relief to the Contractor, similarly to the other books.

9.2.5.8 Valuation of Claims – quantum

'*Quantum*' is a Latin word meaning 'how much'. In the legal environment, this is used to represent the amount of money owed to a Party, as agreed/determined by the CA, decided by the DAAB, or awarded by arbitration or court judgment.

FIDIC Contracts are governed by principles of **contract law**, where the law provides for **compensation of loss or damage** suffered by one Party caused by:

- reasons attributable to the other Party or
- risks allocated to the other Party.

Besides Claims under the Contract, Claims may be made in connection with the Contract based on the governing law.

Limitation of liability is usually stipulated for practical commercial reasons. Sub-Clause 1.15 (Red/Yellow/Emerald)/1.14 (Silver Book), discussed in Chapter 13, reflects those commercial practices, providing reasonable risk allocation.

The FIDIC GCs expressly state when a Party will be entitled to compensation of Costs incurred. 'Cost' as defined under Sub-Clause 1.1.19 (Red/Yellow/Emerald)/1.16 (Silver Book) includes all expenditure reasonably incurred (or to be incurred) by the Contractor in performing the Contract, whether on or off the Site, including taxes, overheads, and similar charges, but does not include profit. The meaning of 'reasonably' in relation to incurred costs will vary from case to case. Therefore, FIDIC strongly recommends that the Contractor keeps contemporaneous records of the costs and the background of why they have been incurred, to be able to demonstrate that it was appropriate and necessary to expend those costs.[55] Sub-Clause 20.2.3 defines 'contemporary records' as records that are prepared or generated at the same time as, or immediately after, the event or circumstance giving rise to the Claim. The claiming Party shall keep such contemporary records as may be necessary to substantiate the Claim. The CA may monitor the Contractor's records and instruct the Contractor to keep additional records.

Besides the contemporary records, Sub-Clause 20.2.4 requires the claiming Party to submit with the fully detailed Claim detailed supporting particulars of the claimed amount. Clearly explained calculations should link the claimed amounts to attached contemporary records, demonstrating causation and legal basis for the entitlement.

9.2.5.9 Employer's Claims for reduction of the Contract Price

Besides any damages for the Contractor's breach of contract, the Employer has express entitlements to reduction of the Contract Price, subject to Sub-Clause 20.2 under:

- Sub-Clause 1.13 (Red/Yellow/Emerald)/1.12 (Silver Book) for additional costs incurred as a result of the Contractor's failure to comply with the applicable Laws;
- Sub-Clause 5.2.2 for cost reasonably incurred in the Review of the resubmitted Contractor's Document previously rejected by the CA for non-compliance;
- Sub-Clause 7.4 for costs incurred due to the Contractor causing delay to specified tests;
- Sub-Clause 7.5 for additional costs incurred due to rejection by the CA of defective Plant/Materials/design/workmanship and consequent retesting;
- Sub-Clause 7.6 for payment to other persons to carry out work due to the Contractor's failure to comply with the CA's instructions;
- Sub-Clause 8.7 for additional costs incurred by the Employer due to the Contractor's revised methods to expedite progress;
- Sub-Clause 8.8 for Delay Damages;
- Sub-Clause 9.2 for costs of Employer's Personnel incurred due to delayed Tests on Completion;
- Sub-Clause 9.3 for additional costs incurred by the Employer due to failure to pass the Tests on Completion and retesting (by reference to Sub-Clause 7.5);
- Sub-Clause 11.4(a) for costs reasonably incurred by the Employer in remedying defects or damage after the Contractor has failed to do so within a reasonable time;
- Sub-Clause 9.4(d)/11.4(b)/(Yellow/Emerald/Silver Book) 12.4 for failure to remedy defects/pass Tests on Completion/Tests after Completion;
- Sub-Clause 11.4(d) for sums paid plus financing charges plus costs if the defect or damage deprives the Employer of substantially the whole benefit of the Works;
- Sub-Clause 11.8 for costs of the search reasonably incurred by the Employer;
- Sub-Clause 11.11 for loss reasonably incurred in case of sale/disposal of remaining items and reinstating/cleaning the Site, when the Contractor fails to comply with its Site clearance obligations within 28 days after the issue of the Performance Certificate;

55 FIDIC, *FIDIC 2017 Contracts Guide* (FIDIC 2022) 18

- Yellow/Silver/Emerald Book Sub-Clause 12.3 for additional costs incurred by the Employer as a result of failure of the Works/Section to pass the Tests after Completion, and retesting, where the Contractor is responsible for remedying the defect or damage causing the failure/ retesting at own cost;
- Sub-Clause 13.6 for adjustments for changes in Laws;
- Sub-Clause 15.4 for costs plus losses and damages in the case of termination for Contractor's default (plus Delay Damages if the Works/Section have not been taken over and the Time for Completion has expired).

Any right of set-off that may exist under the governing law would be subject to Sub-Clause 20.2. In order to set off or deduct any amounts from payments due to the Contractor, the Employer shall submit a Claim under Sub-Clause 20.2. The CA then agrees or determines the amount due. Deducting of an amount by the Employer from an IPC issued by the Engineer would be a breach of contract.

9.2.5.10 Contractor's Claims for additional payment

The Contractor has general duty to mitigate its losses.[56]

Costs are classified into the following broad headings:

(a) direct costs (labour, equipment, materials, and sub-contracted work); and
(b) indirect costs (on-Site overheads and head office overheads), whether time-related or otherwise.[57]

Where the Contractor is entitled under a Sub-Clause of the FIDIC Conditions to payment of Cost, it shall be added to the Contract Price. The Contractor may be entitled to Cost Plus Profit when the Employer is in default. In the case of neutral events where the risk is allocated to the Employer who is not culpable, the compensation includes only Costs.

The effects of the governing law should not be overlooked. In most cases, the governing law will permit compensation only for Costs actually incurred and substantiated with accounting documents.

The grounds for the Contractor's express entitlement to additional payment subject to Sub-Clause 20.2 are presented in Table 9.1.

Besides the provisions listed in Table 9.1 entitling the Contractor to payment of compensation under the Contract, the Contractor may be entitled to payment of:

- damages for the Employer's breach of contract subject to Sub-Clause 20.2:
  - in connection with the relevant obligation under the Contract (e.g. the CA instructs the Contractor not to commence after the expiry of the 21-day Review Period, and the Contractor incurs Cost);
  - at law (e.g. in some jurisdictions, for the Employer's failure to comply with the mandatory obligations to appoint an authorised third party and/or the designer under the Red Book for construction supervision);
- financing charges for delayed payment under Sub-Clause 14.8, without a Claim under Sub-Clause 20.2.

56 'SCL Delay & Disruption Protocol' (SCL 2017) 7
57 'SCL Delay & Disruption Protocol' (SCL 2017) 15

Table 9.1 Provisions entitling the Contractor to claim compensation under Sub-Clause 20.2

Sub-Clause	*Cause of incurred Cost*	*Entitlement*
Red 1.9	Delayed Drawings or instructions	Cost Plus Profit
Yellow/Emerald 1.9	Errors in the Employer's Requirements	Cost Plus Profit
1.13 Red/Yellow/Emerald, 1.12 Silver	Employer's delay or failure to obtain any permit/ licence/approval	Cost Plus Profit
2.1	Delayed access to/ possession of the Site	Cost Plus Profit
4.6	Instruction for Unforeseeable co-operation	Cost Plus Profit
Red/Yellow/Emerald 4.7	Errors in items of reference	Cost Plus Profit
Red/Yellow/Emerald 4.12	Unforeseeable physical conditions	Cost
4.15	Changes to the access route by the Employer or a third party	Cost
4.23	Instructions for dealing with archaeological and geological findings	Cost
7.4	Delayed tests	Cost Plus Profit
7.6	Remedial work attributable to any act by the Employer's Personnel	Cost Plus Profit
8.10	Instruction for suspension and resumption of work	Cost Plus Profit
10.2	Taking over/ using part of the Works before TOC	Cost Plus Profit
10.3	Prevention from carrying out the Tests on Completion	Cost Plus Profit
11.7	Unreasonable delay in permitting access after TOC	Cost Plus Profit
11.8	Search for the cause of any defect, to be remedied at the cost of the Employer	Cost Plus Profit
Yellow/Emerald/ Silver 12.2	Unreasonably delayed by the Employer Tests after Completion	Cost Plus Profit
Yellow/Emerald/ Silver 12.4	Unreasonably delayed access for Tests after Completion	Cost Plus Profit
13.3.2	Contractor's proposal does not become an instructed Variation	Cost
13.6	Changes in Laws after the Base Date	Cost
16.1	Suspension (reducing the rate of works) by the Contractor for a material breach of the Employer's obligations	Cost Plus Profit
16.2.2	Costs incurred after the Contractor's Sub-Clause 16.2.1 Notice of Termination	Cost Plus Profit
16.3	Instructed work after termination	Cost Plus Profit
16.4	Termination for Employer's default	Loss of profit/ other losses and damages
17.2	Instructed rectification of damage to the Works, Goods, or Contractor's Documents in case of Employer's risk	Cost Plus Profit
18.4	Consequences of an Exceptional Event	Cost in case of war or man-made events in the Country

9.2.5.11 Prolongation and disruption Costs

Delay and disruption claims are often greatly exaggerated both as to the extent of delay caused by the Employer's breach and in quantification. The basis for calculation is often excessively theoretical, ignoring the principles that damages are to compensate for actual loss and must be proved. Hence, the compensation shall be the costs that are actually incurred in the period when the delay events impacted on the progress. The Contractor has to prove that:

- **but for** the delay and disruptive events, the cost to the Contractor would have been less than it actually turned out to be, and
- the difference arose **as the result of delay and disruption.**

The 'SCL Delay and Disruption Protocol' defines **prolongation** as 'extended duration of the works during which time-related costs are incurred as a result of a delay'.[58] Entitlement to an EOT does not automatically lead to entitlement to compensation (and vice versa). Where Employer delay to completion and Contractor delay to completion are **concurrent,** then the Contractor should only recover compensation **if it is able to separate the additional costs caused by the Employer delay** from those caused by the Contractor delay. If the Contractor would have incurred the additional costs in any event as a result of Contractor delay, the Contractor will not be entitled to recover those additional costs.[59]

The 'SCL Delay & Disruption Protocol' defines **disruption** as 'disturbance, hindrance or interruption to a Contractor's normal working methods, resulting in lower productivity or efficiency in the execution of particular work activities'.[60] Compensation may be recovered for disruption **only to the extent that the contract permits or there is an available cause of action at law.** The objective of a disruption analysis is to demonstrate the loss of productivity and hence additional loss and expense due to the disruption events for which the Employer is responsible.

Delay and disruption are inherently interrelated. A loss of productivity can lead to delay, and if the impacted activities are on the critical path, to critical delay. Hence, the Contractor may rely upon a disruption analysis to support a critical delay claim in addition to its delay analysis. Equally, delay can lead to disruption. If the Contractor has less time to carry out work activities (absent an EOT for the critical path activities), it is possible that acceleration measures being implemented will lead to those tasks being carried out with a lower productivity than planned and hence, at greater cost. Duplication in claimed entitlement for both delay and disruption has to be avoided.[61]

The practice of contractors to submit global claims without attempting to substantiate cause and effect is discouraged. The case law recommends that the Contractor demonstrate on a balance of probabilities that:

1. Events occurred that entitle it to loss and expense;
2. Those events caused delay and/or disruption; and
3. Such delay or disruption caused it to incur loss and/or expense (or loss and damage).[62]

58 'SCL Delay & Disruption Protocol' (SCL 2017) 70
59 'SCL Delay & Disruption Protocol' (SCL 2017) 7
60 'SCL Delay & Disruption Protocol' (SCL 2017) 9
61 'SCL Delay & Disruption Protocol' (SCL 2017) 10
62 *Walter Lilly & Co Ltd v DMW Developments Ltd [2012] EWHC 1773 (TCC)* [486 (a)]

A careful analysis must establish the entitlement, if any, when the Contractor is faced with:

- partial delay;
- disruption situation (which does not delay the completion date of the project);
- concurrent causes.

Disruption analysis may be used to calculate on-Site overheads. It is focused on the increase in resources required to compensate for productivity loss. Causation is reviewed under the balance of probabilities rule. Productivity-based and cost-based methods are used for disruption analysis. While the productivity-based methods measure the loss of productivity in the utilised resources and then price that loss, the cost-based methods seek more directly to ascertain the difference between actual cost and planned cost without first measuring productivity losses.[63]

Figure 8.2 demonstrates an example with prolongation and disruption costs incurred as a result of delayed access to part M of the Site. The disruption costs start to be incurred when the special machine completing the reshaping of the waste on part K cannot continue its work on part M, as planned. Initially, part M is not on the critical path, but with the continuing effect of the delayed access, it becomes on the critical path, and prolongation costs will be incurred in the period after the initial Time for Completion of 360 days (for extended Performance Security, insurance, etc.).

9.2.5.12 On-Site overheads

On-Site overheads are typically those incurred by the Contractor on Site for the performance of its obligations, like Site offices, local management, secretarial and assistance services, warehouses and storage areas, security, and car pool. They are easily identifiable as costs specifically incurred for the execution of a project. The conventional approach calculates daily time-related site overheads multiplied by the compensable delay from the delay analysis. This approach may be criticised, since the calculation does not consider:

- causation;
- the degree of culpability/suspension;
- double recovery/duplication with Variations;
- concurrency.

9.2.5.13 Off-Site overheads

Off-Site overheads are typically head office overheads, representing the contribution that each project is meant to generate to the financing of the head/supervising office functions of the Contractor, which are required to run a business: general management, business development, human resources, financing and accounting, etc. They are not easily identifiable as costs specifically incurred for the execution of a project, because they have a horizontal nature, serving all activities carried out by the Contractor under all of its projects.[64]

63 'SCL Delay & Disruption Protocol' (SCL 2017) 46

64 FIDIC, *Short Form of Contract (Green Book)* (FIDIC 2021) Guidance Notes 25

The SCL Delay and Disruption Protocol distinguishes two categories of overhead costs:[65]

- **Dedicated overhead costs:** substantiated by specific records like staff time sheet bookings, together with any staff travel expenses, directly or indirectly relating to the Employer's risk event, and
- **Unabsorbed overhead costs:** incurred regardless of the Contractor's volume of work, the retained records should include those relating to rent, rates, heating, lighting, directors' salaries, wages of support staff, pension fund contributions, and auditors' fees.

If the Contractor intends to rely on the application of a formula for the assessment of lost profits and unabsorbed head office overheads, it will first need to produce evidence that it was unable to undertake other work that was available to it because of the Employer's delay.[66] These records may include:

- the Contractor's business plans prior to the Employer's delay;
- the Contractor's tendering history and records of acceptance or rejection of tender opportunities depending upon resource availability;
- minutes of meetings to review future tendering opportunities and staff availability;
- records that support the inputs into the formula used, in particular the Contractor's company accounts for the periods immediately preceding and succeeding the Employer's delay as well as for the period when the Employer's delay occurred.

The parties might consider agreeing relevant rates in the contract, rather than requiring proof of actual costs or loss for certain eventualities. Thus, the claiming party does not need to produce proof of actual cost or loss, and the paying party benefits from a pre-agreed rate.[67]

9.2.5.14 Prolongation costs (PGS) as liquidated damages under the Green Book 2021

Due to the difficulties of proving causation and substantiation of Prolongation Costs (**PGC**) often leading to disputes, the Second Edition of the Green Book sets a formula in the Contract Data Sub-Clause 1.1.35 for their calculation. The PGC is defined as the 'on-Site and off-Site overheads associated with a compensable EOT, as stated in the Contract Data'. The PGC is payable for the period when delay on the critical path was suffered at a daily rate, which is calculated in accordance with a formula that takes into consideration

65 'SCL Delay & Disruption Protocol' (SCL 2017) 55

66 SCL 2017; in *Walter Lilly & Co Ltd v DMW Developments Ltd [2012] EWHC 1773 (TCC)*, Akenhead J held that the use of a formula, supported by relevant factual evidence of opportunities foregone, was a legitimate way of establishing this head of loss. However, in order to prove a claim for loss of profit and head office overheads caused by delay, it had to be shown on the balance of probabilities that but for the delay, the Contractor would have secured other work, which would have contributed to such overheads and/or generated profit, and the loss would not have been sustained [540–543]

67 'SCL Delay & Disruption Protocol' (SCL 2017) 16

the average 'weight' of both on-Site and off-Site overheads per day. The progress of the Works is a Gaussian ('bell curve') function of time, and accordingly, the amount of overheads incurred by the Contractor is not a fixed but a fluctuating amount over time. Several work progress stages are considered, referring to the certified value of the Works, V, as a percentage of the Contract Price stated in the Contract Agreement:

- start of a project: **V is below 33 per cent** when the Contractor progressively mobilises its resources;
- peak production period: **V is greater than 33 per cent but lower than 66 per cent,** resources are mobilised and working at full capacity, and the overheads are at their peak; and
- before completion: **V is greater than 66 per cent,** demobilisation progressively starts to occur, and overheads are accordingly going down.

W is the average Weight of the on-Site and off-Site overheads per day.

W = 20 per cent of the Contract Price stated in the Contract Agreement, divided by the Time for Completion for the Works.

PGC are calculated taking into account the work progress:

- PGP = 25 per cent of W per day of EOT, when V is equal to zero;
- PGP = 60 per cent of W per day of EOT, when V is greater than zero but lower than 33 per cent;
- PGP = 125 per cent of W per day of EOT, when V is in the range 33–66 per cent;
- PGP = 60 per cent of W per day of EOT, when V is greater than 66 per cent.

The Green Book stipulates that 'Prolongation Cost shall be the only compensation due from the Employer to the Contractor for an EOT resulting from a compensable delay'. Therefore, Sub-Clause 1.1.35 acts as a liquidated damages provision for ease of use by the Parties. However, 'this provision shall not affect Contractor's compensation rights for other Costs (if any), such as disruption Costs'.[68]

9.2.6 Payment procedure

9.2.6.1 Advance payment

Sub-Clause 14.2 gives the procedure for advance payment if an amount is stated in the Contract Data. The rate of deduction for the repayments should be checked to ensure that repayment is achieved before the Contractor's completion of the Works. The figures in the GCs are based on the assumption that the advance payment is less than 22 per cent of the Contract Price stated in the Contract Agreement. The acceptable form of Advance Payment Guarantee should be included in the tender documents.

---

68 FIDIC, *Green Book* (FIDIC 2021) Contract Data Sub-Clause 1.1.35

9.2.6.2 Interim payments

The Contractor shall submit a Statement to the CA after the end of the period of payment stated in the Contract Data (if not stated, after the end of each month), showing in detail the amounts to which the Contractor considers that it is entitled, with:

- supporting documents, which shall include sufficient detail, and
- the Progress Report under Sub-Clause 4.20.

Pursuant to Sub-Clause 14.3, the Statement shall include the estimated contract value of the executed Works (including Variations) and the Contractor's Documents produced, and any amounts to be added/deducted under the Contract, deducting from the sum the amounts previously paid.

No amount will be paid to the Contractor until:

- the Employer has received the Performance Security; and
- the Contractor has appointed the Contractor's Representative.

After receiving a Sub-Clause 14.3 Statement from the Contractor, the CA must check it. Pursuant to Sub-Clause 14.6, within 28 days the Engineer shall issue the IPC under the Red/Emerald/Yellow Book, while the Employer shall give the Notice of interim payment under the Silver Book.

The IPC (Silver Book: Notice) shall state the amount that the CA fairly considers to be due, including any additions and/or deductions due, with detailed supporting particulars giving the reasons for difference between the certified (notified) amount and the Sub-Clause 14.3 Statement.

The Employer shall pay to the Contractor within the timeline stipulated under Sub-Clause 14.7. A flowchart in the beginning of each FIDIC book presents the typical sequences of payment events.

9.2.6.3 Withholding amounts

The CA may withhold an IPC (Silver Book: interim payment) in an amount that would (after retention and other deductions) be less than the minimum amount stated in the Contract Data, giving a Notice to the Contractor. The CA has to be careful when using the right to withhold an interim payment, considering the difficulties the Contractor meets and the possible supply-chain disruption.

An IPC (Silver Book: interim payment) shall not be withheld for any other reason, although:

(c) if anything supplied or work done by the Contractor is not in accordance with the Contract, **the estimated cost of rectification/replacement may be withheld until it has been completed;**

(d) if the Contractor fails to perform any work/service/obligation in accordance with the Contract, **the value of this work or obligation may be withheld until it has been performed;** in this event, the CA shall promptly give a Notice to the Contractor describing the failure with detailed supporting particulars of the value withheld; and/or

(e) if the CA finds any significant error or discrepancy in the Statement or supporting documents, the amount for payment may take account of the extent to which this error or discrepancy has prevented or prejudiced proper investigation of the amounts in the Statement until such error or discrepancy is corrected in a subsequent Statement.

The CA may in any IPC (Silver Book: interim payment) make any correction or modification that should properly have been made previously. A Payment Certificate/payment shall not be deemed to indicate acceptance, approval, consent, or Notice of No-objection to any Contractor's Document or to (any part of) the Works.

If the Contractor considers that an IPC (interim payment) does not include any amounts to which the Contractor is entitled:

- these amounts shall be identified in the next Statement;
- the CA shall then make any correction/modification that should properly be made in the next IPC (interim payment);
- to the extent that the Contractor is not satisfied that this next IPC (interim payment) includes the identified amounts, and the identified amounts do not concern a matter for which the CA is already carrying out the agreement/determination procedure, the Contractor may, by giving a Notice, refer this matter for agreement or determination.

9.2.6.4 Schedule of Payments

If the Contract includes a Schedule of Payments specifying the instalments in which the Contract Price will be paid, then:

(a) the instalments quoted in the Schedule of Payments shall be treated as the estimated contract values for the purposes of payment applications;
(b) Sub-Clause 14.5 [Plant and Materials intended for the Works] shall not apply;
(c) if these instalments are not defined by reference to the actual progress of the Works, and the actual progress differs from that on which the Schedule of Payments was based, then the CA may agree or determine revised instalments.

It is better for the amounts to be based on the actual progress achieved by the Contractor in executing the Works, which necessitates careful definition of the payment milestones. Disagreements may arise when the work required for a payment milestone is nearly achieved, but the balance of the work, albeit minor, cannot be completed until some months later. The figures inserted by a tenderer in the Schedule of Payments may be compared with the tender programme in order to assess whether they are reasonably consistent with each other. In order to secure the Contractor's cash flow and prevent disputes, it is recommended:

- to define payment milestones related to clear scope of work;
- to define conditions to be met to achieve milestone payments (e.g. No-objection to Contractor's design, successful tests);
- to link the milestones with achieved stages (e.g. foundations, roof, cabling and ducts, finishing works, fittings, Tests on Completion, etc);
- the CA to be flexible in case of impediments, estimating percentage for partial payment of the milestone value.

9.2.6.5 Retention Money

When the Contract Data specifies a rate for retention, in each Sub-Clause 14.3 Statement, the specified percentage shall be deducted until the amount retained reaches the specified limit of Retention Money from:

- the estimated contract value of the Works (including Variations);
- any amounts to be added/deducted for changes in Laws/Cost;
- other additions/deductions due under the Contract or otherwise.

Sub-Clause 14.9 stipulates the release of the Retention Money. The Contractor shall include in a Statement:

- the first half of the Retention Money after the issue of the TOC for the Works or
- the relevant percentage of the first half of the Retention Money following the TOC for a Section;
- the relevant percentage of the second half of the Retention Money after the expiry date of the DNP for a Section;
- the second half of the Retention Money after the latest of the expiry dates of the DNPs.

The Engineer shall certify the amount in the next IPC/Final Payment Certificate (**FPC**). Under the Silver Book, the Employer shall release the amount in the next interim payment. The CA shall be entitled to withhold the estimated cost of work remaining to be executed after taking over and expiry of the DNP.

9.2.6.6 Payment at completion

Within 84 days after the Date of Completion of the Works, the Contractor shall submit to the CA a Statement at completion under Sub-Clause 14.10 with supporting documents showing:

(a) the value of all work done up to the Date of Completion;
(b) any further sums that the Contractor considers to be due; and
(c) an estimate of any other amounts that the Contractor considers due after the Date of Completion of the Works, under the Contract or otherwise to be shown separately, and include estimated amounts for:
    (i) Claims for which it has submitted a Sub-Clause 20.2 Notice;
    (ii) any matter referred to the DAAB under Sub-Clause 21.4; and
    (iii) any matter for which a Notice of Dissatisfaction (NOD) has been given under SC 21.4.

The CA shall then proceed in accordance with Sub-Clause 14.6.

9.2.6.7 Final payment

The Sub-Clause 14.11.1 draft final Statement (**FS**) submitted within 56 days after the issue of the Sub-Clause 11.9 Performance Certificate shall show in detail, with supporting documents:

(i) the value of all work done in accordance with the Contract;
(ii) any further sums that the Contractor considers to be due at the date of the issue of the Performance Certificate; and
(iii) an estimate of any other amounts that the Contractor considers due after the issue of the Performance Certificate, under the Contract or otherwise, including estimated amounts for the matters described in 9.2.6.6 (c) above.

Figure 9.4 presents the FS procedure. If there are no amounts in the draft FS under (iii) above, the Contractor shall prepare and submit to the CA the FS as agreed. If there are amounts under (iii) above, and it becomes evident that an agreement cannot be reached, the Contractor shall submit to the CA a 'Partially Agreed FS' identifying separately the agreed amounts, the estimated amounts, and the disagreed amounts.

When submitting the (Partially Agreed) FS, the Contractor shall submit a discharge under Sub-Clause 14.12, which confirms that the total of such Statement represents full and final settlement of all moneys due to the Contractor under or in connection with the Contract. It may state that it is subject to any payment in respect of any Dispute, full payment of the amounts in the FS, and the return of the Performance Security. If the Contractor fails to submit the discharge, it shall be deemed to have been submitted and become effective upon payment and receipt of the Performance Security. The discharge shall not affect either Party's liability or entitlement in respect of any Dispute in DAAB/ arbitration.

Within 28 days after receiving the (Partially Agreed) FS and the discharge, the Engineer shall issue to the Employer (with a copy to the Contractor) the FPC (the Silver Book Employer shall give a Notice to the Contractor with supporting particulars) stating:

(a) the amount fairly considered finally due; and
(b) the balance (if any) due.

If the Contractor has not submitted a draft FS within 56 days after the PC, the CA shall request the Contractor to do so. If the Contractor fails to submit it thereafter within 28 days, within a further 28 days, the Engineer shall issue the FPC for such amount as the Engineer fairly considers to be due (the Silver Book: Employer shall give a Notice to the

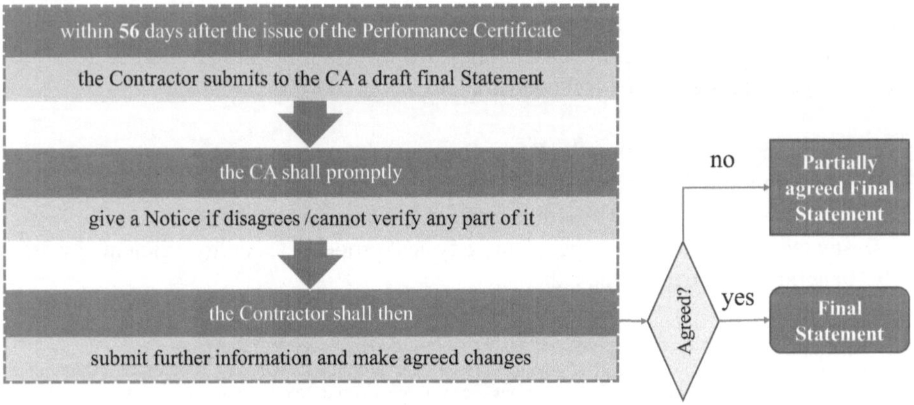

Figure 9.4 Final Statement under Sub-Clause 14.11

Contractor stating the Final Payment). If the Contractor has submitted a Partially Agreed FS or is deemed to have submitted it, the Engineer shall issue an IPC (the Employer shall make an interim payment).

Sub-Clause 14.14 stipulates that the Employer shall not be liable to the Contractor, except to the extent that the Contractor has included an amount in the (Partially Agreed) FS and (except for matters arising after the TOC) the Statement at Completion. Unless the Contractor makes a Claim under SC 20.2 in respect of an amount included in the FPC (Silver Book: Final Payment) within 56 days of receiving it, the Contractor shall be deemed to have accepted the amounts as correct. The Employer shall then be liable only to return the Performance Security. This shall not limit its liability under the indemnification obligations, or in the case of fraud, gross negligence, deliberate default, or reckless misconduct by the Employer.

9.2.7 Innovative Project Cost Management

9.2.7.1 Performance metrics and Earned Value Management (EVM)

The CA may apply the international standard ISO 21508:2018[69] to control the project performance. It defines EVM as a method integrating project scope, actual cost, budget, and schedule for assessment of progress and performance. The Sub-Clause 8.3 Programme shall integrate the time and cost baselines following the WBS. Instructed Variations shall be integrated in the revised Programme as well. Tracking the progress for the monthly report may show the planned versus the actual performance status.

9.2.7.2 Two Stage Open Book

Projects with high degrees of uncertainty or not yet fully defined scope may not benefit from detailed cost calculations. Two Stage Open Book enables innovations with early involvement of the supply chain. This delivery method may be used with the Red Book or with the other books where the Contractor is responsible for the design. It starts with a conditional contract for the design phase, leading to a lump sum contract after the design is approved and the budget fixed.[70]

9.2.7.3 BIM tools

The use of BIM tools for quantity take-offs, estimating, and budgeting is rapidly expanding.[71] By having a BIM model, time and money are saved for the preparation of accurate BQs and budget. The steps to be followed are:

- importing the model into the BQ software;
- importing an existing database or creating a new cost database, inserting each item in the work section with a code and a description, attaching to the items any technical document, specifying physical properties, energy efficiency/ waste generation/ CO_2 emissions, etc.;

69 *ISO 21508:2018 Earned Value Management in Project and Programme Management*

70 King's College London, 'Using Two Stage Open Book and Supply Chain Collaboration' (Queen's Printer and Controller of HMSO 2014) 2 <https://www.gov.uk/government/collections/government-construction> accessed 24 August 2022

71 PMI Inc., *Construction Extension to the PMBOK® Guide* (PMI 2016) 67

- setting up the measurement rules, connecting the items from the BIM model with the cost database (this can be done by clicking on the element on the 3D model and then introducing a formula, e.g. for calculation of surface or volume);
- customising the general parameters for pricing the BQ and creating the budget, such as adding a percentage for contingencies, profit and overheads, and value added tax (**VAT**).

The software applications can generate reports with cost breakdown structure, CO_2 emissions, BQs, and budget.

9.2.7.4 Blockchain implementation

Immediate opportunities for blockchain implementation could include automation and acceleration of payment, automation of accounting, and improved PM. Examining the scope for implementing blockchain solutions in the construction industry's cash flow systems must take into account the legal context and challenges to this. Any solution that does not sit within this framework would not be legal and therefore, could not be implemented.[72]

A technological solution using distributed ledger and blockchain elements could innovate project cash management in FIDIC Contracts:

- automating payments to the Contractor and down the supply chain to Subcontractors and suppliers upon certification by the Engineer under the Red/Emerald/Yellow Book or validation by the Employer under the Silver Book;
- reducing the time for payment and automating the invoicing, thus reducing the supply-chain insolvency risk;
- automating accounting and improving cash management;
- automating and making transparent the cost monitoring and control, facilitating auditability and account reporting.

9.2.8 Example Notices

Appendix 1 includes:

- an example for Contractor's request for reasonable evidence for financial arrangements under Sub-Clause 2.4 of the Yellow Book;
- examples of Engineer's Notices related to measurement under Sub-Clause 12.1 of the Red Book 2017.

72 Financing & Procurement Working Group. Construction Blockchain Consortium, *Blockchain & Construction Cash Flow*, White Paper 1 (CBC 2020) 25 <https://static1.squarespace.com/static/58b6047520099e545622d498/t/5fdb6089ad5a0604f7feaf5e/1608212649913/CBC2020-WP1_Cashflow.pdf> accessed 18 August 2022

CHAPTER 10

Project Quality Management with FIDIC Contracts

10.1 Project Quality Management (QM) processes

The international standard ISO 9000 defines quality as 'the degree to which a set of inherent characteristics of an object fulfils requirements'. The requirements may be stated, generally implied, or obligatory. 'Generally implied' refers to custom or common practice.[1]

According to the PMBOK® Guide, Project QM includes the processes for planning, managing, and controlling project and product quality requirements in order to meet stakeholders' objectives. It also supports continuous process improvement activities.[2]

A specialised quality management system to address the unique characteristics of a specific project is stipulated under Sub-Clause 4.9 of the FIDIC Red/Yellow/Emerald/Silver Books.

Prevention is preferred over inspection. It is better to design quality into deliverables than to find quality issues during inspection.[3]

1 *ISO 9000:2015 Quality Management Systems – Fundamentals and Vocabulary* 3.6.2 and 3.6.4 <https://www.iso.org/obp/ui/#iso:std:iso:9000:ed-4:v1:en:term:3.8.6> accessed 14 September 2022

2 PMI Inc., *PMBOK® Guide* (PMI sixth edition 2017) 271

3 PMI Inc., *PMBOK® Guide* (PMI sixth edition 2017) 274

DOI: 10.4324/9781003272892-10

10.2 Quality management in the service agreements

Quality design and design documentation are critical to the success of any project.[4] The consultant is involved in the quality planning, specifying the requirements for the Works in the tender documents. Appointed as the FIDIC Engineer or Employer's Representative, the Consultant has an important role in the review of the Contractor's Documents and in the quality control (**QC**) of construction. As discussed in Chapter 5, FIDIC advocates for quality-based selection of the Consultant.

Sub-Clause 3.3 of the White Book specifies the standard of care as reasonable skill, care, and diligence to be expected from a consultant experienced in the provision of such services for projects of similar size, nature, and complexity. Without extending the obligation of the Consultant beyond that, the Consultant shall perform the Services with a view to satisfying any function and purpose that may be described in Appendix 1 [Scope of Services]. The Consultant shall comply with all regulations, statutes, ordinances, and other forms of standards, codes of practice, and legislation applicable to the Services and the Agreement.

Sub-Clause 3.3 of the Sub-Consultancy Agreement provides back-to-back obligations for the Sub-Consultant.

10.3 QM planning in FIDIC Contracts for Works

The PMBOK® Guide defines the QM planning as the process of identifying quality requirements/standards for the project and its deliverables, and documenting how the project will demonstrate compliance with them.[5] The contract documents specify the quality standards, including the applicable statutory and legislative quality requirements/ technical quality codes/ standards, adding the specific requirements of the Employer.

10.3.1 Stated requirements

The FIDIC General Conditions (GCs) give detailed provisions for QM, including procedures for verification of the Contractor's compliance with the requirements. Sub-Clause 4.1 expressly requires the Contractor to execute the Works in accordance with the Contract.

The Red Book Drawings and Specification, and the Yellow/Emerald/Silver Book Employer's Requirements, specify the quality requirements. The Employer's Requirements by definition describe the purpose(s) for which the Works are intended, and specify Key Personnel (if any) and the scope, design, and other performance, technical, and evaluation criteria for the Works. They should state the particular requirements for the completed Works, including functional requirements, quality, and scope. Drafters of the tender documents must remember that if any matters are not referred to or covered, the Contractor may well be relieved of any responsibility in respect of such matters.[6]

Under the Yellow/Emerald Books, the Contractor may be entitled to extension of time (EOT)/Costs Plus Profit in the case of Errors in the Employer's Requirements, following

4 FIDIC, *FIDIC Definition of Services Guidelines. Infrastructure Construction / Civil Works* (2016) <https://fidic.org/sites/default/files/_DOS_guide_civil_works_1.pdf> accessed 2 May 2022

5 PMI Inc., *PMBOK® Guide* (PMI sixth edition 2017) 271

6 FIDIC *Yellow Book* (FIDIC 2022) GPPC 10

the procedure under Sub-Clauses 1.9 and 5.1 (described in subsection 13.8.2). Under the Silver Book, the Contractor shall be deemed to have scrutinised, prior to the Base Date, the Employer's Requirements (including design criteria and calculations, if any). The Silver Book Employer shall not be responsible for any error, inaccuracy, or omission of any kind in the Employer's Requirements, with some exceptions listed in subsection 13.8.2.

Promises for higher quality contained in the Yellow/Emerald Book Contractor's Proposal/Silver Book Tender will override the minimum requirements of the Employer.

10.3.2 Obligatory requirements

Sometimes, mandatory requirements from local/international regulations may differ from those in the Specification/Employer's Requirements, and the higher (more conservative) requirements become the law of the contract. For example, the *EU Regulation No 305/2011*[7] is mandatory for construction works in EU countries. Under the *EU Public Procurement Directive*,[8] the Member States must accept designs to the **Eurocodes**. The Eurocodes are a series of ten European Standards, providing a common approach for the design of buildings and other civil engineering works and construction products. The Eurocodes are the standard technical specification for all public work contracts in the EU and European Free Trade Association (**EFTA**) countries. When a designer proposes an alternative design, it must demonstrate that it is technically equivalent to a design solution by the Eurocodes. The EC Recommendation of 11 December 2003 on the implementation and use of Eurocodes for construction works and structural construction products is not binding. Voluntary application of standards is one of the founding principles of European Standardisation. Eurocodes become mandatory after they are translated and adapted, and the National Annexes are published and introduced in the local legislation. In the different regulatory environments, the national regulations either refer to standards, making compliance with them compulsory, or directly introduce a set of design rules.[9] Pressures from international clients and contractors, as well as other stakeholders like the insurance industry, have led to the more rapid application of Eurocodes for private construction.[10]

FIDIC Contracts have several provisions for compliance with the obligatory requirements, no matter whether or not they are stated under the Contract:

- Sub-Clause 1.13 (Red/Yellow/Emerald)/1.12 (Silver Book): the Contractor and the Employer shall, in performing the Contract, comply with all applicable Laws, including by definition all national (or state or provincial) legislation, statutes, acts, decrees, rules, ordinances, orders, treaties, international law and other laws, and regulations and by-laws of any legally constituted public authority;

7 *Regulation (EU) No 305/2011 of the European Parliament and of the Council of 9 March 2011 laying down harmonised conditions for the marketing of construction products* <https://eur-lex.europa.eu/legal-content/EN/TXT/?uri=CELEX:32011R0305> accessed 14 September 2022

8 *Directive 2014/24/EU of the European Parliament and of the Council of 26 February 2014 on public procurement and repealing Directive 2004/18/EC Text with EEA relevance*

9 Joint Research Centre, Technical report 'The Implementation of the Eurocodes in the National Regulatory Framework' (EU Science Hub 2019) <https://ec.europa.eu/jrc> accessed 14 September 2022

10 https://eurocodes.jrc.ec.europa.eu/en-eurocodes-about-en-eurocodes/are-en-eurocodes-mandatory#technical-equivalence> accessed 25 February 2023

- Sub-Clause 4.1 (Red)/5.1 (Yellow/Emerald/Silver Books): the Contractor's design and the Contractor's Documents will comply with the technical standards stated in the Contract and Laws (in force when the Works are taken over under Clause 10);
- Sub-Clause 5.1 (Yellow/Emerald/Silver Books): the designers must be qualified and entitled under applicable Laws to design the Works;
- Sub-Clause 4.4.1(b) (Red)/5.2(b) (Yellow/Emerald/Silver Books): the Contractor's Documents shall comprise, besides those specified in the Employer's Requirements, the documents required to satisfy all permits, permissions, licences, and other regulatory approvals that are the Contractor's responsibility;
- Sub-Clause 5.4 (Yellow/Emerald/Silver Books): the Contractor's Documents, the execution of the Works, and the completed Works shall comply with the Country's technical standards, building, construction, and environmental Laws, Laws applicable to the product being produced from the Works, and other standards specified in the Employer's Requirements, applicable to the Works, or defined by applicable Laws, in force when the Works or Section or Part are taken over under Clause 10;
- Sub-Clause 7.4 requires that all apparatus, equipment, and instruments for the tests shall be calibrated in accordance with the standards specified in the Specification/Employer's Requirements or defined by applicable Laws.

The FIDIC 2017 Contracts Guide advises the Parties, and especially the Contractor, to familiarise themselves with all Laws that are applicable to the Works, since:

- ignorance of the Laws is no excuse for failing to abide by them; and
- if there is an ambiguity between the Laws and the Contract, any and all provisions of the Laws that are of a mandatory nature will overrule the Contract provisions.[11]

10.3.3 Generally implied requirements

Sub-Clause 7.1 specifies the overall general requirements for execution of the Works. Besides those stated in the Contract, the requirements include performance:

- in a proper workmanlike and careful manner;
- in accordance with recognised good practice; and
- with properly equipped facilities and non-hazardous Materials, except as otherwise specified in the Contract.

The Contract cannot specify the requirements for each operation, the length, material, and diameter of each nail. The custom or common practice will provide the requirements where no details are specified. Sub-Clause 6.9 obliging the Contractor's Personnel to be appropriately qualified, skilled, experienced, and competent in their respective trades or occupations ensures compliance with the generally implied requirements.

11 FIDIC, *The FIDIC 2017 Contracts Guide* (FIDIC 2022) 42

10.3.4 QM and Compliance Verification Systems

Sub-Clause 4.9 obliges the Contractor to plan the QM, elaborating two systems specifically prepared for the project:

- a QM System to demonstrate compliance with the requirements of the Contract;
- a Compliance Verification System to demonstrate that the design, Materials, Plant, work, and workmanship comply in all respects with the Contract.

FIDIC Contracts do not contain an express requirement for the QM System to comply with ISO 9001, but it may be included in the Specification/Employer's Requirements. The tender documents often have a requirement for the eligible tenderers to be ISO certified. As the Construction Extension comments, the lack of ISO QM system does not mean the implemented system is ineffective, and having an ISO-compliant one does not guarantee a quality-compliant product.[12] Sub-Clause 4.9.1 requires the QM System to be in accordance with the details stated in the Specification/Employer's Requirements and to include the Contractor's procedures:

(a) to ensure that all Notices and other communications, Contractor's Documents, as-built records, operation and maintenance manuals, and contemporary records can be traced, with full certainty, to the Works, Goods, work, workmanship, or test to which they relate;
(b) to ensure proper coordination and management of interfaces between the stages of execution of the Works, and between Subcontractors; and
(c) for the submission of Contractor's Documents to the CA for Review.

Under the Red/Yellow/Emerald Book Sub-Clause 4.9.1, the Engineer may Review the QM System within 21 days for compliance with the Contract. Then, the Contractor shall revise the QM System to rectify any notified non-compliance within 14 days. Sub-Clause 4.9.1 also sets out requirements for regular audits of the QM System:

- internal audits at least once every 6 months; the Contractor shall submit to the CA a report with the results within 7 days of the audit, including, where appropriate, the proposed measures to improve/rectify the QM System and/or its implementation;
- if the Contractor (each member of the JV) is required by its QM certification to be subject to external audit, the Contractor shall immediately give a Notice to the CA describing any failing identified in any external audit.

The Compliance Verification System shall plan the method for reporting the results of all inspections and tests carried out by the Contractor. It is a new QM tool introduced in the 2017 Edition of the Red/Yellow/Silver Books, strengthening the position that quality assurance (QA) is the Contractor's responsibility, but the CA is entitled to receive evidence of the Contractor's quality control (QC).

12 PMI Inc., *Construction Extension to the PMBOK® Guide* (PMI 2016) 74

In order to secure the implementation of both systems, the Contractor should plan to allocate personnel to QA tasks:

- designers qualified, experienced, and competent in the disciplines of the design under Sub-Clause 5.1 (Yellow/Emerald/Silver Books);
- sufficient superintendents to plan, arrange, direct, manage, inspect, test, and monitor the execution of the Works with adequate knowledge of the operations to be carried out for the satisfactory and safe execution of the Works pursuant to Sub-Clause 6.8;
- dedicated Document Control, Quality Manager, Building Information Modelling (BIM) Coordinator, and Principal Designer may be appointed as Key Personnel under Sub-Clause 6.12.

10.4 QM and QC in FIDIC Works Contracts

10.4.1 Quality terms

The PMBOK® Guide defines QM as the process of translating the QM plan into executable quality activities and QC as the process of monitoring and recording the results of executing the QM activities to assess performance and ensure the project outputs are complete, correct, and meet the expectations.[13] QM uses the QC process data and results to reflect the overall quality status of the project to the stakeholders. QM includes all the QA activities, and is also concerned with the product design aspects and process improvements.[14]

Sub-Clause 4.9.1 empowers the CA at any time to give a Notice to the Contractor stating the extent to which the Contractor is failing to correctly implement the QM System to the Contractor's activities under the Contract. After receiving this Notice, the Contractor shall immediately remedy such failure.

10.4.2 QA/QC of the Contractor's Documents

Sub-Clause 4.4.1 (Red)/5.2 (Yellow/Emerald/Silver Books) regulates the QM of the Contractor's Documents (including Contractor's design, as-built documents, and O&M Manuals):

- the Employer's Personnel shall have the right to inspect the preparation of all these documents, wherever they are being prepared (except under the Silver Book, where the Employer's control is limited);
- Contractor's Documents specified under the Contract shall be submitted to the CA for Review.

The Review by definition includes examination and consideration by the CA of a Contractor's submission in order to assess whether (and to what extent) it complies with

13 PMI Inc., *PMBOK® Guide* (PMI sixth edition 2017) 271
14 PMI Inc., *PMBOK® Guide* (PMI sixth edition 2017) 288–289

the Contract and/or with the Contractor's obligations under or in connection with the Contract. The compliance assessment procedure includes the following steps:

- the Contractor submits the Contractor's Document with a Notice stating that it is ready for Review and that it complies with the Contract;
- the CA shall give a Notice to the Contractor within 21 days:
  - of No-objection (which may include comments concerning minor matters that will not substantially affect the Works), or
  - that the Contractor's Document fails (to the extent stated) to comply with the Contract, with reasons;
- after receiving a Notice for non-compliance, the Contractor shall revise the Contractor's Document and resubmit it for Review.

If the CA gives no Notice within this period of 21 days, the CA shall be deemed to have given a Notice of No-objection to the Contractor's Document. However, Sub-Clause 5.8 of the Yellow/Emerald/Silver Books stipulates that if errors or other defects are found in the Contractor's Documents, they and the Works shall be corrected even where a Notice of No-objection has been (deemed to be) given.

Figure 10.1 presents these design QA/QC procedures. The Review procedure is repeated in the case of changes or rectification of errors/omissions during construction.

The Employer has the following remedies in the case of design non-compliance with the Contract:

- resubmissions until compliance is achieved;
- construction of the relevant works shall not commence until a Notice of No-objection is given;
- the Contractor is not entitled to EOT for any delay caused by any such revision and resubmission and/or by subsequent Review;
- the Employer may be entitled to Delay Damages under Sub-Clause 8.8 if the taking over is delayed;
- all corrections and resubmissions shall be at the Contractor's risk and cost;

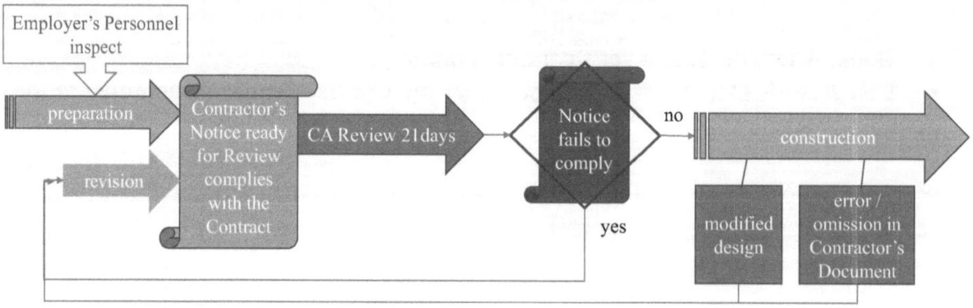

Figure 10.1 Design QA/QC

- the Employer shall be entitled, subject to Sub-Clause 20.2, to payment by the Contractor of the costs reasonably incurred for multiple Reviews under Sub-Clause 7.5;
- the estimated cost of revision and resubmission may be withheld from the amounts due for payment under Sub-Clause 14.6.2 until compliance is achieved;
- rejection under Sub-Clause 7.5 if the Contractor fails to comply with the CA's instructions, and employing another designer at the Contractor's cost (but the Contractor shall have no responsibility for this design);
- non-compliance with the Sub-Clause 7.5 Notice of rejection/ Sub-Clause 15.1 Notice to Correct may lead to termination under Sub-Clause 15.2.

10.4.3 QA/QC during the execution of the Works

The daily QA/QC during the execution of the Works shall guarantee compliance with the Contract. The relevant provisions are under:

- Sub-Clause 4.1(c) (Red)/5.2 (Yellow/Emerald/Silver Books) for compliance with the reviewed design:
    - where the Contractor is responsible for the design, construction may start only after the relevant Notice of No-objection is given (or deemed to be given);
    - construction of such part shall be in accordance with the reviewed Contractor's Documents;
    - the Contractor may modify any design or Contractor's Documents that have previously been submitted for Review, by giving a Notice to the Engineer with reasons; if the Contractor has commenced construction of the relevant part of the Works, work on this part shall be suspended for the Review process and may resume only after a Notice of No-objection is (deemed to have been) given for the revised documents (see Figure 10.1);
- Sub-Clause 5.6 (Yellow/Emerald/Silver Books)/4.4 (Red Book when specified) for 'as-built' records of the execution of the Works, showing the exact as-built locations, sizes, and details of the work as executed by the Contractor to be kept on the Site;
- Sub-Clause 7.1 for QC criteria;
- Sub-Clause 7.2 for samples;
- Sub-Clause 7.3 for examination, inspection, measurement, or testing by the Employer's Personnel of:
    - Materials and Plant during production, manufacture, and construction (at the Site and elsewhere), and
    - works before being covered up;
- Sub-Clause 7.4 for testing by the Contractor.

It is important to check the compliance of the Plant and Materials with the requirements before incorporation in the Works. The QM and Compliance Verification Systems shall ensure that the Contractor verifies compliance in selecting the Materials, provides samples under Sub-Clause

7.2, and then secures proper storage, transportation, and handling pursuant to Sub-clause 4.16, ensuring traceability of the Plant and Materials. Depending on the specific requirements, the CA usually verifies the QC of Materials and Plant in several stages:

- checking of samples supplied under Sub-Clause 7.2, visiting key manufacturing facilities and places from which natural Materials are obtained under Sub-Clause 7.3;
- checking of QM documents, inspection reports, test results, and compliance verification documentation (including certificates of Materials) to be provided with each Progress Report under Sub-Clause 4.20;
- inspecting delivered Materials under Sub-Clause 7.3 for compliance with the verified documents.

As recommended in Chapter 6, the pre-commencement meeting may be used to agree detailed procedures for inspection under Sub-Clause 7.3, to be integrated in the QM System submitted within 28 days of the Commencement Date. Depending on the location and size of different parts of the Site and the availability of the CA's assistants and other Employer's Personnel with expertise in different key disciplines, the Parties should agree with the CA the minimum period for the submission of any Notice of request for inspection (**RFI**). While a 24-hour Notice may be acceptable in general, 2–3-day Notice may be agreed for special works where the experts are not allocated to the Site (such as lifts or special systems). It is important that the template for the RFI includes a list of the appendices to be attached, such as extracts of Drawings, Certificates for materials and tests, acceptance/compliance documents, and any mandatory form under the local Laws to be filled in. Thus, receiving the RFI on the previous day by email, the inspector has all necessary documents on his/her mobile device and is ready for the inspection. On major projects, inspections need a rigorous planning and Contractor's QC:

- to be done in time so as not to incur critical delay;
- the Contractor's superintendents to check readiness and compliance in advance, especially when Subcontractors are involved;
- the Contractor to provide safe access;
- the daily planning to be reasonable, considering the number of inspections and tests to be done, their location, and the availability of representatives of the Parties qualified in the relevant disciplines.

If the Employer's Personnel do not attend at the time stated in the Contractor's Notice (or such time as may be agreed with the Contractor), the Contractor may proceed with covering up, putting out of sight, or packaging for storage or transport. If the Contractor fails to give the RFI Notice in accordance with Sub-Clause 7.3, the Contractor shall, if and when required by the CA, uncover the work and thereafter reinstate and make it good, all at the Contractor's risk and cost.

The CA shall allocate document control duties to an assistant, including:

- registering RFI and distribution to assigned inspectors;

- follow up of inspections, archiving compliance documents;
- follow up of non-conformity reports (**NCRs**) prepared by the inspectors until compliance is achieved.

Sub-Clause 7.4 sets out the procedure for all tests specified in the Contract, other than the Tests after Completion, covered by Clause 12 of the Yellow/Emerald/Silver Books. The Contractor shall promptly forward to the CA duly certified reports of the tests. When the specified tests have been passed, the CA shall endorse the Contractor's test certificate or issue a test certificate to the Contractor. The CA shall be deemed to have accepted the readings as accurate if after being notified, he/she has not attended the test.

Sub-Clause 7.5 sets out the following action plan in case of non-conformities found when checking any Plant, Materials, design, or workmanship:

- the CA gives a Notice to the Contractor describing the defects;
- the Contractor promptly submits a proposal for remedial work;
- the CA may Review the proposal, and may give a Notice within 14 days stating the extent to which the proposed work, if carried out, would not result in the Plant, Materials, design, or workmanship complying with the Contract;
- the Contractor shall then promptly submit a revised proposal;
- if the CA gives a Notice of No-objection or is silent, the Contractor shall remedy the defects;
- after that, if the CA requires retesting, the tests shall be repeated in accordance with Sub-Clause 7.4 at the Contractor's risk and cost.

If the Contractor fails to promptly submit a (revised) proposal for remedial work, or fails to carry out the proposed remedial work to which the CA has (deemed to have) given a Notice of No-objection, the CA may:

- instruct the Contractor under Sub-Clause 7.6 to:
  - repair/remedy (if necessary, off the Site)/ remove from the Site and replace any Plant or Materials that are not in accordance with the Contract;
  - repair/remedy/remove/re-execute any other work not in accordance with the Contract;
- reject the design/Plant/Materials/workmanship by giving a Notice to the Contractor, with reasons, in which case the Employer may (at its sole discretion) have the work carried out by others (including any retesting) at the Contractor's cost, but the Contractor shall have no responsibility for this work.

The Employer has the following remedies under the Contract in case of defects:

- remedial works and retesting until compliance is achieved under Sub-Clauses 7.5, 7.6, and 9.3;
- the Contractor is not entitled to EOT for any delay caused by any such remedial works and retesting, and the Employer may be entitled to Delay Damages under Sub-Clause 8.8, if the taking over is delayed;

- all remedial works/retesting shall be at the Contractor's risk and cost;
- under Sub-Clause 7.5, the Employer shall be entitled subject to Sub-Clause 20.2 to payment by the Contractor of the additional costs incurred due to the rejection and retesting;
- under Sub-Clause 7.6, the Employer shall be entitled subject to Sub-Clause 20.2 to payment by the Contractor of the cost for employing other persons to carry out the work where the Contractor fails to comply with the CA's instruction;
- the estimated cost of remedial works and retesting may be withheld from the amounts due for payment until compliance is achieved under Sub-Clause 14.6.2;
- non-compliance with the Notice of rejection under Sub-Clause 7.5/ instruction under Sub-Clause 7.6 may lead to termination under Sub-Clause 15.2.

10.4.4 QA/QC for taking over of the Works or Section

The QA/QC before taking over by the Employer is covered by Clause 9 [Tests on Completion] and Clause 10 [Employer's Taking Over]. Chapter 6 describes the procedure for taking over and the conditions for a taking-over certificate (TOC):

- Review under Sub-Clause 5.2.2 of the as-built records submitted under Sub-Clause 5.6 and the O&M Manuals submitted under Sub-Clause 5.7;
- inspection of the completed Works under Sub-Clause 7.3;
- testing under Sub-Clause 7.4 and passing the Tests on Completion under Clause 9.

Figure 10.2 presents the Sub-Clause 9.1 QC procedure for the Tests on Completion under the Yellow/Emerald/Silver Books:

- the Contractor shall submit to the CA, at least 42 days in advance, a detailed test programme;

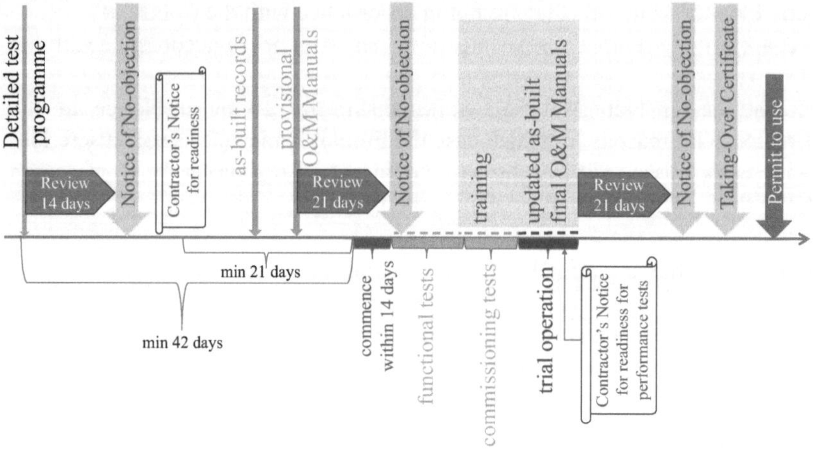

Figure 10.2 Tests on Completion

- the Engineer may Review it within 14 days and may give a Notice to the Contractor stating the extent to which it does not comply with the Contract;
- the Contractor shall revise the test programme within 14 days to rectify such non-compliance;
- the Contractor may commence the Tests on Completion:
  - after the CA gives (or is deemed to have given) a Notice of No-objection;
  - within 14 days after the date notified by the Contractor at least 21 days in advance, or other date instructed by the CA;
  - after submission of the as-built records under Sub-Clause 5.6; and
  - after submission of the provisional O&M Manuals under Sub-Clause 5.7;
- the Contractor shall proceed in accordance with the Contractor's test programme in the following sequence:
  - pre-commissioning tests including inspections and ('dry' or 'cold') functional tests to demonstrate that each item of the Works or Section can safely undertake the next stage;
  - commissioning tests, which shall include the specified operational tests to demonstrate that the Works or Section can be operated safely and as specified in the Employer's Requirements, under all available operating conditions; and
  - trial operation (to the extent possible under available operating conditions), which shall demonstrate that the Works or Section perform reliably and in accordance with the Contract; when the Works or Section are operating under stable conditions, the Contractor shall give a Notice that they are ready for any other Tests on Completion, including performance tests to demonstrate whether the Works or Section comply with the performance criteria specified in the Employer's Requirements and with the Schedule of Performance Guarantees;
- when the Works or Section have, in the Contractor's opinion, passed each stage of the Tests on Completion, the Contractor shall submit a certified report of the results of these tests to the CA;
- the CA shall Review each such report and within 14 days may give a Notice to the Contractor stating the extent to which the results of the tests do not comply with the Contract;
- if the CA does not give such a Notice after receiving the results of the tests, the CA shall be deemed to have given a Notice of No-objection; considering the results, the CA shall make allowances for the effect of any use of the Works by the Employer on the performance or other characteristics of the Works;
- if during the Tests on Completion any error/defect is found in the provisional O&M Manuals, the Contractor shall promptly rectify it at the Contractor's risk and cost and resubmit the final O&M Manuals for Review before taking over;
- in the case of adjustments leading to modification of the Works during the Tests on Completion, the Contractor shall update the as-built records and resubmit them to the CA for Review;
- the CA shall Review the final O&M Manuals and updated as-built records and issue Notices of No-objection before taking over;
- the Contractor shall carry out the training in accordance with Sub-Clause 5.5;

- when compliance with the Employer's Requirements is confirmed, the CA shall issue the TOC;
- the mandatory procedures under the Laws of the Country shall be followed.

Figure 10.3 presents the procedures under Sub-Clause 9.2 and the remedies of the Parties in case of delayed Tests on Completion:

- if the Contractor has given a Notice under Sub-Clause 9.1 for readiness for Tests on Completion, and these tests are unduly delayed for more than 14 days by the Employer's Personnel or by a cause for which the Employer is responsible, Sub-Clause 10.3 [Interference with Tests on Completion] shall apply:
  - the Contractor shall give a Notice to the CA describing such prevention;
  - the Employer shall be deemed to have taken over the Works or Section on the date when the Tests on Completion would otherwise have been completed; and
  - the CA shall immediately issue a TOC for the Works or Section;
  - the Contractor shall carry out the Tests on Completion as soon as practicable before the expiry date of the DNP, after the CA gives a Notice at least 14 days in advance of the possible date;
  - if the Contractor suffers delay and/or incurs Cost as a result of the prevention, the Contractor may be entitled subject to Sub-Clause 20.2 to an EOT and/or payment of Cost Plus Profit.
- if the Contractor unduly delays the Tests on Completion:
  - the CA may, by giving a Notice to the Contractor, require the Contractor to carry out the tests within 21 days;
  - the Contractor shall carry out the tests within this period, giving a prior Notice to the CA of not less than 7 days fixing the dates;
  - if the Contractor fails to carry them out within this period of 21 days, after a second Notice given by the CA, the Employer's Personnel may proceed with the tests, and the Contractor may attend and witness them;

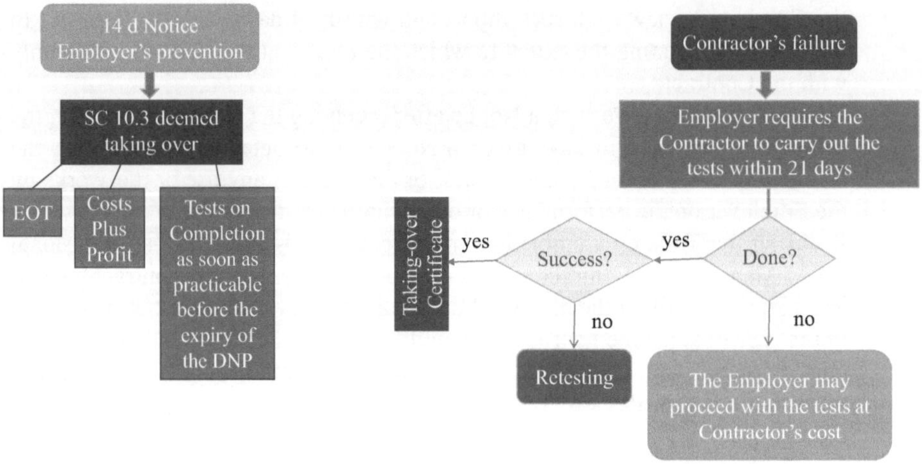

Figure 10.3 Delayed Tests on Completion

- within 28 days of completion, the CA shall send a copy of the test results to the Contractor; whether or not the Contractor attends, tests shall be deemed to have been carried out in the presence of the Contractor, and the results shall be accepted as accurate; and
- if the Employer incurs additional costs as a result of such testing, the Employer shall be entitled subject to Sub-Clause 20.2 to payment by the Contractor of the costs reasonably incurred.

Sub-Clause 9.4 sets out the Employer's remedies in case of the Contractor's failure to pass the Tests on Completion:

- retesting under Sub-Clause 9.3;
- rejection of the Works or Section if the effect of the failure is to deprive the Employer of substantially the whole benefit of the Works, in which event the Employer shall be entitled to terminate the Contract with immediate effect and, subject to Sub-Clause 20.2, to recover from the Contractor all sums paid for the Works, plus financing charges and any costs incurred in dismantling the same, and clearing the Site;
- rejection of the Section if the effect of the failure is that the Section cannot be used for its intended purposes under the Contract, requesting the CA to treat it as an omission, instructed under Sub-Clause 13.3.1; or
- acceptance of the Works or Section requesting the CA to issue a TOC reducing the Contract Price subject to Sub-Clause 20.2 in full satisfaction of this failure.

10.4.5 QA/QC after taking over of the Works or Section

Any minor outstanding work and defects that will not substantially affect the safe use of the Works or Section for their intended purposes found at the inspections before taking over shall be listed in the TOC. The punch list part of the TOC shall specify the dates for completion of the outstanding works and remedying of the specified defects. The QA/QC shall continue during the DNP, and after that before the Performance Certificate is issued, until compliance with the requirements is established, examining the:

- completed work that is outstanding on the relevant Date of Completion;
- executed work required to remedy defects or damage:
  - specified in the TOC;
  - of which a Notice is given to the Contractor by (or on behalf of) the Employer on or before the expiry date of the DNP;
  - found during the final inspection promptly after the expiry of the DNP;
- Contractor's Documents, including the updated as-built records.

Sub-Clause 11.1 sets out the action plan in case a defect appears or damage occurs during the relevant DNP:

- a Notice shall be given to the Contractor by (or on behalf of) the Employer;
- the Contractor and the Employer's Personnel shall jointly inspect the defect or damage;

- the Contractor shall then submit a proposal for necessary remedial work;
- the CA may Review the proposal, and may give a Notice within 14 days stating the extent to which the proposed work, if carried out, would not result in the Plant/Materials/design/workmanship complying with the Contract;
- then, the Contractor shall promptly submit a revised proposal;
- if the CA gives a Notice of No-objection or is silent, the Contractor shall remedy the defects;
- after that, if the CA requires retesting, the tests shall be repeated in accordance with Sub-Clause 7.4 at the Contractor's risk and cost.

The Employer shall be entitled to extension of the DNP under Sub-Clause 11.3 subject to Sub-Clause 20.2, **if and to the extent that the Works, Section, Part, or a major item of Plant cannot be used** after taking over for the intended purposes by reason of a defect or damage that is attributable to the Contractor.

Sub-Clause 11.4 sets out the procedure in case of the Contractor's failure to remedy defects or damages:

- a Notice shall be given to the Contractor by (or on behalf of) the Employer, fixing a date that allows the Contractor reasonable time to remedy the defect or damage;
- if the Contractor fails to remedy the defect or damage by the date stated in the Notice, and this remedial work was to be executed at the cost of the Contractor under Sub-Clause 11.2, the Employer may (at the Employer's sole discretion):
  (a) carry out the work or have the work carried out by others at the Contractor's cost, but the Contractor shall have no responsibility for this work; the Employer shall be entitled, subject to Sub-Clause 20.2, to payment by the Contractor of the costs reasonably incurred;
  (b) accept the damaged or defective work, reducing the Contract Price in full satisfaction of this failure only by an amount appropriate to cover the reduced value to the Employer as a result of this failure;
  (c) require the CA to treat any part of the Works that cannot be used for its intended purpose(s) under the Contract by reason of this failure as an omission, as if such omission had been instructed under Sub-Clause 13.3.1; or
  (d) terminate the Contract as a whole with immediate effect if the defect or damage deprives the Employer of substantially the whole benefit of the Works; the Employer shall then be entitled, subject to Sub-Clause 20.2, to recover from the Contractor all sums paid for the Works, plus financing charges and any costs incurred in dismantling the same, clearing the Site, and returning Plant and Materials to the Contractor.

Clause 12 stipulates the procedure for Tests after Completion, if specified in the Employer's Requirements of the Yellow/Emerald/Silver Books. The Employer shall be entitled, subject to Sub-Clause 20.2, to payment of Performance Damages (set out in the Schedule of Performance Guarantees) by the Contractor in full satisfaction of the failure to pass the Tests after Completion.

Only the Performance Certificate shall be deemed to constitute acceptance of the Works according to Sub-Clause 11.9. The CA shall issue it within 28 days of the expiry of the latest DNP, but the issue may be delayed until all defects found during the final inspection are remedied, the Tests after Completion are successful, and the CA gives a Notice of No-objection to the updated Contractor's Documents.

Pursuant to Sub-Clause 11.10, after the issue of the Performance Certificate, each Party shall remain liable for the fulfilment of any obligation that remains unperformed at that time. As discussed in Chapter 6, the Contractor may be liable for latent defects under the local Laws for some years after the Performance Certificate is issued.

10.5 Continuous quality improvement

The PDCA cycle is the basis for quality improvement as defined by Shewhart and modified by Deming.[15] The international standard ISO 9001:2015 describes the PDCA cycle for continuous quality improvement as follows:[16]

- **Plan:** establish the objectives of the system and its processes, and the resources needed to deliver results in accordance with the requirements and the organisation's policies, and identify and address risks and opportunities;
- **Do:** implement what was planned;
- **Check:** monitor and (where applicable) measure processes and the resulting products and services against policies, objectives, requirements, and planned activities, and report the results;
- **Act:** take actions to improve performance, as necessary.

Figure 10.4 links the PDCA quality improvement cycle with the FIDIC Clauses. The quality of the Works is the Contractor's responsibility pursuant to Sub-Clause 4.1 to be planned under Sub-Clause 4.9. ISO certified or not, the Contractor has to elaborate and apply

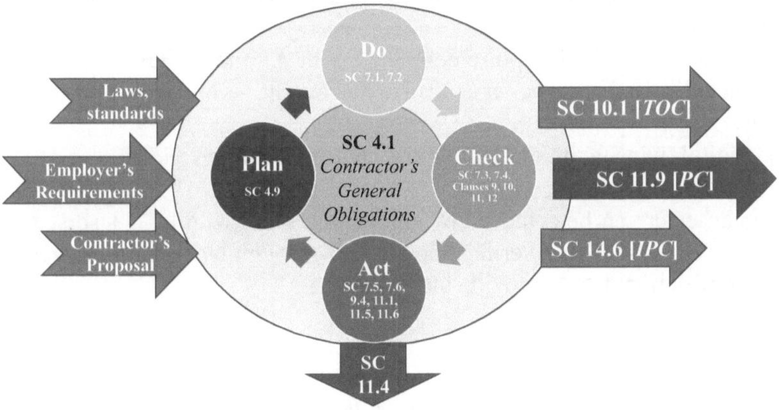

Figure 10.4 PDCA cycle in FIDIC Works Contracts

15 PMI Inc., *PMBOK® Guide* (PMI 2017) 275
16 *ISO 9001:2015 Quality Management Systems – Requirements*

specific QM and Compliance Verification Systems. The CA and the other Employer's Personnel check the quality following the procedures under Clauses 7, 9, 10, and 11. In the case of compliance, payments may be certified under Sub-Clause 14.6, the TOC shall be issued after completion of the Works under Sub-Clause 10.1, and the Performance Certificate shall be issued under Sub-Clause 11.9 after the expiry of all DNPs and remedying of all patent defects. In the case of non-compliance with the requirements, the provisions of Sub-Clauses 7.5, 7.6, 9.4, and (after taking over) Clause 11 shall apply.

In addition to PDCA, quality improvement initiatives such as Total Quality Management (TQM), Six Sigma, and **Lean Six Sigma** may improve both the quality of PM and the quality of the end product, service, or result.[17] While TQM focuses on improving the process, Six Sigma looks for causes of defects and where appropriate, amends processes. The Lean Six Sigma creates a new tool combining the focus on waste elimination of Lean with the Six Sigma methods to solve quality issues.

10.6 Innovative tools for QM

Supply chain collaboration, standardisation, and digitalisation provide numerous opportunities for quality improvement. Long-term relationships based on partnership and co-operation with the supply chain enhance the ability to create value. All tools discussed in Chapter 3 may be used for efficient QA/QC, e.g.:

- clash detection in BIM enables early resolution of issues;
- BIM analytical module enables acoustic and energy efficiency simulations and reports compliance with the contract requirements for different types of premises, using the architectural module and the characteristics of construction products from the available producers' libraries;
- digital fabrication and off-site manufacturing guarantee high quality;
- AI and IoT enable easy transfer of data from BIM to construction mechanisation and robots securing performance in compliance with the requirements and minimising waste;
- drones, sensors, and reality capture combined with AI and digital twin enable 24/7 QC, including safer and better control in narrow spaces and in other difficult conditions;
- BIM and digital twins facilitate QA/QC and continuous improvement.

A reduction in human error can be achieved by automating tasks with IoT sensors, AI, and smart contracts. Certification and verification of tasks through blockchain-protected interfaces feeding an iContract would provide increased QA.[18]

10.7 Example Notices

Appendix 1 includes Notice of the Contractor under Sub-Clause 7.3 of the Yellow Book for readiness for inspection and reply of the Engineer that inspection is not required.

17 PMI Inc., *PMBOK® Guide* (PMI sixth edition 2017) 275

18 Alan McNamara, 'Automating the Chaos: Intelligent Construction Contracts' Chapter in *Smart Cities and Construction Technologies* (IntechOpen 2020) 10

CHAPTER 11

Project Resource Management with FIDIC Contracts

11.1 Project Resource Management processes

According to the PMBOK® Guide, Project Resource Management includes processes to identify, acquire, and manage the resources needed for the successful completion of the project.[1]

Construction projects feature unique characteristics that affect how resources are managed.[2] This chapter links the PMBOK with the FBOK and suggests which innovative tools may support the Project Resource Management with FIDIC Contracts. While the PMBOK® Guide focuses more on human resources management, in construction, the supply chain management, the transportation, storing, and handling of plant and materials, and the smart management of construction equipment are also of primary importance.

The PMBOK® Guide refers to the team or personnel as human resources. According to its Construction Extension, the human resources include the team managing the project and the team performing the construction. FIDIC refers to the human resources as 'personnel'. The PMBOK calls the other resources physical resources, while the FIDIC Works Contracts refer to them as 'Goods'. The definition of 'Goods' includes Contractor's Equipment, Materials, Plant, and Temporary Works.

1 PMI Inc., *PMBOK®* (PMI sixth edition 2017) 307
2 PMI Inc., *Construction Extension to the PMBOK® Guide* (PMI 2016) 81

DOI: 10.4324/9781003272892-11

Recent research by Kusimo[3] revealed seven key factors affecting resource management:

- poor understanding of project complexity at the tender stage;
- incomplete Site survey;
- little visibility of resource profile at the tender stage;
- changes to the project schedule;
- running multiple projects in parallel;
- shortage of specialised resources;
- poor management of Employer-specific authorisations/certifications.

11.2 Project Resource Management using the FIDIC Suite for Services

11.2.1 Management of human resources

The FIDIC Suite for Services includes model agreements for acquiring resources in different ways:

- consulting companies may unite their resources using the Model Joint Venture Agreement;
- the Sub-Consultancy Agreement may be used where a Consultant appointed under the White Book:
  - lacks specific expertise and engages a Sub-Consultant to undertake part of the Services, or
  - cannot obtain mandatory licences/authorisations and signs the Agreement with a local company authorised for design or supervision/technical control, or
  - needs a local consultant for feasibility studies, permit management, etc.;
- a foreign Consultant may enter into a contract with a local Representative using the Model Representative Agreement.

Depending on the scope of services specified in Appendix 1 of the White Book, the Consultant shall provide enough experienced, qualified experts in order to complete the Services exercising reasonable skill, care, and diligence, as expected from a consultant experienced in the provision of similar services (Sub-Clause 3.3 [Standard of Care]). The key personnel proposed by the Consultant to work in the Country[4] shall be subject to acceptance by the Client with regard to their qualifications and experience. Personnel included in the Consultant's proposal as part of the Agreement shall be deemed to be accepted by the Client.

The White Book Sub-Clause 1.1.8 defines 'Consultant's Representative' as the person referred to in the Particular Conditions or appointed and notified to the Client for the administration of the Agreement. Sub-Clause 3.6 obliges the Consultant:

3 Habeeb Kusimo and others, Optimisation of Resource Management in Construction Projects: a Big Data Approach (*World Journal of Science, Technology and Sustainable Development* 2019)

4 Sub-Clause 1.1.9 of the *White Book* defines 'Country' as the country named in the Particular Conditions, or where no country is mentioned, the country of the Project site (or main site)

- to notify the Client of the powers and authority delegated to the Consultant's Representative;
- to designate an individual to liaise with the Client's Representative in the Country, if required by the Client.

Sub-Clause 3.7 provides the procedure for replacement of personnel provided by the Consultant. Any replacement shall have suitable qualifications and experience.

Sub-Clause 3.9 applies where the Consultant is acting as the CA under a Works Contract. Chapter 6 discusses the delegation of authority to the CA vested to act as the Red/Yellow/Emerald Book Engineer/ Silver Book Employer's Representative under the Works Contract. Further, Chapter 6 explains the procedure under Clause 3 of those FIDIC Books for delegation of authority by the CA to his/her assistants with the relevant limitations. The Consultant must appoint an assistant to act as a coordinator or document control. Chapter 6 outlines the importance of team building, communications, and training. A FIDIC accredited in-house training is recommended at the beginning of the project.

Sub-Clause 3.3.3 stipulates that the Consultant shall comply with all regulations, statutes, ordinances, and other forms of standards, codes of practice, and legislation applicable to the Services and the Agreement. The key personnel needed for design services have to comply with the design authorisation regime in the Country or rely on authorised local designers. If the CA also has building control obligations in the scope of its services, the local regulations may have specific requirements for licensing/registering of the company or natural persons.

11.2.2 Case study – legislated requirements for design services

For example, for the design of an office building in Bulgaria, the key personnel authorised under the *Chambers of Engineers and Architects Act*[5] shall include as a minimum an architect, a structural engineer, a heating, ventilation, and air conditioning (HVAC) engineer, a water supply and sewage engineer, a geodetic survey engineer, a geology engineer, an electrical engineer, a landscaping engineer, a fire safety engineer, and a mechanical engineer. The Consultant appointed by the Client[6] for the mandatory compliance assessment of this design and/or construction supervision shall be registered as a legal person with a listed team of qualified experts in different disciplines. The Consultant's personnel allocated to review the design shall include an architect and engineers with the relevant specialisation selected from the list plus a lawyer responsible for reviewing the compliance with the *Property Act* and other regulations for connections to infrastructure, easements, etc.

5 *Chambers of Engineers and Architects in the Investment Design Act* (State Gazette 20 2003)
6 In compliance with the mandatory requirements of the *Spatial Development Act* (State Gazette 1 2001)

11.2.3 Management of other resources

Appendix 2 of the White Book may list equipment and facilities to be supplied by the Client for the execution of the Services.

The equipment supplied by the Consultant depends on the services, but in all cases, the personnel need mobile phones, computers, and other devices for communications and virtual team work. Personal protection equipment and cars are important for contract administration and supervision services. Consultants doing surveys need special equipment like Global Positioning System (**GPS**) and total stations. Special equipment for testing is needed for QC services. Drones, scanners, sensors, and cameras, combined with AI, facilitate the surveys, control, and reporting. BIM needs investment in training and office infrastructure, including computers, monitors, and software.

11.2.4 Innovative human resource management

The project manager should lead and manage the team, taking into account the location of the team members, the cultural differences, and the politics. FIDIC member associations in more than 100 countries include thousands of consulting companies with more than 1 million employees. All of them are required to comply with the FIDIC Code of Ethics,[7] including the obligations to:

- maintain knowledge and skills at levels consistent with development in technology, legislation, and management;
- apply due skill, care, and diligence in the services rendered to the client;
- perform services only when competent to perform them.

As a leader, the project manager is responsible for proactively developing team skills and competencies while retaining and improving team satisfaction and motivation,[8] while at the same time ensuring that all team members comply with the FIDIC Code of Ethics.

The PM style is moving away from command and control towards a more collaborative and supportive management approach, delegating decision making to the team members. Investing in personal emotional intelligence brings efficiency and reduces staff turnover. Project managers must truly understand the talents, values, and potential of themselves and their project teams, and manage their own emotions and the emotions of others.[9] They need to resolve conflicts in a timely manner, maintain mutual trust and motivation, and continually monitor team performance, taking preventive actions.

Globalisation, availability of communication technology, and the COVID pandemic have promoted virtual teams and working from home. This allows resource levelling, flexibility, efficiency, and the use of special expertise from a distant location. Working with BIM enables sharing the same digital space and access to the latest information.

---

7 FIDIC Code of Ethics (FIDIC) <https://fidic.org/sites/default/files/fidic_codeofethics_201510.pdf> accessed 3 October 2022

8 PMI Inc., *PMBOK® Guide* (PMI sixth edition 2017) 309

9 Christine Casper, 'Using Emotional Intelligence to Improve Project Performance' (PMI 2002) <https://www.pmi.org/learning/library/emotional-intelligence-improve-project-performance-1019> accessed 3 October 2022

Using platforms like MS Teams, the Consultant's personnel may co-author documents working at the same time in different countries, store files, and keep a team calendar. Virtual meetings, forums for discussion, and chats facilitate the exchange of knowledge and deal with the feeling of isolation. When managing virtual teams, the project manager must take into account time zones, connectivity, and cultural differences.

11.3 Project Resource Management using the FIDIC Works Contracts

11.3.1 Resource breakdown structure

The PMBOK® Guide defines the resource breakdown structure as a hierarchical representation of resources by category and type. Figure 11.1 links each resource category with the FIDIC provisions. The Clause numbers are from the Yellow Book, but the Red/Silver/Emerald/Gold/Green Books use the same terminology:

- the Contractor's Personnel include all personnel working on the project for the Contractor, including the Contractor's Representative, Key Personnel, staff, labour, other employees of the Contractor and each Subcontractor, and any other personnel assisting the Contractor in the execution of the Works;
- the Key Personnel as defined under Sub-Clause 1.1.48 include those positions (if any) other than the Contractor's Representative stated in the Employer's Requirements (in Figure 11.1, the Design Manager); Sub-Clause 6.12 sets out the requirements in relation to the Key Personnel;
- Sub-Clause 6.8 sets out the requirements related to the Contractor's superintendence, performed in Figure 11.1 by the Site Managers;
- the Goods include all physical resources used for the Works;
- the Contractor's Equipment includes all apparatus, equipment, machinery, construction plant, vehicles, and other items required by the Contractor for the execution of the Works, including all items of the Subcontractors, but excluding Temporary Works, Plant, Materials, and any other things intended to form part of the Permanent Works;
- the Plant includes the apparatus, equipment, machinery, and vehicles intended to form part of the Permanent Works;
- the Materials include anything, except Plant, that is used by the Contractor in the Permanent Works or to be supplied by the Contractor directly to the Employer (such as spare parts);
- the Temporary Works include physical resources of every kind (other than Contractor's Equipment) required on Site for the execution of the Works, which do not form part of the Permanent Works, such as formwork, temporary railings and pathways, and supporting structures.

The Contractor is responsible for the resource management of all these resources, including the Subcontractors' resources. Sub-Clause 4.1 obliges the Contractor to provide the Plant (and spare parts, if any), all Contractor's Personnel, Goods, and consumables required to fulfil its obligations under the Contract.

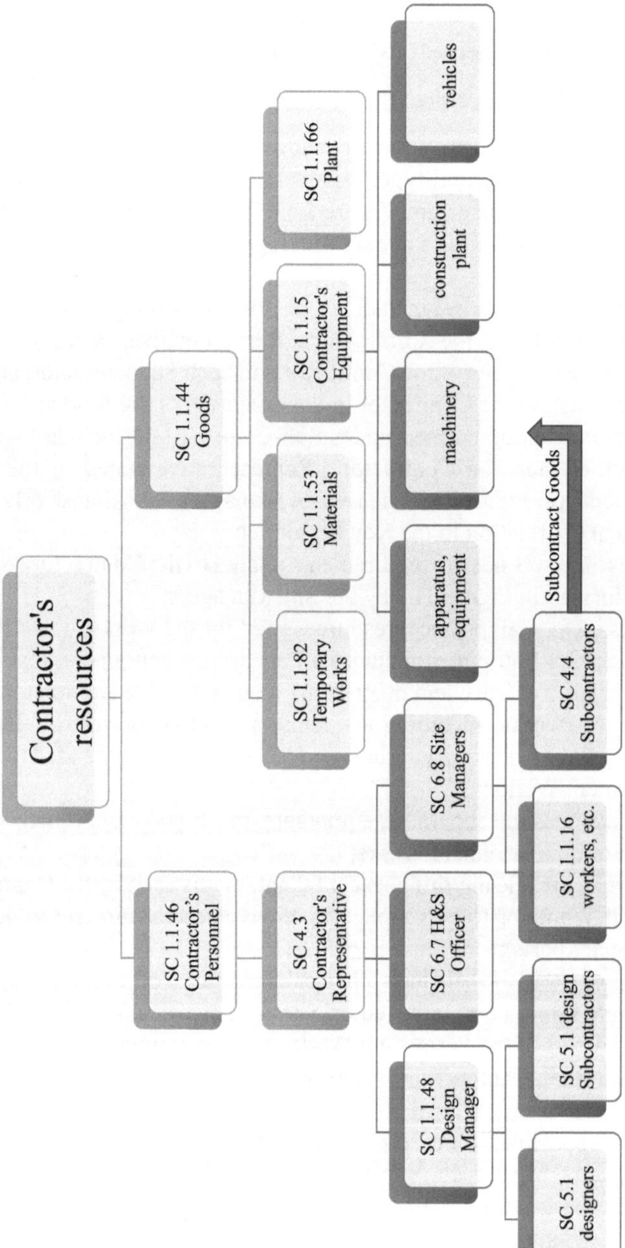

Figure 11.1 Resource breakdown structure under the FIDIC Yellow Book

11.3.2 Resource calendars and records

A resource calendar identifies the working days, shifts, working hours, and availability of resources. The information may be at activity or project level.[10]

Sub-Clause 6.5 [Working Hours] stipulates that no work shall be carried out on the Site on locally recognised days of rest, or outside the normal working hours stated in the Contract Data, unless the work is unavoidable or necessary for the protection of life or property, or for the safety of the Works, or when the CA gives consent. The working hours shall be integrated in the calendar for the Programme under Sub-Clause 8.3, which shall include the dates of all locally recognised days of rest and holiday periods and the key delivery dates of Plant and Materials.

While the GCs require details about the resources to be presented in the supporting report, Chapter 8 recommends the Programme to include a resource-loaded schedule, which may address several of the key factors for resource management listed in the beginning of this chapter, and provide a baseline for the calculation of prolongation and disruption costs.

Sub-Clause 6.10 sets out the Contractor's obligation to include in each progress report under Sub-Clause 4.20 records of:

(a) occupations and actual working hours of each class of Contractor's Personnel;
(b) the type and actual working hours of each item of the Contractor's Equipment;
(c) the types of Temporary Works used;
(d) the types of Plant installed in the Permanent Works; and
(e) the quantities and types of Materials used

for each work activity shown in the Programme, at each work location, and for each day of work. This requirement is extremely burdensome, but the provision allows other arrangements to be proposed by the Contractor and agreed by the CA. The Site Diaries kept at different parts of the Site may be convenient to record this information, if the resources are linked to the work activity codes in the Programme.

Drones and 3D cameras with AI, and a PMIS, facilitate the preparation of the records and may automate the reporting.

11.3.3 Management of human resources

Chapter 4 discusses the role of the Contractor's Representative as project manager of the Contractor and the procedure for his/her appointment under Sub-Clause 4.3. Chapter 6 discusses the Contractor's Representative's authority and delegation of authority to Key Personnel. The Contractor must appoint Key Personnel, suitably competent and experienced, and fluent in the language for communications. Under Sub-Clause 6.8, the Contractor shall provide a sufficient number of competent persons for the superintendence to plan, arrange, direct, manage, inspect, test, and monitor the execution of the Works, with adequate knowledge of the operations to be carried out and of the language for communications.

10 PMI Inc., *PMBOK® Guide* (PMI 2017) 323

Sub-Clause 5.1 [General Design Obligations] of the Yellow/Emerald/Silver Books contains the requirements for the selection of designers, who have to:

(a) be engineers or other professionals, qualified, experienced, and competent in the disciplines of the design for which they are responsible;
(b) comply with the criteria stated in the Employer's Requirements; and
(c) be qualified and entitled under applicable Laws to design the Works.

In addition, under the Yellow/Emerald Book Sub-Clause 5.1, the Contractor:

- shall submit to the Engineer for consent the name, address, detailed particulars, and relevant experience of each proposed designer/design Subcontractor;
- warrants that the Contractor, the Contractor's designers, and design Subcontractors have the experience, capability, and competence necessary for the design;
- undertakes that the designers and design Subcontractors shall be available to attend discussions with the Engineer and/or the Employer at all reasonable times (on or off the Site) until the issue of the Performance Certificate.

No similar requirements are included in the Silver Book, where the Contractor accepts total responsibility, and the Employer should not influence the choice of designers or discuss design aspects with the Contractor's designers.

Clause 6 [Staff and Labour] supplements the Contractor's obligations under Sub-Clauses 4.3 regarding the resource management of the Contractor's Personnel, obliging the Contractor to:

- make arrangements for the engagement of all Contractor's Personnel and for their payment, accommodation, feeding, transport, and welfare (Sub-Clause 6.1);
- pay rates of wages, and observe conditions of labour, that comply with all applicable Laws and are not lower than those established for the trade or industry where the work is carried out (Sub-Clause 6.2);
- comply with all the relevant labour Laws applicable to the Contractor's Personnel, including Laws relating to their employment (including wages and working hours), health, safety, welfare, immigration and emigration, and allow them all their legal rights (Sub-Clause 6.4);
- provide and maintain all necessary accommodation and welfare facilities for the Contractor's Personnel (Sub-Clause 6.6);
- allocate to the project Contractor's Personnel who are appropriately qualified, skilled, experienced, and competent in their respective trades or occupations (Sub-Clause 6.9);
- take all necessary precautions to maintain the health and safety of the Contractor's Personnel, ensuring, in collaboration with local health authorities, availability of medical services at all times at the Site and at any accommodation for Contractor's and Employer's Personnel (Sub-Clause 6.7);
- make suitable arrangements for all necessary welfare and hygiene requirements and for the prevention of epidemics (Sub-Clause 6.7);

- appoint a qualified, experienced, and competent health and safety officer at the Site, responsible for maintaining health, safety, and protection against accidents (Sub-Clause 6.7);
- take all necessary precautions to prevent any unlawful, riotous, or disorderly conduct by or among the Contractor's Personnel, and to preserve peace and protection of persons and property on and near the Site (Sub-Clause 6.11);
- base all Key Personnel at the Site (or at the location of the Works off the Site) for the whole time that the Works are being executed and in case of absence, temporarily appoint a suitable replacement, subject to the CA's prior consent (Sub-Clause 6.12).

Sub-Clause 6.3 prohibits each Party from recruiting staff and labour from among the other Party's personnel.

11.3.4 Subcontractors

Pursuant to Sub-Clause 5.1 (Red)/4.4 (Yellow/Emerald/Silver Books), the Contractor is responsible for the management and coordination of all Subcontractors. The Contractor's Key Personnel interact with the site managers of the Subcontractors to plan all activities. Complying with Sub-Clause 6.8, the Contractor shall provide superintendence to monitor, direct, and verify the execution of the Works carried out by the Subcontractors.

Sub-Clause 5.1 (Red)/4.4 (Yellow/Emerald Books) obliges the Contractor to obtain the Engineer's consent to all proposed Subcontractors not named in the Contract, except for the suppliers of Materials. The Engineer has 14 days to respond. The Contractor shall then give a Notice to the Engineer at least 28 days before the commencement of each Subcontractor's work, and of the commencement of such work on the Site.

Under the Silver Book, the Employer has limited control, and Sub-Clause 4.4 does not require prior consent; it requires Contractor's Notices to the Employer **only when specified in the Contract Data** at least 28 days before the appointment of a Subcontractor and then before commencement of its work.

Sub-Clause 5.2 (Red)/4.5 (Yellow/Emerald/Silver Books) deals with the nominated Subcontractors, named in the Contract, or instructed under Sub-Clause 13.4, including procedures for objection to the nomination and payment to nominated Subcontractors.

The FIDIC Subcontract forms[11] transfer the Contractor's responsibilities related to the resource management back-to-back to the Subcontractors. For example, in the Conditions of Subcontract for Plant and Design-Build:

- similarly to the 'Goods' in the Main Contract, the 'Subcontract Goods' under Sub-Clause 1.1.22 include the Subcontractor's Equipment, Subcontract Plant, the materials for the Subcontract Works, and the Subcontractor's Temporary Works;

11 FIDIC, *Conditions of Subcontract for Construction* (FIDIC 2011) and *Conditions of Subcontract for Plant and Design-Build* (FIDIC 2019)

- Sub-Clause 6.2 stipulates that the Contractor shall appoint a Contractor's Subcontract Representative authorised to act on the Contractor's behalf under the Subcontract;
- Sub-Clause 6.3 stipulates that the Subcontractor shall appoint a Subcontractor's Representative authorised to act on the Subcontractor's behalf under the Subcontract;
- under Sub-Clause 7.5, the Subcontractor takes responsibility for all Subcontractor's Equipment and cannot move major items from the Site without the Contractor's consent;
- the Subcontract Plant is included in the definition of Plant under the Main Contract.

11.3.5 Management of physical resources (Goods)

The resource management planning considers:

- the project size and type, e.g. infrastructure projects require extensive investment in equipment;[12]
- the project location: availability of resources, access roads, etc.;
- requirements of the financing institution;[13]
- connections to utilities.

The Construction Extension outlines the following important points for the resource management of the Goods:

- stocking quantities management;
- expiration dates (important for materials like cement and paint);
- special conditions for handling and storage of materials, including hazardous materials;
- waste management;
- planned machinery maintenance;
- permanent equipment preservation.[14]

Besides the Contractor's general obligations under Sub-Clause 4.1 to provide the Goods and consumables required to fulfil its obligations under the Contract, the other relevant provisions include:

- **Sub-Clause 4.15 [Access Route]** allocating to the Contractor the responsibility for the suitability/availability of the access routes to the Site, their repair and maintenance, obtaining permits from the authorities, and providing signage (to the extent that non-suitability or non-availability arises as a result of changes

12 FIDIC & ITA, *Emerald Book* (2019) Notes

13 FIDIC, *Yellow Book* (FIDIC 2022) GPPC 34: Example Sub-Clause for Goods from Eligible Source Countries

14 PMI Inc., *Construction Extension to the PMBOK® Guide* (PMI 2016) 84

to that access route by the Employer or a third party after the Base Date, the Contractor may be entitled subject to Sub-Clause 20.2 to EOT and/or payment of Cost incurred);

- **Sub-Clause 4.16 [Transport of Goods]** obliging the Contractor to give 21 days' Notice to the CA for the delivery date of any Plant, or a major item of other Goods, to the Site and to be responsible for packing, loading, transporting, receiving, unloading, storing, and protecting all Goods, customs clearance, permits, fees and charges related to the import, transport and handling of all Goods, including all obligations necessary for their delivery to the Site; and to indemnify the Employer in respect of third-party claims arising from their import, transport, and handling.
- **Sub-Clause 4.17 [Contractor's Equipment]** stipulating that Contractor's Equipment brought on to the Site shall be deemed to be exclusively intended for the Works, and the Contractor shall not remove from the Site any major items without the CA's consent; in addition to any Notice given under Sub-Clause 4.16, the Contractor shall give a Notice to the CA within 7 days of the delivery date of any major item, identifying whether it is owned, rented, or leased;
- **Sub-Clause 4.19 [Temporary Utilities]** allocating to the Contractor the responsibility for providing all temporary utilities, including electricity, gas, telecommunications, water, and any other services the Contractor may require for the execution of the Works;
- **Sub-Clause 4.22 [Contractor's Operations on Site]** obliging the Contractor to:
  - store properly or remove from the Site any Contractor's Equipment (subject to Sub-Clause 4.17) and/or surplus materials;
  - promptly clear away and remove from the Site the Temporary Works that are no longer required; and
  - leave the Site and the Works after the issue of a TOC in a clean and safe condition, retaining during the DNP only Goods necessary to fulfil obligations under the Contract;
- **Sub-Clause 7.1 [Manner of Execution]** obliging the Contractor to carry out the manufacture, supply, installation, testing, and commissioning and/or repair of Plant, and the production, manufacture, supply, and testing of Materials:
  - in the manner specified in the Contract;
  - in a proper workmanlike and careful manner, in accordance with recognised good practice; and
  - with properly equipped facilities and non-hazardous Materials, except as otherwise specified in the Contract;
- **Sub-Clause 7.2 [Samples]** requiring the Contractor to submit to the CA for consent specified or standard samples of Materials, and relevant information, prior to using the Materials in or for the Works;
- **Sub-Clauses 7.3 [Inspections], 7.4 [Tests], 7.5 [Defects and Rejection], 7.6 [Remedial Works]** giving the QC procedures;
- **Sub-Clause 7.7 [Ownership of Plant and Materials]** stipulating that, to the extent consistent with the mandatory requirements of the Laws of the Country, the Employer becomes the owner of each and every item of Plant and Materials

when delivered to the Site, or if earlier, when the Employer pays for that item under Sub-Clause 8.11 or Sub-Clause 14.5;

- **Sub-Clauses 7.8 [Royalties]** setting out the Contractor's obligation to pay all royalties, rents, and other payments for natural Materials obtained from outside the Site, and for the disposal of material from demolitions and excavations and of other surplus material;
- **Sub-Clauses 17.1 [Responsibility for Care of the Works]** allocating the responsibility for the Goods to the Contractor from the Commencement Date until the issue of the TOC, when responsibility for the care of the Works shall pass to the Employer;
- **Sub-Clause 17.3 [Intellectual and Industrial Property Rights]** dealing with indemnifications of the Employer in case of third-party claims for infringement of any patent, trademark, etc., which may arise in relation to the use of Contractor's Equipment.

Sub-Clause 2.6 [Employer-Supplied Materials and Employer's Equipment] enables the Employer to make available materials and/or equipment to the Contractor in accordance with the details, times, arrangements, rates, and prices stated in the Specification/ Employer's Requirements. The Contractor shall be responsible for each item of Employer's Equipment while any of the Contractor's Personnel is operating it, driving it, directing it, using it, or in control of it.

Sub-Clause 4.19 [Temporary Utilities] includes an option for the Employer to provide utilities for the Contractor's use. The Contractor shall be entitled to use, for the purposes of the Works, the utilities on the Site for which details and prices are given in the Specification/Employer's Requirements. The Contractor shall, at the Contractor's risk and cost, provide any apparatus necessary for the Contractor's use of these services and for measuring the quantities consumed and reimbursement of the Employer.

11.3.6 Innovative resource management

AI-powered resource monitoring solutions use the IoT and video analytics to facilitate resource management. AI and IoT:

- save time for monitoring and enable improvements in gathering information and reporting;
- save time for finding tools and materials on the Site;
- automate activities, thus reducing the need for human resources;
- reduce waste, transferring exact updated information from BIM to the Contractor's Equipment;
- improve quality with fewer resources spent;
- enable data-driven decision making;
- facilitate selection of sustainable Materials and Plant;
- improve collaboration and transparency.

Recent research by Kusimo identifies that most of the problems in UK resource management are due to poor data management processes and the practice of having data in silos.

Overcoming the challenge requires the adoption of Big Data approaches to allow the integration of large and different forms of data.[15]

Resource and sequence optimisation tools leveraging advanced analytics in generative scheduling enable the optimal construction sequence to be found by comparing millions of options. Re-sequencing the programme and reallocating the teams may reduce required labour, eliminate structural overtime, and accelerate construction.[16]

Off-site manufacturing reduces the need for labour, eliminates waste, and secures a controlled environment with proper health and safety conditions.

Proof of source of Plant and Materials can be achieved through digital technology with the potential to provide better record keeping across the entire supply chain for a project through a traceable, immutable digital capture of actions and transactions. Tracking of Goods and services throughout the supply chain, offering near real-time data about components in the construction to the BIM model, provides updates on the as-built state that can be achieved by a complete digital ledger solution.[17]

11.4 Recent challenges in resource management

Persistent scarcity of skilled labour and changing logistics operations are major issues in construction. Requirements for sustainability and safety at work are increasing. New, lighter-weight materials enable improved logistics and modern architectural solutions.[18] Recent extraordinary events such as COVID-19 and the war in Ukraine caused disruption in resource management, leading to schedule delays and cost overruns.

11.4.1 COVID-19

COVID-19 was first reported in China at the end of 2019. In view of its exponential growth, it was declared a world pandemic by the **World Health Organization (WHO)** on 11 March 2020. As China is considered the world's factory, the pandemic's disruptions to supply chains around the world started there before spreading elsewhere with significant negative effects on the efficiency of operations and the supply chain.[19] Emergency legislation enacted in different countries during the COVID-19 waves imposed measures impacting the movement of personnel and goods. Lockdowns, border crossing restrictions, and

---

15 Habeeb Kusimo and others, Optimisation of Resource Management in Construction Projects: a Big Data Approach (*World Journal of Science, Technology and Sustainable Development* 2019)

16 McKinsey & Company, 'Generative Scheduling: Saving Time and Money in Capital Projects' (2022) <https://www.mckinsey.com/capabilities/operations/our-insights/generative-scheduling-saving-time-and-money-in-capital-projects> accessed 9 October 2022

17 Alan McNamara, 'Automating the Chaos: Intelligent Construction Contracts' Chapter in *Smart Cities and Construction Technologies* (IntechOpen 2020) 11

18 McKinsey & Company, 'The Next Normal in Construction' (2020) <https://www.mckinsey.com/~/media/McKinsey/Industries/Capital%20Projects%20and%20Infrastructure/Our%20Insights/The%20next%20normal%20in%20construction/executive-summary_the-next-normal-in-construction.pdf> accessed 8 October 2022

19 Maciel Queiroz, Dmitri Ivanov, and others, 'Impacts of Epidemic Outbreaks on Supply Chains: Mapping a Research Agenda amid the COVID-19 Pandemic through a Structured Literature Review' (Springer 2020) <https://link.springer.com/article/10.1007/s10479-020-03685-7> accessed 18 October 2020

quarantine of the personnel led to suspension of the works on many FIDIC projects. The sites that remained open faced disrupted supply chains and operational restrictions.[20]

Ninety per cent of the world's goods are transported across the seas. Due to the disruption, many containers were stuck in the wrong places, and demands for shipping spiralled. Staff shortage additionally caused shipping delays. Shipping prices escalated, leading to an increase in the price of timber and steel. Construction materials and machines were in short supply, as the supply chains also encountered inadequate staff numbers and materials due to the lockdown policies. The blocked transportation in towns and villages made the delivery of materials more difficult. The lockdowns in China in 2022 caused further disruption in the supply chains.

Both migrant and domestic construction workers were unable to reach jobsites and needed to adhere to new on-site protocols reducing productivity.[21] For situations when workers were unable to return due to strict lockdown policies, new workers had to be hired and trained, or available staff were required to undertake multiple jobs with suitable measures to meet the urgent demand for human resources. PM teams had to coordinate with suppliers and transportation departments to seek an early recovery of material supply. They also had to optimise resource allocation plans and mitigate the impact of material shortage.[22]

Figure 11.2 links the impact of the pandemic on projects using FIDIC Works Contracts with the GCs providing relief to the Contractor. COVID-19 was a unique, broad-scale, long-term, and extraordinary event creating long-term challenges for parties around the world, which caused significant disruption to global construction. Following the Claims procedure, the Contractor may be entitled to EOT under Sub-Clause 18.4 to the extent that the Contractor is prevented from performing its obligation due to the COVID impact. Since the WHO declared that the pandemic originated in bats, it falls in the ambit of Sub-Clause 18.1 (f) as a natural event, entitling the Contractor under Sub-Clause 18.4 only to EOT. If WHO had discovered after the investigation that it had been spread by a laboratory, the entitlement under Sub-Clause 18.4 might have included, besides EOT, payment of Cost.

Emergency acts and decrees imposed rules for quarantine, travel restrictions, and lockdowns. Sub-Clause 13.6 may entitle the Contractor to EOT and Costs, but only as a result of the changes in the Laws of the Country in which the Site is located, while on many projects, contractors encountered difficulties due to the emergency legislation in the country of the major suppliers.

Under Sub-Clause 8.5 (d) of the Red/Yellow/Emerald Books, the Contractor shall be entitled, subject to Sub-Clause 20.2, to an EOT if and to the extent that completion for the purposes of Sub-Clause 10.1 is or will be delayed by Unforeseeable shortages in the

---

20 The impact of COVID-19 on the execution of construction projects and the operation of construction contracts for 51 jurisdictions around the world is outlined in the books edited by Donald Charrett and published by Informa Law from Routledge: *FIDIC Contracts in Asia Pacific* (2022); *FIDIC Contracts in Europe* (2023); *FIDIC Contracts in the Americas* (2023); *FIDIC Contracts in Africa and the Middle East* (2024).

21 McKinsey & Company, 'How Companies and Leaders Can Reset for Growth beyond Coronavirus' (2020) <https://www.mckinsey.com/capabilities/operations/our-insights/how-construction-can-emerge-stronger-after-coronavirus> accessed 18 October 2022

22 Zhimin Wang and others, Risk Identification and Responses of Tunnel Construction Management during the COVID-19 Pandemic (*Hindawi Advances in Civil Engineering* 2020) <https://www.hindawi.com/journals/ace/2020/6620539/> accessed 18 October 2022

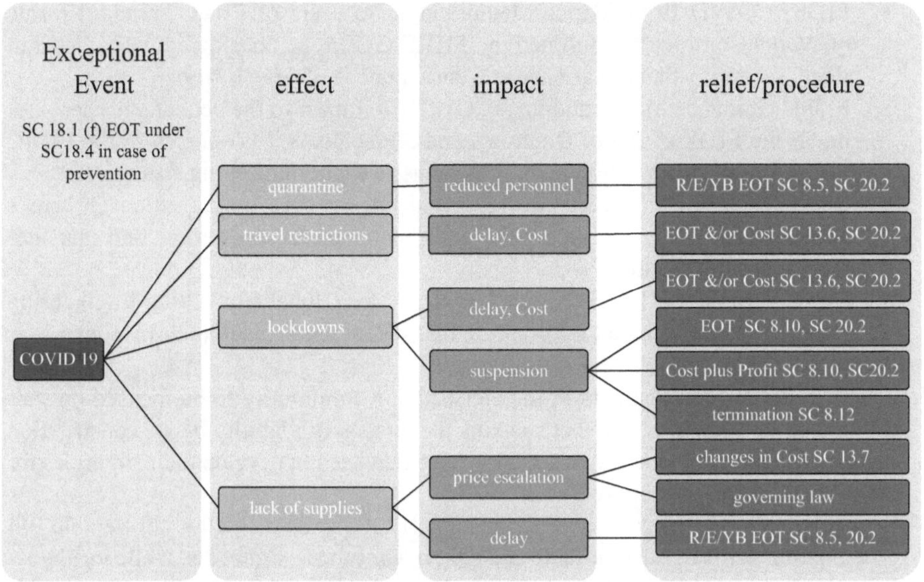

Figure 11.2 COVID-19 impact and FIDIC provisions providing relief to Contractors

availability of personnel or Goods (or Employer-Supplied Materials, if any) caused by epidemic or governmental actions. This risk is allocated to the Contractor under the Silver Book. However, in the case of quarantined key personnel of the Employer, Sub-Clause 8.5 (e) may entitle the Contractor to an EOT under all Books if the Contractor is prevented from continuing the Works due to delayed inspections and tests.

A lockdown may lead to suspension under Sub-Clause 8.9. Sub-Clause 8.10 entitles the Contractor to EOT and/or payment of Cost plus Profit if the Contractor suffers delay and/or incurs Cost complying with the instruction for suspension/resuming the work, subject to Sub-Clause 20.2. Under Sub-Clause 8.11, the Contractor may be entitled also to payment for Plant and Materials not delivered to the Site during the suspension. If the suspension continues for more than 84 days, it may entitle the Contractor to termination, following the procedure under Sub-Clause 8.12.

The price escalation may result in significant adjustment of the Contract Price under Sub-Clause 13.7 where a Schedule of cost indexation is included in the Contract. The governing law may provide additional relief.

FIDIC responded quickly to the new challenges, creating a COVID page[23] publishing guidance to the industry on the pandemic's effect on contracts and projects. Alongside this, FIDIC developed a series of webinar programmes providing an opportunity for industry leaders and stakeholders to discuss how they were addressing the COVID-19 challenge. Besides the newsletters and bulletins identifying the best practices for the response to the pandemic, FIDIC published on its dedicated webpage the following freely downloadable guidance:

23 <https://fidic.org/COVID-19> accessed 22 October 2022

- FIDIC COVID-19 Guidance Memorandum to users of FIDIC standard forms of Works Contract:[24] it outlined the FIDIC GCs to be considered with regard to likely scenarios that are arising as a consequence of COVID-19;
- FIDIC Guidance Memorandum – COVID-19 impact to the consultant's services under the FIDIC Suite of Contracts and Agreements:[25] it referred to the FIDIC White Book and JV Agreement provisions relevant in different scenarios.
- FIDIC COVID-19: On-site working and project team organisation Guidance Memorandum:[26] it identified risks and recommended preventive and management measures at site.
- Coronavirus (COVID-19): FIDIC Guidance for Global Consulting Engineering Businesses:[27] it included information that will assist firms in their response.

FIDIC recommended all members of the construction community to be focused on successful project delivery in a way that sustains the long-term viability of the construction community. All players in the construction value chain need to develop their strategies for dealing with or leading disruption.

The special circumstances caused by the COVID-19 pandemic called for flexible PM and coordination skills to raise suitable and effective response strategies, while local governments made substantial contributions to solving the difficulties. Dealing with inadequate resources and tight schedule pressure, optimal resource allocation and collaboration are vital in catching up with the construction progress.

One of the lessons learned from the pandemic is that contractors must have reliable suppliers and enough storage facilities, because the lean approach for 'just in time' delivery is risky. The trend is to select agile and resilient suppliers.

Big Data analytics plays a vital role in overcoming supply-chain disruptions and enhancing the sustainability of supply chains.[28] Digital transformation is a key to smarter risk management. An intelligent supply-chain planning system provides the advantage of being able to more accurately predict, sense, and react to the opportunities hidden in with the risks. When disruptions occur, it enables companies to make rapid, data-driven decisions.[29]

The COVID-19 crisis accelerated change that had already started to occur at scale related to digitalisation and off-site manufacturing:

---

24 <https://fidic.org/sites/default/files/COVID%2019%20Guidance%20Memorandum%20-%20PDF.pdf> accessed 22 October 2022

25 <https://fidic.org/sites/default/files/FIDIC%20COVID%2019%20Guidance%20Memorandum%20for%20Consultant_Services_270821_2.pdf> accessed 5 May 2023

26 <https://fidic.org/sites/default/files/FIDIC%20Guidance%20Memorandum_RLQC.pdf> accessed 5 May 2023

27 <https://mcusercontent.com/b2d4de85982c2d1149046b241/files/cf491425-3f2d-4d7a-b2f1-edcd5df25941/FIDIC_Sector_Guidance_Final.pdf?fbclid=IwAR24Kg_x-W8m4EO9DIgvZ4UabrZf4uqMk8jtH93_dR57eJtVR05uPB9bkgY> accessed 5 May 2023

28 Tariq Al-Shbail and others, Big Data Analytics for Supply Chain Sustainability: Amid the Outbreak of the COVID-19 Pandemic (*World Customs Journal* 15 2020) 15 <https://worldcustomsjournal.org/Archives/Volume%2015%2C%20Number%202%20(Sep%202021)/1922%2001%20WCJ%20v15n2%20Shbail%20et%20al.pdf> accessed 22 October 2022

29 GEP, '3 Definitive Characteristics of an Agile and Resilient Supply Chain' (2021 *Supply Chain Software Blogs*) <https://www.gep.com/blog/technology/3-definitive-characteristics-of-an-agile-and-resilient-supply-chain> accessed 5 October 2022

- designers and administrators are shifting to remote ways of working;
- the adoption of BIM is accelerating worldwide;
- contractors are using 4D and 5D simulation to replan projects and reoptimise schedules;
- integrated digital-twin solutions are being developed to be used end to end, from project concept to commissioning;
- contractors are looking to online channels for monitoring their employees' well-being through apps, for ordering construction materials, for managing scarce resources more accurately, and for maintaining cash flow with minimal manning at site offices;
- e-commerce models minimise physical interactions with sales teams working remotely using digital tools;
- more players are investing in automation of on-site and back-office processes.

Recent research presents data that the response to the disrupted supply chains leads to the consolidation of the contractors and suppliers to establish economies of scale and support investment in IT, talent, research, and technology. Industry players are already starting to vertically integrate to increase efficiency and as a route to standardisation and control of design and execution.[30]

11.4.2 The war in Ukraine

Another Exceptional Event, the war in Ukraine, is disrupting some projects where contractors had relied on Goods from Ukraine/Russia or Ukrainian personnel. The conflict has contributed to global energy inflation, pushing up prices of the three types of fossil fuels, which together accounted for 83.1 per cent of the world's primary energy in 2020.[31] Price escalation of materials and transportation, and the embargo restrictions, have created logistic and financial problems threatening the completion and viability of projects. The impossibility of acquiring the resources as planned due to the war may entitle contractors under Sub-Clause 18.4 to EOT and compensation of the incurred losses subject to Sub-Clause 20.2. The embargo preventing import from Russia may be treated as changes in the Laws of the Country and entitle the Contractor to similar relief under Sub-Clause 13.6.

These overlapping Exceptional Events led to escalation of the prices of timber, steel, oil, and derivatives, which may impact the Costs in construction much more than the agreed adjustments for changes in Cost under Sub-Clause 13.7. The governing law may provide additional relief in some civil law countries. The Parties need to collaborate to find acceptable solutions for the benefit of the project, following the advice of the FIDIC

30 McKinsey & Company, 'How Construction Can Emerge Stronger after Coronavirus' (2020) https://www.mckinsey.com/capabilities/operations/our-insights/how-construction-can-emerge-stronger-after-coronavirus> accessed 22 October 2022

31 Børge Brende and Bob Sternfels, 'Resilience for Sustainable, Inclusive Growth' (McKinsey & Company 2022) <https://www.mckinsey.com/capabilities/risk-and-resilience/our-insights/resilience-for-sustainable-inclusive-growth?stcr=525844586EF14145AA121F3F7901DA2E&cid=other-eml-alt-mip-mck&hlkid=e9f1023b56e04dc5ad5b68ba865380a0&hctky=13345602&hdpid=307d9623-80a7-44f4-8602-2386fc7d887b> accessed 24 October 2022

Memorandums.'[32] If the projects are no longer financially viable, or the Employer cannot find additional financing to cover the Contract Price adjustment due, termination under Sub-Clause 15.5 [Termination for Employer's Convenience] or Sub-Clause 18.5 [Optional Termination] may follow. Alternatively, the Parties may be discharged from further performance due to the impossibility of fulfilling their contractual obligations under Sub-Clause 18.6 [Release from Performance under the Law].

11.5 Example Notice

Appendix 1 includes Contractor's Notice of delivery of Contractor's Equipment under Sub-Clause 4.17 of the Yellow Book.

32 *FIDIC Contracts Guidance on the Effects of Inflation, Unavailability of Goods and Labour Following the Global COVID-19 Pandemic and the War in Ukraine* (FIDIC 2023) <https://www.fidic.org/sites/default/files/Guidance%20Memo%20-%20War%20memorandum_170323_final.pdf> accessed 26 March 2023

CHAPTER 12

Project Communications Management with FIDIC Contracts

12.1 Project Communications Management processes

According to the PMBOK® Guide, Project Communications Management includes the planning, managing, and monitoring processes necessary to ensure that the information needs of the project and its stakeholders are met through the development of artefacts and implementation of activities designed to achieve effective information exchange.[1]

Communications describe the possible means by which the information can be sent or received, either through communication activities, such as meetings and presentations, or through artefacts, such as emails, social media, project reports, or project documentation.

Project communications should be managed proactively to ensure that the information being communicated is appropriately generated, received, and understood. Communications can have a serious effect on the cost and schedule of a project and lead to Claims. Project communications are critical to minimise the impact of risks and prevent disputes.

Major complex projects need an agreed centralised management of project communications between the main project stakeholders to ensure effective and efficient access to and distribution of project information.[2] The leading organisation should centralise document storage, control, and distribution, establishing proper access and authorisation rules.

Project communications may be:

- internal between the project participants or external;
- formal (notices, reports, records, etc.) or informal (emails, chats, messages, *ad hoc* informal discussions, etc.);
- written or oral.

1 PMI Inc., *PMBOK® Guide* (PMI sixth edition 2017) 359
2 PMI Inc., *Construction Extension to the PMBOK® Guide* (PMI 2016) 95

DOI: 10.4324/9781003272892-12

12.2 Improving communications in construction

Communication has been identified as the most important requisite of success, and lack of it is perhaps the most significant factor in human failure.[3]

The message itself and how it is transmitted are influenced by the sender's current emotional state, knowledge, background, personality, culture, and biases.[4] Diversity in construction multiplies the influences that may distort the interpretation of the message by the receiver. People from different nationalities, religions, and cultures with different experience and expertise (including language skills) gather for a project in a new environment. Smooth and transparent communications are crucial for building the team and efficient PM. The famous instruction of Elon Musk to his employees targets efficient PM: 'Excessive use of made up acronyms is a significant impediment to communication. No one can actually remember all these acronyms and people don't want to seem dumb in a meeting, so they just sit there in ignorance. This is particularly tough on new employees.'[5] His statement is fully applicable to the Site-specific jargon in major international projects causing misunderstanding and alienation of new team members. This author has endeavoured to ensure that all acronyms and defined terms are defined where first used, and included in the Glossary.

Improved communications would lead to less delay and lower expense. All stakeholders' satisfaction about both the process and the quality of the product could improve if they communicated in a better way.[6]

The PMBOK recommends using the 5Cs of written communication:

- correct grammar and spelling;
- concise expression and elimination of excess words;
- clear purpose and expression directed to the needs of the reader;
- coherent logical flow of ideas;
- controlling flow of words and ideas.[7]

12.3 Project Communications Management using the FIDIC Suite for Services

Sub-Clause 1.1.19 of the White Book defines 'Notice' as a written communication identified as a Notice and issued in accordance with the provisions of Sub-Clause 1.3. Sub-Clause 1.1.26 defines 'Variation Notice' similarly. Pursuant to Sub-Clause 1.3, other formal communications may include, without limitation, approvals, consents, instructions, and decisions. Sub-Clause 1.3 [Notices and other Communications] gives the rules for the formal communication:

3 Nael and Lydia Bunni, *Risk and Insurance in Construction* (Routledge 2022) #3.3.32/3.6.27 in *Sutcliffe v Chippendale and Edmonson (1971), 18 BLR, 149*, an architect was held liable for certifying payment of defective work, caused by a failure of communication between the architect and the quantity surveyor

4 PMI Inc., *PMBOK® Guide* (PMI 2017) 373

5 Ashlee Vance, *Elon Musk. Tesla, SpaceX, and the Quest for a Fantastic Future* (Ecco 2015) 239

6 Mieke Hoezen and others, 'The Problem of Communication in Construction' (2006) <https://www.researchgate.net/publication/254858936_The_problem_of_communication_in_construction> accessed 25 October 2022

7 PMI Inc., *PMBOK® Guide* (PMI sixth edition 2017) 363

(i) to be identified as such with reference to the Clause under which it is issued;
(ii) to be in writing (meaning 'handwritten, type-written, printed or electronically made and resulting in a permanent uneditable record' under Sub-Clause 1.2);
(iii) to be delivered by hand against receipt, sent by mail or courier, or transmitted by any form of agreed system of electronic transmission stated in the Particular Conditions;
(iv) to be delivered, sent, or transmitted to the address for the recipient's communications as stated in the Particular Conditions/notified by the recipient or if the recipient has not stated otherwise sent to the address from which the request was issued;
(v) not to be unreasonably withheld or delayed.

Sub-Clause 1.4 adds one more rule: the language for all communications shall be the ruling language stated in the Particular Conditions, or where no language is stated, then all communications shall be in the language in which the Agreement (or most of it) is written.

Clause 1 sets out additional rules in relation to external communications:

- Sub-Clause 1.1.6 defines 'Confidential Information' as 'all information specifically identified by the disclosing Party as confidential at the time of disclosure, or information that a reasonable person would consider from the nature of the said information and circumstances to be confidential';
- Sub-Clause 1.8 stipulates that from the signing of the Agreement until 2 years after the completion of the Services (or the termination of the Agreement), neither Party shall disclose or cause or permit their employees, professional advisers, agents, or sub-consultants to disclose to third parties any Confidential Information without the prior written consent of the other Party (except where the information is public knowledge/ not acquired from the disclosing party/ independently developed/ required to be disclosed by the authorities);
- Sub-Clause 1.9 stipulates that during the period with confidentiality restrictions under Sub-Clause 1.8, publication shall be subject to approval of the Client, and the Consultant may use material and information relating to the Services and the Project for commercial tendering purposes.

12.4 Project Communications Management using FIDIC Works Contracts

12.4.1 Rules for formal communications

The primary purpose of Sub-Clause 1.3 is to provide an arrangement for formal communications under the Contract to be effectively dispatched and received as agreed under the Contract. Sub-Clause 1.3 differentiates between giving of a Notice (including a Notice of Dissatisfaction [NOD]/Notice of Claim) and the issuing, providing, sending, submitting, or transmitting of another type of communication. Some of the mandatory rules are similar to those under the White Book, but there are differences and additional requirements for the formal communications:

(i) to be in writing (the meaning is the same as under the White Book);
(ii) a Notice to be identified as a Notice, another form of communication to be identified as such;

(iii) unlike the rule of the White Book, due to the serious effect of the validity of a Notice, the requirement to include reference to the Clause under which the communication is issued is applicable for other communications where appropriate, but not for Notices;

(iv) to be a paper-original signed by the authorised representative or an electronic original generated from any of the systems of electronic transmission stated in the Contract Data (if not stated, acceptable to the CA), transmitted by the electronic address uniquely assigned to each of such authorised representatives, or both, as stated;

(v) to be delivered by hand/ sent by mail or courier against receipt, or transmitted by any form of agreed system of electronic transmission;

(vi) to be delivered, sent, or transmitted to the address for the recipient's communications as stated in the Contract Data or later notified by the recipient;

(vii) to have effect when (deemed to have been) received at the recipient's current address; an electronically transmitted communication is deemed to have been received on the day after transmission, provided no non-delivery notification was received by the sender;

(viii) not to be unreasonably withheld or delayed;

(ix) when a Notice or NOD or certificate is issued by a Party or the CA, a copy to be sent to the CA or the other Party, as the case may be; all other communications to be copied to the Parties and/or the CA as stated in the Contract.

The FIDIC 2017 Contracts Guide points out that the

> effectiveness (or otherwise) of a formal communication, which is sent to an address or by a method of transmission other than as required under this Sub-Clause, may depend on such matters as the recipient's subsequent actions, the consequences of the communication and the law governing the Contract.[8]

Sub-Clause 1.4 adds one more rule: the language for communications shall be stated in the Contract Data. If the language for communications is not stated in the Contract Data, it will be the ruling language (as under the White Book), which may create difficulties for the numerous key personnel of the Parties and for the permitting, and lead to disagreements and/or Disputes.

Sub-Clause 1.12 sets out the confidentiality rules for both Parties.

12.4.2 Requirements for Notices

'Notice' has the same definition as under the White Book, but no requirement for reference to the provisions of the Contract, which may apply to the other communications. There is no specific mandatory form. This facilitates notification and risk mitigation, enabling early recording and monitoring of the effect of the events/circumstances.

Other chapters of this book list Notices in relation to PM procedures. Appendix 1 also includes Notices to demonstrate how the CA and the Parties can apply the FIDIC GCs to the facts.

8 FIDIC, *The FIDIC 2017Contracts Guide* (FIDIC 2022) 84

The Notice provisions concern the communication of events/issues/circumstances to the project Parties and the CA, the knowledge of which is necessary for effective PM. Compliance with the Notice provisions:

- is a matter of good contract administration practice;
- is an early warning enabling risk mitigation and scrutiny of records;
- may trigger or give rise to additional entitlements.

Since many contractual entitlements are conditional upon a Notice being given, the relevant provisions shall be strictly complied with to avoid uncertainty. Failing to comply with the contractual Notice provisions may result in the loss of entitlements, e.g. under Sub-Clause 20.2.1.

Besides the mandatory requirements under Sub-Clauses 1.3 and 1.4, the following rules are applicable for the drafting of Notices:

(a) the Notice shall be given within the stipulated timeline, e.g.:
- Notice of an Exceptional Event under Sub-Clause 18.2 (see Figure 12.1): within 14 days after the claiming Party became aware, or should have become aware, of the Exceptional Event or circumstance; in the case of a continuing event like COVID-19 or war, under Sub-Clause 18.3, the affected Party shall give further Notices describing the effect every 28 days after the first Notice and immediately give a Notice when it ceases to be affected;
- Notice under Sub-Clause 4.23: as soon as practicable after discovery of archaeological/geological findings to enable prompt investigation;
- Notice of Claim under SC 20.2.1: as soon as practicable, but no later than 28 days after the claiming Party became aware, or should have become aware, of the event or circumstance (in case of failure, the entitlement may be lost);

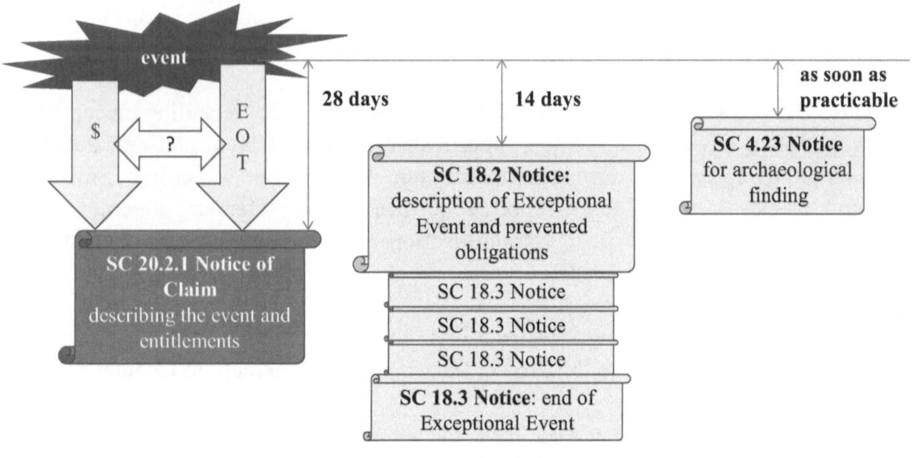

Figure 12.1 Examples of Notices under different provisions

- Notice stating the Commencement Date under Sub-Clause 8.1: not less than 14 days before the Commencement Date, to give enough time to the Contractor for mobilisation;
- Notice for a TOC under Sub-Clause 10.1: not earlier than 14 days before estimated readiness (the Contractor is under pressure, trying to escape the Delay Damages, and its opinion is usually not realistic but rather, optimistic);
- Notice under Sub-Clause 16.1: not less than 21 days before suspension by the Contractor, to give enough time to the Employer to remedy the default;

(b) the specified time for compliance with the Notice must be reasonable in all circumstances prevailing at the time when it is given;[9]

(c) the contents of the Notice shall comply with the relevant provision of the GCs as amended by the Particular Conditions; using the language of the Contract is highly recommended.

Figure 12.1 presents examples for the timeline and contents of different Notices:

- The Sub-Clause 18.2 Notice of prevention caused by an Exceptional Event is to be given within 14 days, whereas the Sub-Clause 20.2.1 Notice of Claim has to be given within 28 days, so arguably, both shall be given separately complying with the rules under Sub-Clause 1.3. Commentators consider that the requirement to give additional notices every 28 days in the case of continuing impact of an Exceptional Event under Sub-Clause 18.3 will often be redundant and may be waived by the Parties. There is no obvious sanction for failing to provide these follow-up Notices apart from any damages that could arise from a general law breach of contract.[10]
- The Contractor shall give the Notice under Sub-Clause 4.23 as soon as practicable after discovery of any archaeological/geological finding of interest to enable prompt investigation. Depending on the CA's instructions for dealing with the finding, the Contractor may give a Notice of Claim within 28 days under Sub-Clause 20.2.1 if it suffers critical delay and/or incurs Cost from complying with the instructions.

12.4.3 Other formal communications

Sub-Clause 1.3 lists other forms of formal communication, including acceptance, acknowledgement, advising, agreement, approval, certificate, Claim, consent, decision, determination, disagreement, discharge, instruction, No-objection, record(s) of meeting, permission, proposal, record, reply, report, request, Review, Statement, statement, submission, or any other similar type of communication. Some of them are commented on at length in other chapters of this book:

- CA's determinations and instructions, reports, records, TOC, Performance Certificate in relation to the Project Integration Management in Chapter 6;

9 *Obrascon (OHL) v Her Majesty's Attorney General for Gibraltar* [2014] EWHC 1028 (TCC) J Akenhead at #318 in relation to the Notice to Correct given under the *Yellow Book* 1999 Sub-Clause 15.1

10 Corbett & Co., *FIDIC 2017: A Practical Legal Guide* (Corbett & Co 2020) 441

- Statements, Payment Certificates, and discharge under Clause 14 in relation to the Project Cost Management in Chapter 9;
- Claims and related Notices shall be discussed in Chapter 16;
- DAAB decisions, and NOD with a DAAB's decision shall be discussed in Chapter 17.

Practical guides include hundreds of draft forms for Notices and other communications, which should be used with care, because they are prepared for a certain edition of a specific FIDIC Book, and the GCs may be amended by the Particular Conditions.[11] The Green Book contains 40 forms for communications.

12.4.4 Meetings

The records of meeting, often called Minutes of Meetings (**MoM**):

- are listed under Sub-Clause 1.3 among the formal contract communications;
- may record intent to deliver formal instructions and Notices, but cannot substitute for them: those communications must be sent separately, complying with the relevant requirements.

Sub-Clause 3.8 (Silver Book 3.6) explicitly requires the CA to keep a record of each management meeting and distribute it to the participants and Employer, identifying the responsibilities in accordance with the Contract for the actions to be taken. The CA is responsible for the meeting management, including preparation of the agenda, invitation of essential participants, dealing with conflicts during the meeting, reaching agreement on timely actions, and recording those. Proper meeting management solves issues effectively and efficiently.

Digital tools can help:

- platforms for remote meetings save time and CO_2, and enable attendance by important participants working in different locations;
- the software for virtual meetings enables easy sharing of documents/seeing issues on the Site, while keeping the discussion focused on the topic;
- in a virtual meeting, the participants cannot interrupt each other, and usually, conflict management is easier;
- the platforms for virtual meetings enable recording, creating a useful record, which may be viewed by invitees who could not attend the meeting;
- visualisation on a big screen of BIM and presentations facilitate the discussion (where face-to-face meetings are preferred to enable people to know each other and be more open in the communication, which leads to brainstorming and finding better solutions);

11 For example, Michael Robinson's books *A Contractor's Guide to the FIDIC Conditions of Contract* (John Wiley & Sons 2011) and *An Employer's and Engineer's Guide to the FIDIC Conditions of Contract* (John Wiley & Sons 2011) give free access to numerous model e-forms for the Red Book 1999, and the *FIDIC 2017: A Practical Legal Guide* (Corbett & Co 2020) includes more than 100 forms for the 2017 Edition of the *Yellow Book*

- for both face-to-face and virtual meetings, writing the MoM during the meeting and sharing on the screen facilitates collaboration; otherwise, when the MoM is distributed after the meeting, managers and lawyers who have not attended the meeting may try to change the record, which creates tension.

The DAAB online meetings and Site visits became the standard during the COVID-19 pandemic, enabling the dispute prevention process to continue despite the travel restrictions. The 2022 Reprint DAAB Rule 3.3 provides for online meetings only in exceptional circumstances, but the DAAB and the Parties may agree on a mixture of face-to-face and online meetings, saving time, costs, and CO_2. For this purpose, they shall include amendments of the procedural rules annexed to the GCs in the DAAB Agreement Clause 2.

12.4.5 PMIS

Project information is managed and distributed using a variety of tools, including electronic PM tools. Factors that influence the technology include whether the team is collocated, the confidentiality of any information that needs to be shared, and resources available to the team members. PMIS can ensure that the project participants can easily retrieve the updated information they need. All documents may be stored on a designated virtual cloud, accessible by all project participants from any point (either from a remote point or from any local network with Internet connection). A system for managing privileged access to the Contract data should be created.

The use of BIM as a single repository of complete project data (generated and coordinated concurrently by all the parties involved in various project phases, from conceptual design to O&M) may provide better understanding of the project and facilitate effective communication, coordination, and collaboration by avoiding errors and omissions, enabling early conflict detection.[12] BIM creates a transparent environment, building trust among the project participants. Transparent communication:

- ensures fast workflow, leading to increased productivity;
- updates the project team, making them feel competent and involved, resulting in higher motivation and commitment;
- prevents misunderstandings and clears up existing ones;
- strengthens relationships and enables better decision making.[13]

12.4.6 Innovative communications management

Besides those used for meetings and PMIS discussed earlier, teams use modern digital tools for informal communication, such as:

- intranets: private networks connected through the web;
- apps like Viber, WhatsApp, etc. for mobile phones, tablets, and computers enabling group chats, video calls, sharing of photos, videos, and files, and creating polls;

12 PMI Inc., *Construction Extension to the PMBOK® Guide* (PMI 2016) 94

13 BlogIn, 'Transparent Team Communication: Why and How to Embrace It' <https://blogin.co/blog/transparent-team-communication-why-and-how-to-embrace-it-69/> accessed 27 October 2022

- digital screens to disseminate information at the Site;
- internal social networks.

MS Teams makes it easy to meet and chat online, discuss in a forum, share files, and work on them simultaneously.

Digital apps using information from drones, video monitoring, and sensors combined with AI may compare the progress on Site with the digital twin, automating reporting and notifications. Generative AI like ChatGPT-4 goes further, providing instant answers to questions and streamlining communication among project participants.

FIDIC Sub-Clause 1.3 enables agreement on systems of electronic transmission. The Particular Conditions may amend the requirement for one paper-original for Contractor's Documents under Sub-Clause 1.8, progress reports under Sub-Clause 4.20, and applications for payment under Clause 14, leaving only the electronic transmission under Sub-Clause 1.3 for all formal communications. Thus, the project may become paperless, contributing to efficient communications with CO_2 savings. Such amendments should consider the availability of digital government in the Country of the Site to ensure compliance with the Laws in relation to approvals, consents, and permits by the authorities. Digital building permitting is already a reality in some municipalities.[14]

The EU Change toolkit for digital building permit (**CHEK**) will support the digitisation of building permit issuing and automated compliance checks.[15] CHEK will do this by providing a kit of both methodological and technological tools to fully digitise building permitting and partially automate compliance checks on building designs. The EC expects an improvement of 60 per cent in efficiency, leading to the uptake of digital permitting by potentially 85 per cent of municipalities in Europe. This requires:

- aligning available digital technologies to the municipal process, enabling new methods and business models;
- development of data Open standards, including BIM, 3D city models, and reciprocal integration (GeoBIM);
- up/re-skilling officers and users;
- improving, aligning, and integrating technology;
- realising and demonstrating scalability.

12.5 Example Notices

Appendix 1 includes example Contractor's Notices of errors in the Employer's Requirements under Sub-Clause 1.9 of the Yellow Book.

14 Since 2013, all building permits in the Republic of North Macedonia have been issued electronically, based on a full electronic procedure: <http://www.bef.uceb.eu/wp-content/uploads/2018/05/06-Macedonian-Chamber.pdf> accessed 28 February 2023

15 <https://chekdbp.eu/> accessed 27 February 2023

CHAPTER 13

Project Risk Management with FIDIC Contracts

13.1 Project Risk Management processes

ISO 31000:2018 defines risk as the **effect of uncertainty on objectives.**[1] The effect is a deviation from the expected. It can be positive, negative, or both, and can address, create, or result in opportunities and threats. Risk is usually expressed in terms of risk sources, potential events, their consequences, and their likelihood. **Likelihood** is defined as the chance of something happening. An event as a risk source can have one or more occurrences, and several causes and **consequences.** An event can also be something that is

1 ISO 31000:2018 *Risk Management – Guidelines*

DOI: 10.4324/9781003272892-13

expected which does not happen, or something that is not expected which does happen. A consequence:

- can be certain or uncertain;
- can have a positive or negative, direct or indirect effect on objectives;
- can be expressed qualitatively or quantitatively;
- can escalate through cascading and cumulative effects.

The COVID-19 waves and their aftermath illustrate this. Chapter 11 discusses their consequences for construction projects using FIDIC contracts.

The PMBOK uses 'probability' instead of 'likelihood' and 'impact' instead of 'consequence'. According to the PMBOK® Guide, the objectives of Project Risk Management are **to increase the probability and/or impact of positive risks and to decrease the probability and/or impact of negative risks, in order to optimise the chances of project success.** The Project Risk Management processes include:[2]

1. Plan Risk Management.
2. Identify Risks.
3. Perform Qualitative Risk Analysis.
4. Perform Quantitative Risk Analysis.
5. Plan Risk Responses.
6. Implement Risk Response.
7. Monitor Risks.

The key concepts defined in the PMBOK for Project Risk Management are applicable to construction projects:

- all projects are risky, since they are unique undertakings with varying degrees of complexity that aim to deliver benefits in a context of constraints and assumptions, while responding to stakeholder expectations that may be conflicting and changing;
- experienced consultants and contractors should choose to take project risk in a controlled and intentional manner in order to create value while balancing risk and reward;
- the effectiveness of Project Risk Management is directly related to project success:
  - unmanaged threats may result in delay, cost overruns, performance shortfall, or loss of reputation;
  - opportunities that are captured can lead to benefits such as reduced time and cost, or improved performance or reputation;
- risk is initially addressed during project planning by shaping the tender strategy;
- risk should also be monitored and managed as the project progresses to ensure that the project stays on track and emergent risks are addressed;
- it is necessary to tailor the Project Risk Management processes considering the project size, complexity, importance, and PM approach (waterfall, agile or hybrid).

Appendix X3 of the Construction Extension lists more than 100 common causes of risks in construction projects.

2 PMI Inc., *PMBOK® Guide* (PMI sixth edition 2017) 395

13.2 Risk identification in construction

Every project faces uncertainty due to different factors, listed at the end of Chapter 2. 'Risk can be managed, minimised, shared, transferred or accepted. It cannot be ignored.'[3]

Identifying risks is an iterative process. For better planning and identification, risks may be grouped under the same risk category or trigger event that might allow several risks to be realised.[4] Risks are often categorised as insurable and non-insurable. Appendix X3 of the Construction Extension presents a generic categorisation by sources of risk, grouping them as technical, construction, external, organisational, PM, and business risks.

Figure 13.1 shows a risk categorisation based upon the negative risk consequences. The first category includes risks that might be insurable, like defective design/materials/workmanship, Acts of God, fire, human error, and failure to take necessary precautions. Those risks may lead to personal injury or material damage. The second category includes risks that are not insurable, like late possession of the Site, Variations, delayed information, and quality defects in construction. The impact of the second category is economic loss, including time loss and financial loss. Contract provisions specify the risk allocation through the rights and obligations of the Parties.[5] In most cases, the contract expressly allocates to the Employer specific risks, leaving the remaining risks to the Contractor. Contractual risk allocation is, however, independent of insurance cover.

13.3 Risk analysis

Risk analysis measures the combined effect of the probability (likelihood) and the impact (consequence) of risk events.

Qualitative Risk Analysis is the process of prioritising individual project risks for further analysis or action by assessing their probability of occurrence and impact, as well as

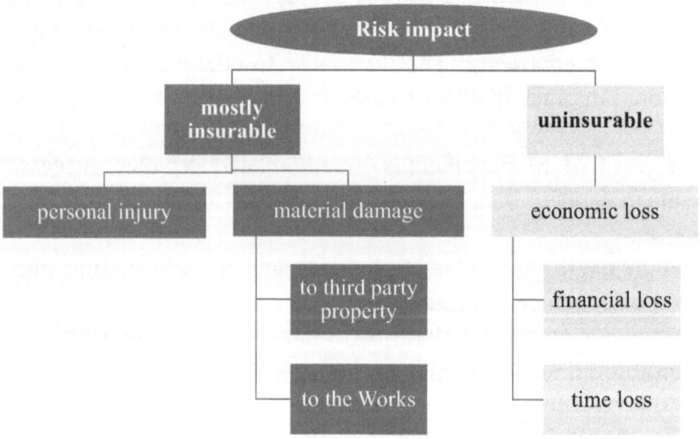

Figure 13.1 Risk categorisation in FIDIC Works Contracts

3 Michael Latham, *Constructing the Team* (HMSO 1994) 3.7
4 *Construction Extension to the PMBOK® Guide* (PMI 2016) 104
5 Nael and Lydia Bunni, *Risk and Insurance in Construction* (Routledge 2022) 4.1.33–4.1.38

other characteristics.[6] Previously identified risks may change, and new risks may appear. The project team may use a rapid brainstorming approach for qualitative risk analysis, based on available information and experience. A matrix similar to the one presented in Figure 13.2 may map the probability of each risk occurrence and its impact on project objectives if that risk occurs.

Figure 13.3 illustrates how risks can be prioritised for further analysis and planning of risk responses using the matrix presented in Figure 13.2. High-priority threats or opportunities may require priority actions and proactive response strategies. Low-priority risks may be placed in the Risk Register as part of the watch list, and a contingency reserve may be added (e.g. Sub-Clause 13.4 Provisional Sums).

The Risk Register should be updated with emerging risks. It may present the priority level or risk score, the nominated risk owner, and the list of potential risk responses. Additional columns may include causes, effects on the objectives, risk triggers, WBS reference or affected activities, current risk status, etc.

The Emerald Book introduces Sub-Clause 1.16 [Contract Risk Register and Contract Risk Management Plan], which may be included in the Special Provisions of the other forms for major works. It obliges the Contractor to:

- complete and maintain the Contract Risk Register;
- prepare and maintain the Contract Risk Management Plan to manage and control the identified risks;
- submit the initial Contract Risk Register and Risk Management Plan to the CA for Sub-Clause 5.2.2 Review within 28 days of the Commencement Date;
- submit the updates at specified regular intervals (3 months if not specified) and promptly after the occurrence of any risk event requiring a revision or update.

Both documents shall comply with the Employer's Requirements and include the Contractor's procedures for regular Contract Risk Management meetings for their review and update, and follow-up actions, which may include:

1	2	3	4	5		
Insignificant	Below the average	Average	Above the average	If occurs, a disaster		
5	10	15	20	25	Has occurred	5
4	8	12	16	20	Above the average	4
3	6	9	12	15	Average (50%)	3
2	4	6	8	10	Below the average	2
1	2	3	4	5	Low	1

probability

impact (consequence)

Figure 13.2 Probability and impact matrix

6 PMI Inc., *PMBOK® Guide* (PMI 2017) 419

Value from the matrix	Priority	Risk
1-4	1	Limited, **acceptable** risk
5-9	2	Essential risk, **attention is** necessary
10-16	3	Major risk, **measures and actions** mitigating the risk are necessary
17-25	4	Suspension of activities until **elimination/mitigation** of the risk

Figure 13.3 Prioritisation of risks

- giving Sub-Clause 8.4 advance warning Notices;
- calling meetings under Sub-Clause 3.8.

Quantitative risk analysis quantifies overall project risk exposure, and it can also provide additional quantitative risk information to support risk response planning.[7] It usually requires specialised software and expertise in the development and interpretation of risk models. It is most likely to be appropriate for large/complex/strategically important projects, or where it is a contractual requirement, or a key stakeholder requires it.[8] It consumes additional time and cost.

13.4 Risk response

Risk responses should be appropriate for the significance of the risk, cost-effective in meeting the challenge, realistic within the project context, agreed upon by the parties involved, and owned by a responsible person. A contingency plan can be developed if the selected strategy is not effective. Secondary risks arising as a result of implementing a risk response should also be identified.[9]

The following strategies may be considered **for dealing with each identified threat**:

- **Escalation**: if the threat is outside the scope, or the response is outside the CA's authority, the CA shall communicate to the Employer that the risk cannot be managed by the project team (e.g. the Contractor may object to Unforeseeable varied work under Sub-Clause 13.1, and the CA may cancel the instruction);
- **Avoidance**: actions to eliminate the threat or protect the project from its impact (e.g. suspension of the Works by the Employer under Sub-Clause 8.9 during a COVID-19 wave);
- **Transfer**: shifting ownership of the threat to a third party (e.g. to the parent company with a Guarantee under Sub-Clause 4.2, to insurers under Clause 19);
- **Mitigation**: actions to reduce the probability of occurrence and/or impact of a threat (e.g. weekly testing and other COVID-19 prevention measures under Sub-Clause 6.7);
- **Acceptance**: no proactive actions, but establishing a contingency reserve and planning actions to handle the threat if it occurs (Site protocols planned the response to COVID-19 under Sub-Clause 6.7).

7 PMI Inc., *PMBOK® Guide* (PMI sixth edition 2017) 428
8 PMI Inc., *PMBOK® Guide* (PMI sixth edition 2017) 429
9 PMI Inc., *PMBOK® Guide* (PMI 2017) 439

Chapter 9 explores the contingency planning under FIDIC contracts, presented in Figure 9.1. Sub-Clause 13.4 Provisional Sums and Sub-Clause 13.5 Daywork Schedules enable timely risk response. The Sub-Clause 4.2 Performance Security and Retention Money accumulated under Clause 14 provide securities in case of Contractor's default.

Risk responses should be planned and implemented not only for individual project risks but also **to address overall project risks**:

- **Avoidance**: where the level of overall project risk is significantly negative and outside the agreed-upon risk threshold: the Variation procedure may be used to omit the high-risk works; in some cases, the Contract may be terminated under Sub-Clause 15.5 (e.g. if it is obvious that the Employer's design has serious defects and the risk to be used is not acceptable), or the Parties may be released from performance under Sub-Clause 18.6 if a risk event arises outside the control of the Parties and makes it impossible or unlawful for them to fulfil their obligations;
- **Exploitation**: where the level of overall project risk is significantly positive and outside the agreed-upon risk threshold, the Sub-Clause 13.2 Value Engineering (VE) procedure or the Variation by request for proposal under Sub-Clause 13.3.2 may be used to add value;
- **Transfer/sharing**: Clause 19 stipulates transfer of risks to insurers; the GCs stipulate the sharing of risks (e.g. Parties share the climatic risks: exceptionally adverse climatic conditions entitle the Contractor to EOT under the Red/Yellow/Emerald Books, but the Contractor bears the prolongation Costs and the risk for foreseeable adverse climatic conditions; see Figure 13.9);
- **Mitigation/enhancement**: the Variation procedure may be used for necessary changes; the Sub-Clause 8.3 Programme may be revised, changing the sequence of works to reduce the critical time impact;
- **Acceptance**: active acceptance includes contingency reserves (e.g. Sub-Clause 13.4 Provisional Sums); passive acceptance involves no proactive actions apart from periodic review of the risk level (done in the monthly reports under Sub-Clause 4.20 and during the management meeting under Sub-Clause 3.8 (Silver Book: 3.6).

In order to select the appropriate risk response strategy, the following may be used:

- **Alternative analysis** to compare different risk response options;
- **Cost-benefit analysis,** where the impact of an individual project risk can be quantified in monetary terms;
- **Multi-criteria decision analysis** to select a preferred option using a decision matrix.

13.5 Risk monitoring

The following data analysis techniques can be used for monitoring risks:

- **Technical performance analysis:** requires quantifiable measures of technical performance to compare actual results against targets; deviations can

indicate the potential impact of threats or opportunities (e.g. the Schedule of Performance Guarantees links the extent of the failure of the Clause 12 Tests after Completion under the Yellow/Emerald/Silver Book with Performance Damages);
- **Reserve analysis:** compares the amount of contingency reserves remaining with the amount of risk remaining in order to determine if the remaining reserve is adequate (e.g. when the accumulated amount of Variations exceeds 30 per cent of the ACA [Silver Book: Contract Price], the Contractor may request and the Employer shall provide within 28 days reasonable evidence under Sub-Clause 2.4 that financial arrangements have been made; the failure to do so entitles the Contractor to suspend work, giving a 21-day Notice, until the Employer demonstrates the availability of additional financing).

The Contractor's Sub-Clause 4.20 progress reports contain information enabling monitoring of emerging risks and contingencies:

- a list of Variations under Clause 13;
- a list of Notices of Claim given by either Party under Sub-Clause 20.2.1;
- details of any hazardous incidents and activities relating to environmental aspects and public relations;
- comparison of actual and planned progress, with details of any events or circumstances that may adversely affect the completion of the Works and the measures adopted to overcome delays.

13.6 Risk allocation in construction contracts

The obligations and rights allocated by the Contract arising as a consequence of a risk event materialising are referred to by the popular term 'risk allocation'.[10]

Besides setting out the scope and cost of the project to be constructed, the purpose of a construction contract is to allocate the risks to which that project is exposed, and to provide a clear statement as to how these risks are to be dealt with and managed.[11] The risk allocation in construction is traditionally based on a sharing between the parties involved, in accordance with the provisions of the contracts for works and services. From the works contract flows another line of risk sharing between the Contractor and subcontractors/suppliers/manufacturers/insurers and others.[12] From the agreement for services flows risk sharing with Sub-Consultants/insurers.

Chapter 5 recommends early appointment of the Consultant as the project manager. During the project preparation, the Employer and the Consultant identify project risks, analyse them, and plan the responses. Since most of the risks cannot be eliminated, the selected contract form allocates the risks to the Parties in line with the

10 Donald Charrett, *FIDIC Contracts in Europe. A Practical Guide to Application* (Informa law from Routledge 2023) 46

11 Nael and Lydia Bunni, *Risk and Insurance in Construction* (Routledge 2022) #1.3.11

12 Nael Bunni, 'Managing Risk' (2005) 5.5

tender strategy. The choice depends on the competence of the Consultant and the risk appetite of the Employer.

According to Max Abrahamson, the rules for allocation of risks in a construction project revolve around the ability of a party to:[13]

(a) control any arrangements that might be required to deal with the hazard or any triggering incident relating to it;
(b) control the risk or influence any of its resultant effects;
(c) perform a task relating to the project, such as obtaining and maintaining insurance cover; and
(d) benefit from the project.

Dr. Bunni notes that if risks are not allocated in a contract, and a dispute arises between the parties regarding to whom a particular risk is allocated, then an arbitrator or a judge would most likely examine the following criteria for risk allocation and determine the dispute accordingly:[14]

1. Which party could best foresee that risk?
2. Which party could best control that risk and its associated hazards?
3. Which party could best bear that risk?
4. Which party most benefits or suffers when that risk eventuates?

Dr. Bunni concludes that in general,

> liabilities arising from the duties and obligations of the parties to a contract should be covered by indemnities given by one party to the other, or provided in one form or another, including the provision of insurance policies. Thus, the logical sequence that emerges is a flow from Risk to Responsibility to Liability to Indemnity to Insurance.[15]

When the risks are allocated, the flow presented in Figure 13.4 will apply. Each Party becomes responsible and/or liable for certain risks. The liability does not necessarily follow the responsibility: one may be responsible for a certain risk but not liable for its consequences[16] (e.g. the Contractor is responsible for the care of the Works under Sub-Clause 17.1, but the Employer is liable for the consequences of certain risks under Sub-Clause 17.2; see Figure 13.8).

Figure 13.4 Risk flow in construction contracts according to Dr. Bunni

13 Max Abrahamson, 'Risk Management' [1983] ICLR 241
14 Nael Bunni, 'Managing Risk' (2005) 5.4
15 Nael Bunni, 'Managing Risk' (2005) 2.7 <https://fidic.org/sites/default/files/5%20bunni_managing_risk_ibc_oct05.pdf> accessed 27 October 2022
16 Nael and Lydia Bunni, *Risk and Insurance in Construction* (Routledge 2022) 4.1.44, 5.1.3

13.7 Project Risk Management using the FIDIC Suite for Services

Among the risks associated with the feasibility study, design, and preparation of tender documents are:[17]

- selection of Consultant and choice of experts;
- form of contract conditions;
- choice of Site;
- adequacy of surveys and investigations;
- adequacy of cost planning;
- Exceptional Events;
- inappropriate choice of design with respect to others or society;
- negligence and lack of care;
- state of the art, codes, and technical knowledge;
- lack of knowledge, inadequate checking;
- work carried out in haste;
- lack of communication;
- failure to take account of foreseeable problems;
- use of untested and unproven techniques;
- inadequate performance of mechanical and electronic equipment;
- lack of safety precautions;
- selection criteria for Contractor;
- choice of nominated Subcontractors.

Most of these risks will be applicable in relation to the services for contract administration and construction supervision. Additional risks are Contractor's inadequate quality control (QC), Parties' adversarial behaviour, extreme climatic conditions, lack of infrastructure, lack of qualified personnel willing to be allocated to the Site, technical complexity and innovations, ground/underground risks, illegal activities, conflicting requirements of stakeholders, delays in permitting/performance, many Variations, and Claims, often combined with lack of funds.

The Task Group drafting the fifth edition of the White Book considered the standard of care to the Client for the Services, endeavouring to achieve a fair balance of risk between the Client and the Consultant.[18] The approach for risk allocation is the following:

- the standard for performance shall be reasonable skill, care, and diligence to be expected from a consultant experienced in the provision of similar services (Sub-Clause 3.3.1);
- without extending the standard of care, the Consultant shall perform the Services with a view to satisfying any function and purpose stated in Appendix 1 [Scope of Services] (Sub-Clause 3.3.2);

17 Nael and Lydia Bunni, *Risk and Insurance in Construction* (Routledge 2022) figures 3.1 and 3.2
18 FIDIC, *White Book* (FIDIC 2017) Foreword

- if the performance does not meet the required standard of care, then the Client should be entitled to appropriate redress against the Consultant, usually covered by professional indemnity (**PI**) insurance;
- the Consultant shall not be liable for defective or inadequate services arising out of unforeseeable or uncontrollable events, so the obligation should be fully insured under any PI insurance policy (which may be difficult in some jurisdictions);
- where there is no evidence of Consultant's fault or breach, it is not a fair or reasonable balance of risk to make the Consultant strictly liable for the outcome of professional services.

The Consultant's performance shall be considered in relation to the existing knowledge in the profession:[19] 'The standard is that of the reasonable average. The law does not require of a professional man that he be a paragon, combining the qualities of polymath and prophet.' Referring to risks related to innovations, 'The law requires even pioneers to be prudent.'[20]

Sub-Clause 4.6 sets out the procedure and stipulates the reliefs in the case of Exceptional Events preventing a Party from performing any of its obligations:

- the Party shall give a Notice within 14 days from when it becomes aware, or should have become aware, of the event or circumstance and shall be excused from performance during the period of prevention;
- the Consultant shall be entitled to an EOT where the Exceptional Event gives rise to a delay;
- the Client shall issue a Variation to the Services under Sub-Clause 5.1 where the risk impact leads to an unavoidable change in the scope of Services.

Besides the Exceptional Events and Variations to the Services, pursuant to Sub-Clause 4.4, any delay, impediment, or prevention attributable to the Client or its other consultants, contractors, or other third parties, as well as any other event or circumstance listed under a specific provision, may entitle the Consultant to an EOT. If any of those circumstances cause the Consultant to incur additional costs, they shall be compensated as Exceptional Costs.

Sub-Clause 1.1.12 defines 'Exceptional Costs' as costs arising out of any necessary work, cost, expense, or delay incurred by the Consultant that is additional to the Services (or Variations) and that is necessarily and unavoidably performed under the Agreement and identified as such. For example, a change to the local legislation may cause the Consultant to incur Exceptional Costs, leading to adjustment of the agreed remuneration under Sub-Clause 1.5.

Clause 9 stipulates the transfer of risk to insurers. The Consultant shall maintain for the duration of the Services:

19 *Eckersley & Others v Binnie & Partners and Others* (1988) 18 ConLR 1

20 *Independent Broadcasting Authority v EMI Electronics and BICC Construction* (1980) 14 BLR 1

- PI insurance and public liability insurance with minimum amount of cover not less than the amount specified in the Particular Conditions, provided that such insurance is available at commercially reasonable rates;
- workers' compensation or employer's liability insurance;
- any other insurances as may be required by the applicable law.[21]

The PI insurance usually covers losses and damages resulting from performance of the Services not complying with the required standard of care, including injury to others and physical damage to property.

TG15B is developing GPs for Service Agreements to be complied with by FIDIC users for fair risk allocation.

13.8 Project Risk Management using FIDIC Works Contracts

13.8.1 Risk allocation approach

Chapter 5 outlines the allocation of design responsibility, price risk, and ground risk in relation to the choice of a FIDIC contract for works. The following subsections address the relevant risk sharing procedures. Figure 5.4 presents the algorithm for risk-based selection of the appropriate FIDIC book. The Particular Conditions should not change the balance of risk allocation provided for in the selected GCs; otherwise, the resulting contract would infringe GP3.

The Red/Yellow/Emerald/Silver Books allocate the risks resulting in economic loss through different provisions (e.g. the risk for delayed right of access to the Site is allocated to the Employer under Sub-Clause 2.1, while the risk for suitability/availability of access routes to the Site is allocated to the Contractor under Sub-Clause 4.15 but will be shared with the Employer in case of changes by the Employer or a third party after the Base Date).

Chapter 8 discusses the entitlement of the Contractor to an EOT under different Sub-Clauses as a result of Employer's risks listed in Table 8.1. The delays caused by Contractor's risks may lead to Delay Damages under Sub-Clause 8.8 and reduction in the Contract Price by the additional costs incurred by the Employer.

Chapter 9 lists the Contractor's risks leading to the Employer's entitlement to reduction of the Contract Price. Table 9.1 includes the Employer's risks leading to the Contractor's entitlement to compensation.

Specific provisions under Clause 17 allocate the risks for bodily injury, material loss or damage, and infringement of intellectual property. Clause 18 deals with Exceptional Events. Clause 19 contains provisions for risk transfer to insurers.

Sub-Clause 4.2 requires the Contractor to provide a Performance Security besides the Retention Money under Clause 14 as contingency for Contractor's risks resulting in

21 For example, Article 171 of the *Bulgarian Spatial Development Act* (State Gazette 2001) requires separate professional liability insurance policies to be maintained for design, technical control, and supervision of design and construction services, with minimum amounts of insurance cover depending on the project category. The mandatory insurance covers liability for bodily injury to other participants in the construction and/or to third parties, as well as for material damage to the property of other participants in the construction or to third parties, caused by the consultant. <https://cpcp.mrrb.government.bg/cms/assets/Laws/SPATIAL%20DEVELOPMENT%20ACT.pdf> accessed 30 October 2022

economic loss for the Employer. Sub-Clauses 13.4 [Provisional Sums] and 13.5 [Daywork] set out procedures for spending contingency funds in the case of Employer's risk events. Sub-Clause 1.15 (Silver Book: 1.14) limits the risk exposure.

The approach of the Gold Book is different: Clause 17 [Risk Allocation] groups the overall risks during the design-build period and then during the operation service period. Additionally, Sub-Clauses 17.1 and 17.3 divide the risks allocated to the Employer to Commercial Risks and Risks of Damage. Sub-Clause 1.1.7 defines 'Commercial Risk' as a risk resulting in financial loss and/or time loss, where insurance is not generally available. Sub-Clause 1.1.67 defines 'Risk of Damage' as a risk resulting in physical loss or damage to the Works or other property belonging to either Party, other than a Commercial Risk. Sub-Clauses 17.2 and 17.4 allocate the remaining risks to the Contractor, subject to the limitation of liability under Sub-Clause 17.8. The indemnification provisions under Sub-Clauses 17.9, 17.10, and 17.11 follow the risk allocation.

The innovative approach of the Green Book is discussed at the end of this chapter.

13.8.2 Design responsibility and liability for related risks

As discussed in Chapter 5, the Employer is responsible for the design under the Red Book, while the Contractor is responsible for the design under the Yellow/Silver/Emerald/Gold Books. In the case of design responsibility, the Contractor accepts a fitness for purpose obligation. The fitness for purpose is a strict obligation, independent of negligence.

Under Sub-Clauses 5.1 and 1.9 of the Yellow/Emerald Books, the Contractor may be entitled to relief in case of errors in the Employer's Requirements, following the procedure. The Employer retains the risk for such errors, being the Party in a better position to manage this risk during the project preparation. The Contractor shall examine the Site and carry out a detailed review of the Employer's Requirements (including the design) in order to find and notify about errors:

1) **during the tender period,** the tenderers have to exercise due care when examining the Site and the Employer's Requirements to discover any errors, fault, or other defect; they have opportunity to ask questions and receive clarifications and also comment on the Employer's Requirements in their proposal;
2) **promptly upon receiving the Notice to commence under Sub-Clause 8.1 within the period stated in the Contract Data** (if not stated, 42 days) calculated from the Commencement Date, the Contractor shall scrutinise the Employer's Requirements (including the Employer's design) and give a Notice to the Engineer describing the error, fault, or defect; the Contractor should have mobilised its designers as early as possible to have enough time for Site surveys and analysis of the Employer's design;
3) **if after the expiry of the second period,** the Contractor finds errors, it shall also give a Notice describing them. Under Sub-Clause 1.3, Notices shall not be unreasonably withheld or delayed.

Receiving a Notice under Sub-Clause 1.9 of the Yellow/Emerald Book during the second and third periods, the Engineer shall proceed under Sub-Clause 3.7 to agree or determine:

(a) whether or not there is an error, fault, or defect;
(b) whether or not (taking account of cost and time) an experienced contractor exercising due care would have discovered it during the first period for a Notice given during the second period or during the first two periods for a Notice given during the third period;
(c) what measures (if any) the Contractor is required to take to rectify the error, fault, or defect.

If an experienced contractor would not have discovered the error during the first two periods:

(i) the Sub-Clause 13.3.1 Variation procedure shall apply to the measures that the Contractor is required to take (if any); and
(ii) if the Contractor suffers delay and/or incurs Cost as a result of the error, the Contractor shall be entitled following the Sub-Clause 20.2 Claims procedure to EOT and/or payment of such Cost Plus Profit.

The objective test compares the appointed Contractor with any experienced contractor who may be eligible under the prequalification criteria. Sometimes, it is difficult to determine fairly whether the error could have been discovered in the first two periods. For the tender period, the author recommends the following tests:

1. Did any of the tenderers discover the error? (The questions and answers usually become part of the Contract documents and can be checked by the Engineer.)
2. Did any of the tenderers discover the error and reflect this finding in its proposal? (This can be checked by the Employer only.)

For the mobilisation period, the test is more difficult. Usually, the tenderers use in-house experts for the preparation of the tender, but they rarely have designers with all key qualifications and experience in similar projects. If a consulting company is not a partner in the Contractor's joint venture (JV) or a named Subcontractor, tenderers rarely engage designers to check the Employer's design during the tender period. When the Notice for the Commencement Date comes, the Contractor starts its procurement for designers, organising competitions and trying to find the best value for money (often seeking the lowest price). In many cases, the first month is lost in negotiations of fees, and there is no time for the experts to scrutinise the design and find the errors in the agreed period. Of course, the objective test should not take this into account but should consider whether qualified designers allocated on the project from the date when the Notice for the Commencement Date was received would have discovered the error. The scope of work, the availability of access, the need to do additional surveys, errors in the cadastre, and other public data are possible considerations.

Under the Silver Book, the Contractor takes all the responsibility for the correctness of the Employer's Requirements, with some exceptions under Sub-Clause 5.1:

a. portions, data, and information that are stated in the Contract as being immutable or the responsibility of the Employer;
b. definitions of intended purposes of the Works or any parts thereof;
c. criteria for the testing and performance of the completed Works; and
d. portions, data, and information that cannot be verified by the Contractor, except as otherwise stated in the Contract.

Figure 13.5 compares the design responsibility and related risk allocation under the Red/Yellow/Silver/Gold/Emerald Books:

- under Sub-Clause 4.1 of the Red Book, the Contractor has limited design risk: the Contractor is responsible for all Contractor's Documents (which may include shop drawings, as-built drawings, and design needed for permits and approvals under Sub-Clause 1.13, like Traffic Relocation Plan, Health and Safety Plan), the design of the Temporary Works, method statements, such design of each item of Plant and Materials as is required for the item to be in accordance with the Contract, and as an exception, design for specified minor parts of the Permanent Work;
- since the Employer is responsible for the design of the Permanent Works under the Red Book, in the case of delayed drawings, the Contractor may be entitled under Sub-Clause 1.9, subject to Sub-Clause 20.2, to EOT and/or payment of Cost Plus Profit;
- under Sub-Clause 5.1 of the Yellow/Gold/Emerald Books, the Contractor is responsible for the design of the Works but may be entitled, subject to Sub-Clause 20.2 (20.1 Gold Book), to EOT and/or payment of Cost Plus Profit in the case of timely notified errors in the Employer's Requirements under Sub-Clauses 5.1 and 1.9 (1.10 Gold Book), and Clause 13 shall apply to any measures that the Contractor is required to take;
- the Contractor accepts the total design responsibility and associated risks under Sub-Clause 5.1 of the Silver Book, with the rare exceptions listed there as Employer's responsibility.

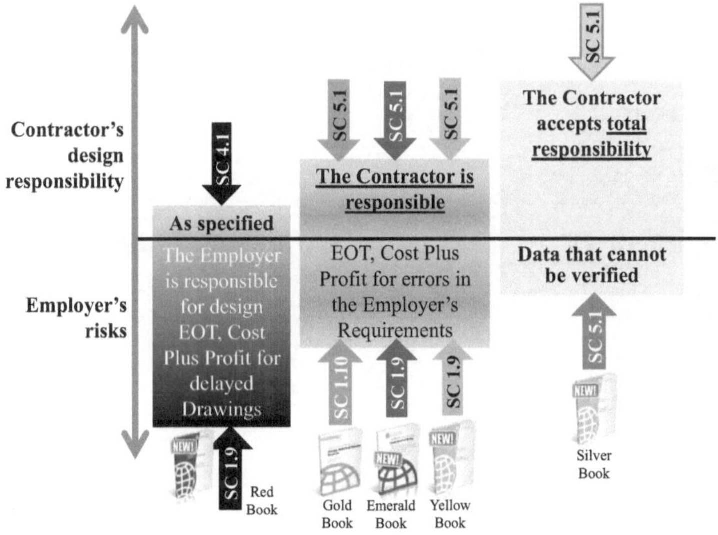

Figure 13.5 Design responsibility under a FIDIC Works Contract

13.8.3 Risk of errors in setting out the Works

The Contractor is responsible for the correct positioning of all parts of the Works. However, under the Red/Yellow/Silver/Gold/Emerald Books, the Employer retains the risk of errors in the items of reference submitted with the Site data under Sub-Clause 2.5 (Red/Yellow/Emerald Books)/4.7(Gold Book). The items of reference include the original survey control points, lines, and levels of reference, which may be specified in the Specification/ on the Drawings (Red Book), Employer's Requirements (Yellow/Emerald/Gold Books) or issued to the Contractor by a Notice from the Engineer. The Contractor is obliged to verify the accuracy of those items of reference before using them and may notify errors under Sub-Clause 4.7. The CA then follows the procedure under Sub-Clause 4.7 to agree or determine the Contractor's relief and measures (if any) for the rectification of the error. The considerations are similar to those under Sub-Clause 1.9 (1.10 Gold Book). The Variation procedure shall apply to any measures that the Contractor is required to take. If there are no measures, but the Contractor suffers delay and/or incurs Cost, following the Sub-Clause 20.2 (20.1 Gold Book) procedure, the Contractor shall be entitled to EOT and/or payment of Cost Plus Profit.

The Contractor accepts the risk for errors in the items of reference and for the correct positioning of the Works under Sub-Clause 4.7 of the Silver Book.

13.8.4 Price risk

The Employer should select the method of valuation following the risk allocation principles formulated by Max Abrahamson and Nael Bunni, retaining the risk in case of uncertainty and innovations.

The risk of price escalation may be shared under Sub-Clause 13.7 using a Schedule of cost indexation. Such adjustment shall not be applied to work valued as Cost Plus Profit.

Figure 13.6 presents the price risk allocation under the Red/Yellow/Emerald/Silver/Gold Books, as stipulated under Sub-Clause 14.1 of the GCs.

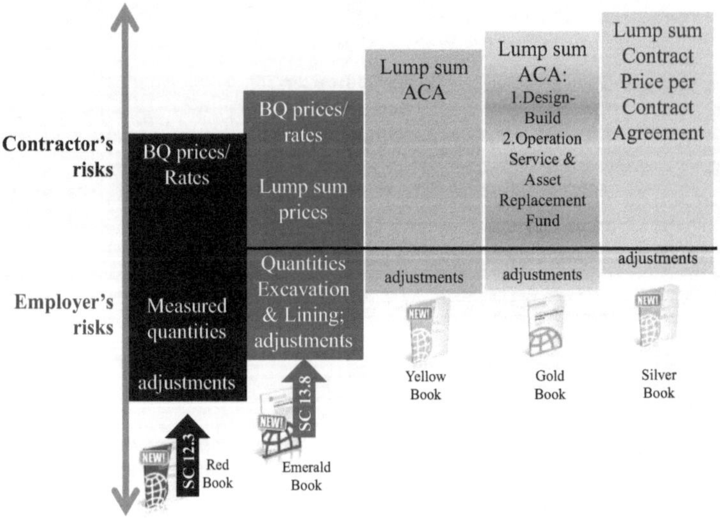

Figure 13.6 Price risk allocation under the FIDIC Works Contracts

Under the Red Book, the Contract Price is calculated by **measuring the as-built quantities** of the Works and valuing them by applying the appropriate rates and prices in the BQs or other Schedule. The Employer carries the risk of changes in quantities, while the Contractor takes the risk for the unit rates and prices.

Under the Emerald Book, the Contract Price is **calculated in two different ways** depending on the type of works. Taking into account that the cost of the Excavation and Lining Works will largely depend on the subsurface physical conditions, their quantities are to be measured and remain the Employer's risk, while the Contractor takes the risk for their rates/unit prices. The Contractor also takes the risk of changes in quantities, and the risk in respect of cost related to other works, paid as a lump sum. The Employer retains some risks (like Unforeseeable physical conditions, errors in Employer's Requirements, etc.), and their impact may lead to adjustments.

Under the Yellow Book, the Contract Price is **the lump sum ACA,** accepted by the Employer based on the Contractor's proposal. However, the Employer keeps some risks (like Unforeseeable physical conditions, errors in Employer's Requirements, etc.), and their impact may lead to adjustments to the ACA, including additions or deductions.

Under the Gold Book, the Contract Price is the **lump sum amount submitted by the Contractor** including amounts for Design-Build of the Works, Operation Service, and Asset Replacement Fund. The Contract Price is adjusted similarly to the Yellow Book as a result of Employer's risks.

Under the Silver Book, the Contract Price is the **lump sum stated in the Contract Agreement.** The Employer has fewer risks leading to price adjustments (e.g. for errors in the Employer's Requirements only where the information could not have been verified during the tender preparation).

However, the Particular Conditions may change the price risk allocation as follows:

- the Red Book may become a lump sum contract when the Employer's design has enough details that the tenderers can take the risk of changes in quantities; private Employers often submit to the tenderers drawings prepared by their designers for office or residential buildings in ACAD to calculate easily the quantities and submit lump sum prices; the tender documents should include a schedule of payment instead of BQs. Some tender documents include BQs for information only as part of the Specification; the Sub-Clause 14.1 GPPC example provision for lump sum contract may be used;
- the Red Book may become a Cost-plus contract for reasons of urgency or where the Employer prefers early involvement of the Contractor, combining the Cost-plus method with some incentives;
- the Yellow Book may become a hybrid contract similar to the Emerald Book, where any part of the Works is to be paid on the basis of measurement; the GPPC proposes example provisions for measurement of specified parts;
- the Silver Book GPPC includes an example provision for interim valuation procedure for projects where the Works consist of only a few different types of operations.

These changes are in line with The FIDIC Golden Principles, when the goal is fair risk allocation.

13.8.5 Ground risk

Figure 13.7 presents the ground risk allocation under the Red/Yellow/Silver/Gold/Emerald Books. Under Sub-Clause 2.5 (4.10: Gold Book), the Employer is under an obligation to provide the Contractor with all relevant information in its possession concerning the ground conditions of the Site.

Under Sub-Clause 4.10 of the Red/Yellow/Emerald Books, the Contractor is responsible for the interpretation of the Site data submitted by the Employer. Additionally, the Contractor shall be deemed to have obtained all necessary information as to ground risks, but its responsibility is limited 'to the extent which was practicable (taking account of cost and time)'. This limitation is related to the tender period and the availability and accessibility of information. 'To the same extent, the Contractor shall be deemed to have inspected and examined the Site' and to have been satisfied before submitting the Tender as to the form and nature of the Site, including sub-surface conditions.

Sub-Clause 4.11 of the Red/Yellow/Emerald Books confirms that the Contractor takes the risk for the correctness and sufficiency of the ACA, based on the sub-surface data, interpretations, necessary information, inspections, and examinations as described in Sub-Clause 4.10. The ACA is the agreed amount, at the time the Contract is entered into, to be paid by the Employer to the Contractor. The ACA may be adjusted, adding the Cost due to Unforeseeable ground conditions under Sub-Clause 4.12. 'Unforeseeable' is defined as 'not reasonably foreseeable by an experienced contractor by the Base Date' (28 days before the latest date for submission of the Tender). The Engineer may take into account possible reductions due to more favourable ground conditions in other parts of the Works than could reasonably have been foreseen by the Base Date. However, the net effect of all additions and reductions shall not result in a net reduction in the Contract Price. The approach of the Gold Book is similar.

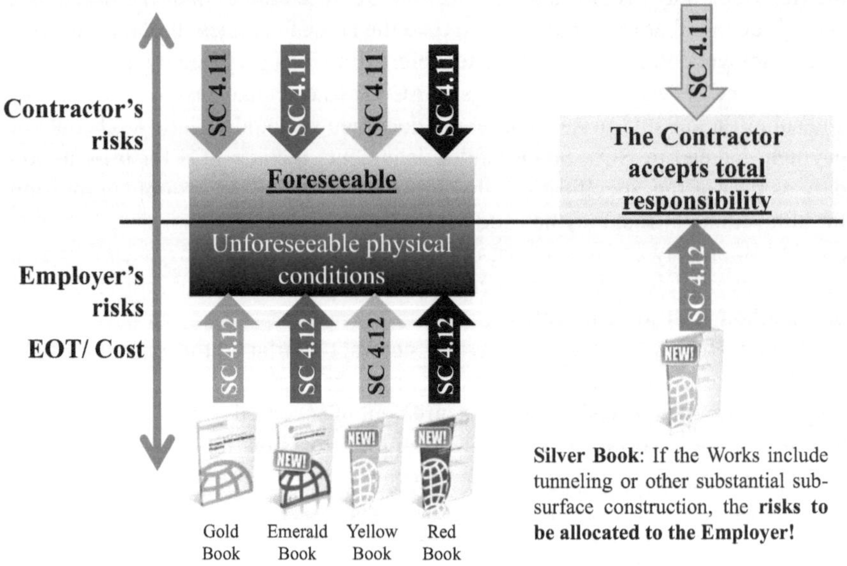

Figure 13.7 Ground risk allocation under a FIDIC Works Contract

The Emerald Book has a ground risk allocation similar to the Red/Yellow/Gold Books, but its approach is more sophisticated. The authors take into account the specific characteristics of the underground works and define the Geotechnical Baseline Report (GBR) as the single source of risk allocation related to subsurface physical conditions. All ground conditions not addressed in the GBR shall be considered Unforeseeable. The risks arising out of the foreseen properties of the ground, obstacles, and adverse reaction to the excavation and ground support processes are assigned to the Contractor, as well as the production rates and cost of performing the Works under the same conditions. Conversely, the risks arising out of unforeseen physical conditions of the ground, obstacles, and adverse reaction to the excavation and ground support processes are allocated to the Employer, warranting EOT and/or reimbursement of cost to the Contractor.[22]

Under the Silver Book, the Contractor is responsible not only for interpreting the Site data under Sub-Clause 4.10 but also for verifying it. The Employer's responsibility for errors in the data is limited to the exceptions specified in Sub-Clause 5.1(a) to (d), referring to information stated to be Employer's responsibility. The ground risk allocation under the Silver Book must be very limited; otherwise, the Tenders will have risk contingencies from zero to 30 per cent, unless several tenderers are prequalified and have enough time for additional Site surveys. The agreed amount at the time the Contract is entered into is the Contract Price as stated in the Contract Agreement. The Contractor accepts total responsibility for having foreseen all ground risks under Sub-Clause 4.12. The Contract Price cannot be adjusted to take into account any Unforeseeable difficulties or costs. However, the opening text of Sub-Clause 4.12, 'Except as otherwise stated in the Particular Conditions', allows the drafters to amend the provision in case of major ground risk, in line with the FIDIC GPs, demonstrated in the following case study.

13.8.6 Case study – Silver Book

An interesting example for fair risk allocation is a hydro power plant in Asia with a Silver Book contract, where Sub-Clause 4.12 [Unforeseeable Physical Conditions] from the Yellow Book replaced the Silver Book Sub-Clause 4.12 [Unforeseeable Difficulties] in the Particular Conditions. The Plant and related processes are leading in the procurement of the hydro power plants, and Employers may select the Silver Book similarly to nuclear/thermal power plant projects. At the same time, the infrastructure for a hydro project is usually built by the same Contractor in steep mountains with difficult access and changing ground conditions. The tenderers cannot price the ground risks properly, so the experienced contractors would have a high contingency, increasing the Contract Price under the Silver Book. By retaining the ground risk, the Employer may save public money if the actual ground conditions are not more adverse than foreseen.

22 FIDIC, *Emerald Book* (FIDIC 2019) Notes

13.8.7 Material damages to the Works, Goods, and Contractor's Documents

Figure 13.8 presents the allocation of the risk for material damage. Sub-Clause 17.1 allocates the responsibility for care of the Works, Goods, and Contractor's Documents to the Contractor from the Commencement Date until the issue of the TOC or date of termination (if the Contract is terminated). After that, responsibility for the care of the works shall pass to the Employer, but Sub-Clauses 17.1/17.2 stipulate that the risk for certain losses and damages to the Works, Goods, and Contractor's Documents is allocated to the Contractor for:

- outstanding work on the Date of Completion until its completion;
- any loss or damage caused by the Contractor after the issue of a TOC; or
- loss/damage occurring after the issue of a TOC, arising from an event occurring before the issue of a TOC, for which the Contractor was liable.

The Contractor's liability for losses and damages following the responsibility for care of the Works, Goods, and Contractor's Documents is limited:

- within the total amount under Sub-Clause 1.15 (Silver Book: 1.14);
- outside the Employer's risk events listed under Sub-Clause 17.2.

The Variation and Claims procedures may be applicable in relation to the risk response under Sub-Clause 17.2. For example, in the case of damage to the Works and stored Materials as a result of an Unforeseeable storm, the Contractor shall rectify the loss and damage to the extent instructed by the CA under Sub-Clause 17.2. The instruction shall be treated as a Variation under Sub-Clause 13.3.1. If the Contractor has not protected the building with temporary waterproofing or has not properly stored the Materials, then the Contractor may share the risk with the Employer, claiming for a proportion of EOT and/or Cost Plus Profit under Sub-Clause 17.2 subject to Sub-Clause 20.2.

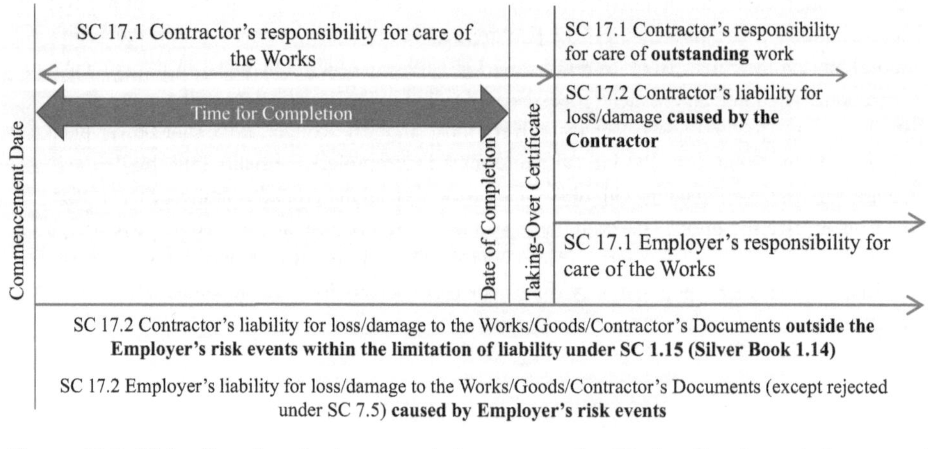

Figure 13.8 Risk allocation for losses and damages to the Works, Goods, and Contractor's Documents

Sub-Clause 17.2 provides a mechanism for dealing with rectification of the losses and damages caused by Employer's risk events. The FIDIC 2017 Contracts Guide explains that the list of Employer's risk events takes account of the extent to which insurance cover is available in many countries, but it is not intended to reflect insurance coverage, since some of the risks are not insurable[23] (like the war risk).

13.8.8 Exceptional Events

The Exceptional Events are included in the list of Employer's risks entitling the Contractor to an EOT and Cost Plus Profit for rectification of damages to the Works, Goods, and Contractor's Documents under Sub-Clause 17.2. Generally, the Exceptional Events are not insurable, but some of them, such as earthquake, are insurable and may be listed in the Contract Data as risks that shall not be excluded from the insurance cover for the Works under Sub-Clause 19.2.1(iv). With identical wording to the definition of Exceptional Event under Sub-Clause 1.13 of the White Book, Sub-Clause 18.1 of the Red/Yellow/Emerald/ Silver Books defines 'Exceptional Event' as an **exceptional**[24] event or circumstance that:

(i) is beyond a Party's control;
(ii) the Party could not reasonably have provided against before entering into the Contract;
(iii) having arisen, such Party could not reasonably have avoided or overcome; and
(iv) is not substantially attributable to the other Party.

Sub-Clause 18.2 stipulates that if a Party is (or will be) prevented from performing any obligation under the Contract (except to pay) as a result of the Exceptional Event, the affected Party shall be excused performance of the prevented obligations during the risk impact, subject to timely notifications. Under Sub-Clause 18.3, each Party has a duty to mitigate the risk, using all reasonable endeavours to minimise any delay in the performance of the Contract.

Sub-Clause 18.1 includes an open list for different groups of Exceptional Events. Subject to Sub-Clause 20.2, the Contractor may be entitled under Sub-Clause 18.4 to an EOT for the critical delay during the time of prevention and/or compensation of incurred Cost resulting from man-made risks in the Country of the Site/ war risk in any country. Sub-Clause 18.4 may provide relief to the Contractor only when performance is prevented, not made more onerous or expensive.

13.8.9 Interaction of risk sharing provisions

The entitlement to an EOT for instructed rectification of damage caused by Exceptional Events under Sub-Clause 17.2 is separate from the entitlement to an EOT for the previous period of prevention under Sub-Clause 18.4. Table 13.1 compares the events, procedures, and Contractor's remedies related to Exceptional Events and Employer's Risks listed in Sub-Clause 17.2.[25]

---

23 *The FIDIC 2017 Contracts Guide* (FIDIC 2022) 465

24 The word 'exceptional', present in the 1999 *Force Majeure* definition but missing in the 2017 Edition and the *Emerald Book*, is added in the 2022 Reprint for clarity

25 The list is non-exhaustive; e.g. the pandemic is not included

Table 13.1 Risk sharing under Sub-Clauses 17.2 and 18.4

	Employer's Risks Sub-Clause 17.2	*Exceptional Events Clause 18*
Result	loss or damage to the Works, Goods, Contractor's Documents	prevents the Parties from performing their obligations
Test	listed events under SC 17.2 Works, Goods, Contractor's Documents not rejected under SC 7.5 for forces of nature: Unforeseeable or against which an experienced Contractor could not reasonably have been expected to have taken adequate preventive precautions	exceptional, beyond a Party's control the Party could not reasonably have provided against before entering into the Contract having arisen, the Party could not reasonably have avoided or overcome is not substantially attributable to the other Party
Event	(a) interference with right of way, light, air, water, or other easement, unavoidable result of the execution of the Works;	(a) war, hostilities, invasion, act of foreign enemies;
	(b) use or occupation by the Employer of any part of the Permanent Works, except as specified;	(b) rebellion, terrorism, revolution, insurrection, military/usurped power, or civil war;
	(c) Employer's design error, omission, or defect, which may be contained in the (Red Book) Specification and Drawings/ (Yellow/Emerald Book) Employer's Requirements, which an experienced Contractor exercising due care would not have discovered before submitting the Tender);	(c) riot, commotion, disorder by persons other than the Contractor's Personnel and other employees of the Contractor and Subcontractors;
	(d) Unforeseeable operation of the forces of nature or against which an experienced Contractor could not reasonably have taken adequate preventive precautions;	(d) strike or lockout not solely involving the Contractor's Personnel and other employees of the Contractor and Subcontractors;
	(e) any Exceptional Event Sub-Clause 18.1 (a)–(f) event sright column here →	(e) encountering munitions of war, explosive materials, ionising radiation, or contamination by radio-activity, except as may be attributable to the Contractor's use of such munitions, explosives, radiation, or radio-activity; or
	(f) any act or default of the Employer's Personnel or Employer's other contractors.	(f) natural catastrophes such as earthquake, tsunami, volcanic activity, hurricane, or typhoon.

(*Continued*)

	Employer's Risks Sub-Clause 17.2	*Exceptional Events Clause 18*
Notice	promptly Notice of Claim under Sub-Clause 20.2 in case of shared liability within 28 days	14 days after awareness under Sub-Clause 18.2 description every 28 days and immediately after the effect ceases under Sub-Clause 18.3 **20.2** Notice of Claim within 28 days under Sub-Clause 18.4
Contractor's entitlement	EOT and/or compensation for the instructed rectification under SC 13.3.1 (Variation by Instruction) proportion of EOT and/or Cost Plus Profit for the instructed rectification in case of shared liability	subject to SC 20.2, EOT, mitigation expected subject to SC 20.2, Cost by reason of man-made events in the Country or war anywhere

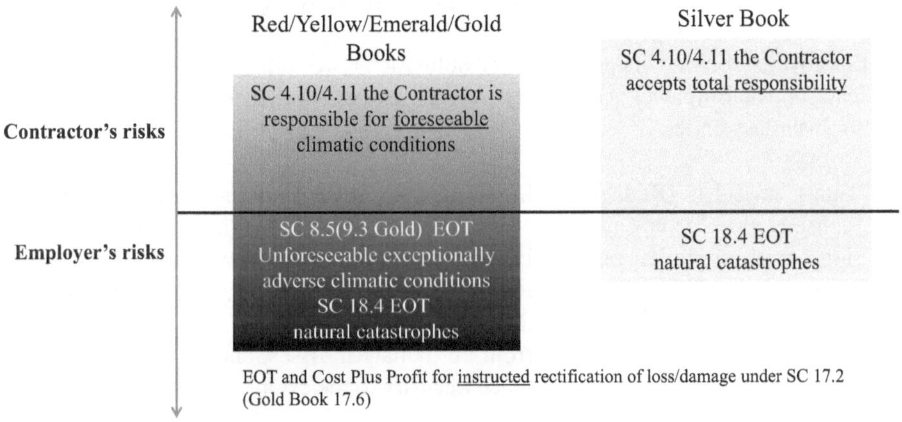

Figure 13.9 Climate risk allocation

Figure 13.9 presents the climate risk sharing between the Parties. Following the logic of Sub-Clause 4.11 for sufficiency of the ACA (Silver Book: Contract Price), the Contractor bears the risk for foreseeable climatic conditions, having based its proposal on the Site data and its interpretation under Sub-Clause 4.10. The Contractor is entitled to an EOT under Sub-Clause 8.5 (Red/Emerald/Yellow)/9.3 (Gold Book) for Unforeseeable exceptionally adverse climatic conditions. The Silver Book Contractor accepts total responsibility for climatic conditions, except for natural catastrophes, where the entitlement is the same as under the Red/Emerald/Yellow/Gold Books. Under Sub-Clause 18.4, the Contractor is entitled to an EOT but no prolongation costs in the case of exceptional climate risks like hurricanes for the period of prevention. The Contractor is entitled to an EOT and Cost Plus Profit for the instructed rectification of the losses/damages to the Works, Goods, and Contractor's Documents resulting from Unforeseeable forces of nature under Sub-Clause 17.2 (Gold Book: 17.6).

13.8.10 Indemnities

An agreement to indemnify means an undertaking to compensate for loss or damage suffered.[26] The FIDIC forms contain significant indemnities by each of the Parties in favour of the other Party. The scope and effect of these indemnities will depend on their treatment under the particular governing law, as well as the particular wording that is used.[27] The FIDIC 2017 Contracts Guide notes that in some legal jurisdictions (including English law), the words 'indemnify' or 'indemnity' as used in the 2017 Books have a specific meaning, entitling the indemnified Party to recover certain losses that may not otherwise be recoverable by law. For example, a Party's indemnity may be construed to permit the other Party to recover losses or damages that might otherwise be considered as indirect or consequential. As such, before entering the Contract, it is recommended that the Parties work with their professional advisers to review this terminology for compliance and consistency with accepted practice under the governing law of the Contract.[28] The purpose of allowing the recovery of loss of profit, etc., under indemnities for personal injury and property damage of third parties is to enable the Party seeking indemnification to recover the claim by the third party in full (including legal fees and expenses) to the extent that the third party's claim includes loss of profit, etc., and to avoid any argument that the recovery of that third party's loss of profit is excluded.[29]

The Employer's specific indemnification obligations are not limited by Sub-Clause 14.14 [Cessation of Employer's Liability]. The Employer shall indemnify and hold the Contractor harmless under:

- Sub-Clause 1.13 (Red/Yellow/Emerald)/1.12 (Silver Book) against and from the consequences of any delay or failure to obtain the planning/zoning or building permit or similar permits/licences/approvals for the Permanent Works, and other permits/licences/approvals described as Employer's responsibility in the Specification/Employer's Requirements;
- Sub-Clause 4.2.2 against and from all damages, losses, and expenses resulting from a claim under the Performance Security to the extent that the Employer was not entitled to make the claim;
- Sub-Clause 17.3 against and from any claim alleging an infringement of intellectual and industrial property rights that is:
    - an unavoidable result of the Contractor's compliance with the Employer's Requirements and/or any Variation; or
    - a result of any Works being used by the Employer for other purposes or if not disclosed before, the Base Date/stated in the Contract in conjunction with anything not supplied by the Contractor;
- Sub-Clause 17.5 against and from third-party claims, damages, losses, and expenses in respect of:

26 Nael and Lydia Bunni, *Risk and Insurance in Construction* (Routledge 2022) 6.1.1

27 Ellis Baker and others, *FIDIC Contracts: Law and Practice* (Informa Law from Routledge 2009) para. 8.136

28 FIDIC, *FIDIC 2017 Contracts Guide* (FIDIC 2022) 105

29 Ellis Baker and others, *FIDIC Contracts: Law and Practice* (Informa Law from Routledge 2009) para. 7.83

  - bodily injury, or loss of or damage to any property other than the Works, attributable to any negligence, wilful act, or breach of the Contract by the Employer; and
  - damage to or loss of any property, real or personal (other than the Works), to the extent that such damage or loss arises out of any Employer's risk event under Sub-Clause 17.2;
- under Sub-Clause 19.1 against all direct losses and claims arising from Employer's failure to comply with any condition of the insurances effected under the Contract.

Additionally, under Sub-Clause 5.2.2 (Red Book)/4.5.1 (Yellow/Silver/Emerald Book), the Employer may agree to indemnify the Contractor against and from the consequences of instructed employment of a nominated Subcontractor, against whom the Contractor has raised reasonable objections.

The Contractor shall indemnify and hold the Employer harmless under:

- Sub-Clause 1.13 (Red/Yellow/Emerald)/1.12 (Silver Book) against and from the consequences of any failure to give all notices, pay all taxes, duties, and fees, and obtain all other permits, permissions, licences, and/or approvals, as required by the Laws in relation to the execution of the Works unless the failure is caused by the Employer's failure to comply with Sub-Clause 2.2 [Assistance];
- Sub-Clause 4.14 against and from all damages, losses, and expenses (including legal fees and expenses) resulting from unnecessary or improper interference with the convenience of the public or with access to or use and occupation of roads and footpaths;
- Sub-Clause 4.16 against and from all damages, losses, and expenses resulting from the import, transport, and handling of all Goods;
- Sub-Clause 17.3 against and from any claim alleging an infringement of intellectual and industrial property rights that arises out of or in relation to the Contractor's execution of the Works or the use of Contractor's Equipment;
- Sub-Clause 17.4:
  - against and from third-party claims, damages, losses, and expenses in respect of:
    - bodily injury due to the Contractor's execution of the Works, unless attributable to any negligence, wilful act, or breach of the Contract by the Employer; and
    - damage to or loss of any property, real or personal (other than the Works), to the extent that such damage or loss arises out of or in the course of or by reason of the Contractor's execution of the Works, and is attributable to any negligence, wilful act, or breach of the Contract by the Contractor;
  - against all negligent acts, errors, or omissions by the Contractor in carrying out the Contractor's design obligations that result in the Works (or Section or Part or major item of Plant, if any), when completed, not being fit for the purpose(s) for which they are intended under Sub-Clause 4.1;
- Sub-Clause 19.1 in respect of Contractor's failure to comply with any condition of the insurances effected under the Contract.

Sub-Clause 17.6 sets out additional rules for risk sharing between the Parties:

- the Contractor's liability under Sub-Clause 17.4 may be reduced proportionately to the extent that Employer's risk events listed in Sub-Clause 17.2 may have contributed to the damage, loss, or injury;
- the Employer's liability under Sub-Clause 17.5 may be reduced proportionately to the extent that Contractor's risk event may have contributed to the damage, loss, or injury.

The Employer and the Contractor shall jointly and severally indemnify and hold the DAAB Member harmless against and from all damages, losses, and expenses resulting from any claim from which he/she is relieved from liability under Sub-Clause 8.1 of the DAAB Agreement.

13.8.11 Limitation of liability

Sub-Clause 1.15 (Silver Book: 1.14) [Limitation of Liability] enables reasonable risk/reward allocation. It contains exclusions to the liability of both Parties for certain types of loss and limitations on the liability of the Contractor, subject to certain exceptions, presented in Figure 13.10.

It is not in the Party's/project's best interest if a Party is forced into an unacceptable financial situation by carrying unlimited liability. It is very difficult for a tenderer to assess its risk exposure if it carries unlimited financial liability for the consequences of

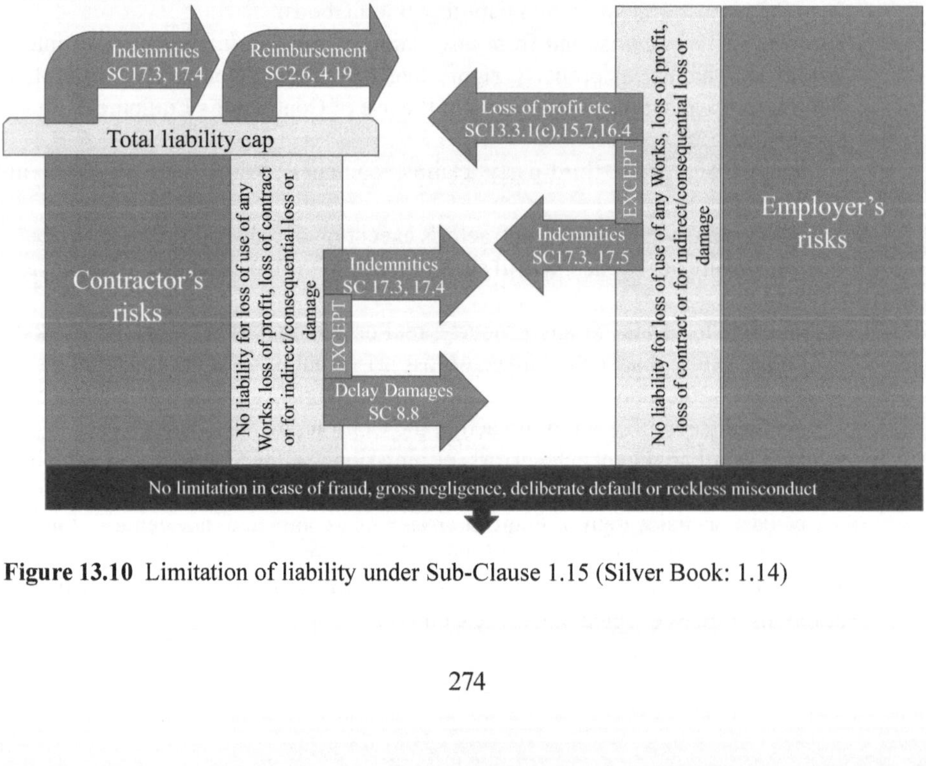

Figure 13.10 Limitation of liability under Sub-Clause 1.15 (Silver Book: 1.14)

certain risk events. Additionally, the Employer might wrongly believe that it is covered for this unlimited risk exposure.[30]

Pursuant to the first paragraph of this Sub-Clause, neither Party shall be liable to the other for:

(i) loss of use of any Works;
(ii) loss of profit;
(iii) loss of any contract; or
(iv) any indirect or consequential loss or damage that may be suffered by the other Party in connection with the Contract.

The exceptions to these general limitation of liability are the following:

- Delay Damages under Sub-Clause 8.8, which are pre-determined liquidated damages that may have been calculated as loss of use of the Works and loss of profit;
- the Contractor may be entitled to any loss of profit or other loss and damage suffered as a result of:
  - omissions instructed under Sub-Clause 13.3.1 (c) where the Parties have agreed the omitted works to be carried out by others;
  - termination under Sub-Clause 15.5 or 16.2;
- indemnities under Sub-Clause 17.3 for intellectual/industrial property rights, for third-party claims for bodily injury under Sub-Clauses 17.4/17.5, and damage to third-party property under Sub-Clause 17.5 arising out of Employer's risk events.

The second paragraph of Sub-Clause 1.15 (Silver Book: 1.14) limits the total liability of the Contractor to the Employer with the following exceptions:

- payments for Employer-supplied Materials, Employer's Equipment, and temporary utilities provided by the Employer under Sub-Clauses 2.6 and 4.19;
- indemnities by the Contractor for third-party claims under Sub-Clauses 17.3 and 17.4.

If the Contract Data does not include a maximum total figure for Contractor liability, the ACA (Silver Book: Contract Price stated in the Contract Agreement) is the default limit.

Sub-Clause 1.15 (Silver Book: 1.14) shall not limit liability in any case of fraud, gross negligence, deliberate default, or reckless misconduct by the defaulting Party.

The impact of the governing law shall be considered as well.

13.8.12 Risk transfer to insurers

The risk transfer to insurers under Clause 19 does not change either Party's obligations or responsibilities under the Contract. The Contractor shall effect and maintain the following insurances with insurers and in terms subject to consent by the Employer:

30 *The FIDIC 2017 Contracts Guide* (FIDIC 2022) 104

- for the **Works and Contractor's Documents, together with Materials and Plant** for incorporation in the Works, in the joint names of the Contractor and the Employer **from the Commencement Date until the date of the issue of the TOC for the Works**;
- for the **Goods** in the joint names of the Contractor and the Employer **from delivery to the Site until they are no longer required for the Works**;
- to the extent that the Contractor is responsible for the design, **PI insurance** against liability arising out of any act, error, or omission by the Contractor in carrying out the Contractor's design obligations **for the period specified in the Contract Data** (if stated in the Contract Data, it may cover the fitness for purpose liability, which may not be possible in some jurisdictions);
- for **injury to persons and damage to property** (other than the Works) arising out of the performance of the Contract and occurring before the issue of the Performance Certificate, other than loss or damage caused by an Exceptional Event; in the joint names of the Contractor and the Employer, to be effected **before the Contractor begins any work on the Site and maintained until the issue of the Performance Certificate;**
- for **injury of employees** of the Contractor, the Employer, and the Engineer (but may exclude liability for acts and negligence of the Employer's Personnel) **during the time they assist in the execution of the Works**;
- other **local mandatory insurances** according to the details in the Contract Data.

Completing the Contract Data, the Employer should be advised by a professional with extensive experience in construction insurance and liability. If the Contractor fails to effect and keep in force any of the listed insurances, the Employer may step in and pay any premium as may be necessary and then recover the paid amounts from the Contractor.

If either the Contractor or the Employer fails to comply with any condition of the insurances effected under the Contract (e.g. timely notification), the Party in default shall indemnify the other Party against all direct losses and claims arising from such failure.

The insurances to be maintained under Sub-Clause 19.2.1 for the Works and under Sub-Clause 19.2.2 for the Goods must be in the joint names of the Parties and cover the cost of loss/damage rectification to the extent that the risks are insurable and except the deductible amounts. If the Contractor fails to comply with the insurance obligations, the Employer may reduce the compensation due under Sub-Clause 17.2 for the instructed rectification with the amount to be covered by the insurance.

The insurance for the Works under Sub-Clause 19.2.1 covers:

- the full replacement value of the Works and Contractor's Documents, together with Materials and Plant for incorporation in the Works;
- loss or damage of any part of the Works as a consequence of failure of elements defectively designed or constructed with defective material or workmanship; and
- an additional amount of 15 per cent of the replacement value (unless other amount is specified in the Contract Data) for additional costs, including professional fees, cost of demolition, and removal of debris.

The insurance cover under Sub-Clause 19.2.1 may exclude uninsurable risks like:

- the cost of making good any defective part of the Works (only the losses/damages caused by its failure to **other** parts of the Works are covered);
- indirect or consequential loss or damage, including reductions in the Contract Price for delay;
- wear and tear, shortages, and pilferages; and
- risks arising from Exceptional Events, unless otherwise stated in the Contract Data.

Before filling in the requirements for insurance in the Contract Data, the drafters of the tender documents should take specialised advice for the insurance cover available on the local market.

13.9 Innovative Risk Management using the FIDIC Suite for Works

13.9.1 Risks related to digitisation

A risk assessment for delivery of asset and project information should be carried out to help understand, communicate, and manage information delivery risks. Its outcome should be taken into account when preparing any information protocol.[31] The Risk Register shall integrate the information management risks, as recommended in ISO 19650-2. Digitalisation requires collaborative risk mitigation in case of gaps in the risk allocation.[32]

13.9.2 Enhancing project resilience to emerging risks

Resilience should be seen as the ability to deal with adversity, withstand shocks, and continuously adapt and accelerate as disruptions and crises arise over time.[33] Emergent risks can be tackled through developing project resilience.[34] This requires the project to have:

- the right level of budget and schedule contingency for emergent risks, in addition to a specific risk budget for known risks (Chapter 9 gives recommendations for allocating Sub-Clause 13.4 Provisional Sums to specific probable risks and separately to unknown risks);
- flexible PM, including strong change management (enabled by the Variation and Claim procedures);
- empowered project team that has clear objectives to get the job done within agreed-upon limits (Chapter 6 comments on the procedures for the authorisation

31 UK BIM Alliance, Centre for Digital Built Britain and BSI, *Information Management According to BS EN ISO 19650. Guidance Part 1: Concepts* (second edition 2019) <https://www.ukbimalliance.org/wp-content/uploads/2019/04/Information-Management-according-to-BS-EN-ISO-19650_-Guidance-Part-1_Concepts_2ndEdition.pdf> accessed 10 April 2022

32 May Winfield, 'Digital and Transformational Change in Construction Projects: The (Liability) Bumps in the Road' (pbctoday 1 March 2023) <https://www.pbctoday.co.uk/news/digital-construction/bim-news/digital-and-transformational-change-construction-projects-liability-bumps-road/121918/> accessed 14 March 2023

33 Børge Brende and Bob Sternfels, 'Resilience for Sustainable, Inclusive Growth' (McKinsey & Company 2022) <https://www.mckinsey.com/capabilities/risk-and-resilience/our-insights/resilience-for-sustainable-inclusive-growth> accessed 6 November 2022

34 PMI Inc., *PMBOK® Guide* (PMI sixth edition 2017) 399

of the CA and Contractor's Representative and further delegation of power and authority; Chapter 8 commends the integration of the Last Planner® System [LPS] enabling agile schedule management);

- resilient supply chain with digital management, using regional sources, planning alternative suppliers, and increasing inventory (discussed in Chapter 11);
- frequent review of early warning signs to identify emerging risks as early as possible (besides the numerous provisions for giving Notices in case of emerging risks, Sub-Clause 8.4 obliges the CA and the Parties to advise each other in advance of any known or probable risk that may impact the project);
- clear input from stakeholders to clarify areas where the project scope/budget/strategy can be adjusted in response to emergent risks (pursuant to Sub-Clause 3.8/3.6, the CA/Contractor's Representative may call a management meeting and invite the other stakeholders to discuss the risk response).

13.9.3 Collaborative risk management

For any team to manage risk effectively, it needs to establish the time and processes necessary to undertake joint risk management after contractor selection and prior to starting on Site.[35] FIDIC is drafting the Collaborative Contract form, but Parties may use the collaborative provisions in the existing books, with some additional tools.

It is always better for the project the CA and the Parties to manage the risks together using:

- BIM (Advisory Notes for FIDIC Users available);
- 4D simulations for risk identification and selection of appropriate risk response;
- advance warning under Sub-Clause 8.4;
- management meetings under Sub-Clause 3.8/3.6 (1.16 Emerald Book Contract Risk Management meetings);
- co-operation under Sub-Clause 4.6;
- VE sharing the benefits under Sub-Clause 13.2;
- Variations and Provisional Sums/Daywork Schedules under Clause 13;
- Schedules of Payment under Sub-Clause 14.4 with pain/gain share adjustments combined with a Maximum Guaranteed Price under Sub-Clause 14.1;
- dispute avoidance procedure applied by the DAAB upon agreement of both Parties under Sub-Clause 21.3;
- agreed liquidated damages as compensation for emerging risks (e.g. Prolongation Costs under Green Book);
- agreed quantifiable risk allocation linked with Site data (e.g. risk allocation under the Emerald Book, discussed in Chapters 5 and 9);
- Risk Registers and Risk Management Plans (the Emerald Book Sub-Clause 1.5 integrates them in the documents forming part of the Contract; Sub-Clause 1.16 may be plugged in the other books);
- sensors combined with AI for instant warning, automated reporting, and establishing the risk impact.

35 David Mosey, *Collaborative Construction Procurement and Improved Value* (John Wiley and Sons Ltd 2019) 348

13.9.4 The innovative approach of the Green Book

The new Green Book tabulates the Employer's Risks and the linked Contractor's entitlements under the Contract, spread in numerous interacting Clauses under the FIDIC books for major works. The related guidance demystifies the risk allocation concepts integrated in the FIDIC Suite for works. The user-friendly approach has the following benefits, leading to effective PM:

- one-page table under Sub-Clause 11.1 catches all Employer's Risks and clearly states when the Contractor is entitled to:
  - EOT only (for shared risks resulting in excusable delay, but no compensation);
  - Cost and/or EOT (only the loss of profit is shared by the Contractor);
  - Cost Plus Profit and/or EOT (risks are fully allocated to the Employer);
- the formula for calculation of the Prolongation Costs (**PGC**) for compensable EOT (discussed in subsection 9.2.5.14) is agreed under the Contract as liquidated damages, saving time for substantiation of incurred Costs and preventing disagreements leading to disputes;
- besides the Delay Damages and PGC, liquidated damages are agreed to facilitate the Parties' agreement in the case of termination under Sub-Clause 10.4.1;
- Sub-Clause 11.2 stipulates that the Employer may have entitlements/relief flowing from the Contractor's risks in the following cases:
  - Cost reduction resulting from changes in the local Laws after the Base Date;
  - costs incurred due to the Contractor's failure to fulfil any obligation;
  - any cause giving an entitlement under or in connection with the Contract;
- the Guidance clearly explains the risk allocation and the method of calculation of PGC and outlines the importance of the governing law, which may change the tabulated risk allocation, especially in case of concurrent risk events;
- Sub-Clause 11.3 summarises the indemnities related to third-party claims related to personal injury, damage to property other than the Works, and infringement of intellectual/industrial property rights;
- a table is provided in Contract Data Sub-Clause 12.1, summarising the insurance requirements;
- the insuring Party provides an insurance certificate endorsed by the insurer, evidencing compliance with the Contract requirements.

FIDIC users may consider integrating some of these innovative features in the Particular Conditions of the contracts for major works.

13.10 Example Notices

Appendix 1 includes examples of Contractor's Notices of an Exceptional Event under Sub-Clause 18.2 and its impact under Sub-Clause 17.2 of the Yellow Book.

CHAPTER 14

Project Stakeholder Management with FIDIC Contracts

14.1 Project Stakeholder Management processes

According to the PMBOK® Guide, Project Stakeholder Management includes the processes required to identify the people, groups, or organisations that could impact or be impacted by the project, to analyse their expectations and impact on the project, and to develop appropriate management strategies for effectively engaging stakeholders in project decisions and execution.[1]

Every project has stakeholders who are impacted by or can impact the project in a positive or negative way. Some stakeholders may have a limited ability to influence the project's work or outcomes; others may have significant influence on the project and its expected outcomes. The key to effective stakeholder engagement is a focus on continuous communication with all stakeholders, including team members, to understand their needs and expectations, address issues as they occur, manage conflicting interests, and foster appropriate stakeholder engagement in project decisions and activities. Academic research and analyses of high-profile project disasters highlight the importance of a structured approach to the identification, prioritisation, and engagement of all stakeholders.

14.2 Project Stakeholder Register

The identification of the stakeholders should start with project initiation and continue through the project's life. A Stakeholder Register may be included in the tender documents for the selection of consultants and contractors, with data about the important stakeholders like:

- project beneficiary/operator;
- authorities giving consents/issuing permits;

1 PMI Inc., *PMBOK® Guide* (PMI sixth edition 2017) 503

DOI: 10.4324/9781003272892-14

- utility operators;
- professional chambers registering/licensing designers, supervisors, contractors;
- suppliers of specific Goods, if limited in number;
- neighbours, non-governmental organisations (NGOs), etc. with significant influence on the project (e.g. in relation to right of way or environmental impact decision), etc.

The Stakeholder Register may include the following data:

- identification and contact details;
- major requirements, expectation, potential impact;
- stakeholder classification (internal/external, impact/influence/power/interest, etc.).

The Stakeholder Register is used with the Road Map recommended in Chapter 5 for efficient permit management. It may be updated with the Risk Register and/or Issue Log, discussed in Chapters 6 and 7.

14.3 Project Stakeholder Management with FIDIC Contracts

Figure 14.1 presents the usual interaction between project stakeholders in a FIDIC Red Book Contract.

The primary stakeholders in a FIDIC Red Book Contract are the Parties and the Engineer appointed by the Employer. The designer is also appointed by the Employer. The White Book may be used for both agreements for services. The Contractor has remedies for delayed design and Engineer's instructions under Sub-Clause 1.9. Besides Clause 2 [Employer], Clause 3 [Engineer], and Clause 4 [Contractor], many other provisions regulate the interaction of the project participants.

The Contractor manages Subcontractors and suppliers. Sub-Clauses 4.4 [Subcontractors] and 4.5 [Nominated Subcontractors]/ Red Book Clause 5 [Subcontracting] set out the

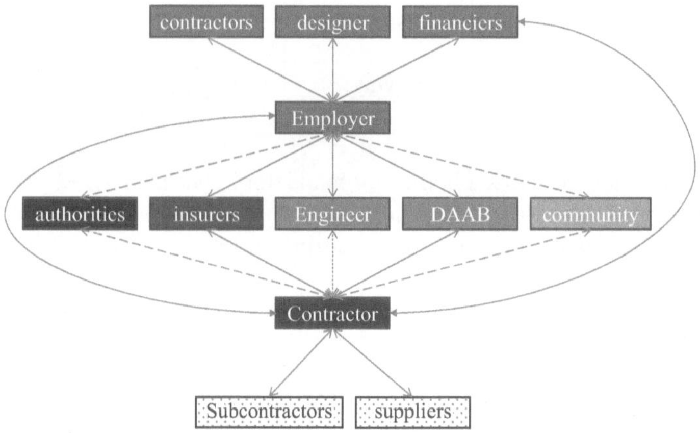

Figure 14.1 Project Stakeholder Management FIDIC Red Book

Contractor's obligations related to the management of Subcontractors. The FIDIC Subcontracts (2011 for Construction, 2019 for Plant and Design-Build, and others in the pipeline) further regulate the management of Subcontractors. For example, pursuant to Sub-Clause 2.3 of the 2019 Subcontract for Plant and Design-Build, the Subcontractor has no obligation to comply with any instruction issued by the Employer/Engineer unless and until it has been confirmed in writing as a Contractor's instruction.

Sub-Clauses 21.1–21.4 include the provisions for the constitution and operation of the DAAB. Each DAAB member signs a DAAB Agreement with both Parties.

Sub-Clause 4.6 [Co-operation] sets out the Contractor's obligation for co-operation with the Employer's Personnel, other Employer's contractors, representatives of authorities, and utility operators. Sub-Clause 2.3 [Employer's Personnel and Other Contractors] obliges the Employer to ensure reciprocal co-operation of the Employer's Personnel and other contractors with the Contractor, supporting its efforts under Sub-Clause 4.6.

Sub-Clause 3.8 (Silver Book: 3.6) [Meetings] enables the Engineer or the Contractor's Representative to call a management meeting and invite other stakeholders such as the Employer's other contractors, representatives of public authorities, utility companies, and Subcontractors.

14.3.1 Financial institutions

The Employer secures financing and is obliged under Sub-Clause 2.4 to provide reasonable evidence that financial arrangements have been made and are being maintained. The requirements of the financial institutions have to be integrated in the tender documents, as discussed in Section 9.1.[2] The Contractor usually relies on commercial banks to obtain a Performance Security under Sub-Clause 4.2/bank guarantees under Sub-Clauses 14.2.1/14.5, and credit lines to leverage its cash flow.

14.3.2 Public authorities, authorised third parties, and private utility companies

Both Parties have interaction with authorities for permit managements. Sub-Clause 1.13 (Silver Book: 1.12) divides between the Parties the responsibility and liability in relation to compliance with the local Laws. The Employer must secure the successful outcome of the permitting process, managing many stakeholders, such as:

- land owners to secure ownership or right of way;
- environmental authority;
- municipalities;
- fire authority and/or other specialised bodies for mandatory consents;
- building control authority/third-party technical control;
- utility companies for relocation of/connection to utilities, etc.

For example, an Air Traffic Service Authority requests the Employer in a project for a skyscraper to reduce its height by 100 meters due to interference into the flight route. The Employer invites top experts to find a technical solution, and they successfully solve the problem.

2 See also *Yellow Book* (FIDIC 2022) GPPC 46

In extreme cases, poor stakeholder management may result in project failure, where the Employer has to terminate the Contract under Sub-Clause 15.5 [Termination for Employer's Convenience] or the Parties are discharged from further performance under Sub-Clause 18.6 [Release from Performance under the Law].

Under Sub-Clause 1.13 (Silver Book: 1.12):

- the Employer interacts with the authorities to obtain the planning, zoning, or building permit or similar permits, permissions, licences, and/or approval for the Permanent Works and others described in the Specification/Employer's Requirements;
- the Contractor interacts with the authorities to obtain all permits/approvals in relation to the execution of the Works (those may be related to the Traffic Plan, Health and Safety Plan, Waste Management Plan, etc.).

Sub-Clause 1.13 (Silver Book: 1.12) (c) sets out the Contractor's obligation to assist the Employer as stated under the Contract or reasonably required by the Employer with the permit management within the Employer's scope. Sub-Clause 2.2 enables the Contractor to request from the Employer reasonable assistance for permit management within the Contractor's scope.

Under Clause 5 of the Yellow/Silver/Emerald Books, where the Contractor is responsible for the design, the Contractor usually interacts with the authorities and utility companies for design consent/approval. The Contractor undertakes that the design, the Contractor's Documents, the execution of the Works, and the completed Works will be in accordance with the Laws of the Country (Sub-Clause 5.3) and the technical standards and other specified standards (Sub-Clause 5.4). The Contractor has to secure the timely involvement of licensed laboratories/ registered third parties/ technical surveillance authorities for certification of high-risk Plant (like portal cranes and high-pressure vessels).

Under Sub-Clause 4.16, the Contractor is responsible for interacting with the customs authorities for the import of the Goods but may request the Employer's assistance under Sub-Clause 2.2.

Sub-Clause 4.19 allocates the responsibility for the provision of temporary utilities to the Contractor. Unless the Employer provides temporary services to the Contractor, the Contractor manages the relations with the utility companies for temporary connections.

Sub-Clause 4.23 contains special provisions related to archaeological/geological findings. The Contractor has obligations to notify and prevent damages to the artefacts, following the instructions of the CA. The Employer is the Party interacting with the relevant authority. Such findings often delay infrastructure and city centre projects, but the Contractor is entitled to an EOT and/or payment of compensation, subject to Sub-Clause 20.2.

Depending on the Laws of the Country, the authorities may be involved in the commencement of the Works under Sub-Clause 8.1 and the commissioning and taking over under Sub-Clause 10.1.

The interaction with the building control that may be performed by a state/local authority or an authorised third party is discussed in Chapters 5, 6, and 11.

The Road Map and the Stakeholder Register are very helpful for the timely identification of the numerous authorities/utility companies and successful permit management.

14.3.3 Community

The procedures for environmental impact assessment, changes to zoning plans, and construction permits are usually announced to interested stakeholders, and their management may be very difficult and decisive for the project outcome. Sometimes, the stakeholders may have conflicting interests or try to suspend the procedure in order to request benefits from the Employer. Projects like nuclear power plants, waste landfills, and refugee centres usually meet strong opposition from the neighbourhood, and the Employers have to manage the concerned stakeholders efficiently during the project preparation and through all site assessment/permitting stages.

Sub-Clause 4.14 restricts the Contractor's interference with the convenience of the public or the use/occupation of roads and footpaths and includes an indemnity provision for third-party claims resulting from unnecessary or improper interference. Sub-Clause 6.5 forbids construction on locally recognised days of rest, or outside the normal working hours stated in the Contract Data, unless agreed otherwise. But, there may be situations when work is unavoidable for the protection of life and property or for the safety of the Works, and in those cases, the community shall be informed and the protests managed.

The Contractor takes the responsibility for keeping unauthorised persons off the Site under Sub-Clause 4.21.

The Employer interacts with third parties to obtain the right of access to the Site in order to give it to the Contractor under Sub-Clause 2.1. The Contractor may negotiate with third parties to obtain special/temporary rights of way and the use of land for any additional facilities outside the Site for the execution of the Works under Sub-Clause 4.13. Sub-Clause 4.22 obliges the Contractor to take the necessary precautions to keep the Contractor's Personnel and the Contractor's Equipment within the Site and the additional areas obtained, acknowledged by the CA as working areas.

The Contractor shall report under Sub-Clause 4.20 (g) any public relations activities.

14.3.4 Labour unions/chambers

The Contractor is responsible for the arrangements for engagement of Contractor's Personnel under Sub-Clause 6.1. These usually include many local people, who may be united in labour unions. The labour unions may negotiate the minimum wages, compensation, working time, etc. Sub-Clause 6.2 is a good starting point for the relations with the labour unions, since it obliges the Contractor to pay rates of wages and observe conditions of labour complying with the applicable Laws and not lower than those established locally.

The Contractor and/or its designers may have obligations to be registered in professional chambers. The procedures enabling design and construction may be long and difficult for foreign persons. The Contractor has to resolve those issues with early identification of the relevant stakeholders and management of the procedures with local support. Sub-Clause 5.1 of the Yellow/Emerald/Silver Book points out that the designers have to be qualified and entitled to design under the local Laws. When this is not possible for foreign designers, local designers may be appointed either to develop the design or to review it for compliance with the local Laws and authorise it.

14.3.5 Insurance providers

Insurance providers are important stakeholders, and the PM team should be aware of their requirements, because restrictions may impact project execution.[3]

Clause 9 of the White Book, Clause 19 of the Red/Yellow/Emerald/Silver Books, and Clause 12 in the Green Book set out the provisions for insurance, discussed in Chapter 13.

14.4 Innovative Project Stakeholder Management with FIDIC Contracts

BIM secures transparency in a collaborative working environment for the primary stakeholders and funders/beneficiaries. 3D/4D visualisation and short videos from drone footage facilitate the presentation of the project to the DAAB and indirect stakeholders at all stages.

Project websites are an important tool for the continuous information of indirect stakeholders. The information may be filtered to comply with the confidentiality requirements. Forums give opportunities to the community to be heard and their concerns to be addressed in a timely manner.

E-Government streamlines the permitting processes and enables transparency, thus facilitating the establishment of any entitlement to EOT under Sub-Clause 8.6 [Delay Caused by Authorities].[4]

14.5 Example Notice

Appendix 1 includes an example Contractor's Notice under Sub-Clause 6.5 of the Yellow Book.

3 *Construction Extension to the PMBOK® Guide* (PMI 2016) 137

4 The service entity for the built environment, Lupapiste, and dmCity, a service aimed at digitalising the land use life cycle, have developed digital solutions allowing municipalities in Finland to act according to the reform of the *Land Use and Building Act*, including digital permitting <https://cloudpermit.com/media-room/esri-finland-dmcity-partners-with-cloudpermit-finland> accessed 27 February 2023

CHAPTER 15

Project Health, Safety, Security, and Environmental Management with FIDIC Contracts

15.1 Project Health, Safety, Security, and Environmental (HSSE) Management Processes

The PMBOK® Guide does not include a section dedicated to Project HSSE Management, but its Construction Extension has one. It presents HSSE considerations for managing, assuring, and controlling construction projects.

Project HSSE processes include all activities of the Employer/Client and the Contractor/Consultant ensuring the project is executed in a manner preventing accidents and avoiding personal injury, fatalities, or property damage, and protecting the environment. These processes should address both **the management of the project and the product of the Project.** The HSSE processes should be integrated with the risk management processes.

15.2 HSSE Management with FIDIC Contracts

Three distinct sets of HSSE requirements should be complied with:

- mandatory requirements;
- Client/Employer requirements;
- requirements of the performing Consultant/Contractor.

 DOI: 10.4324/9781003272892-15

The first two sets are considered requirements of the Contract. The third set is voluntary, but it often also becomes part of the Contract, to the extent that those commitments are included explicitly or by reference in the Contract (e.g. in the Yellow/Emerald/Gold Book Contractor's Proposal/ Silver Book Tender).

15.2.1 Requirements for compliance with HSSE Laws in the Works Contracts

Chapter 10 lists the provisions setting out general responsibilities for compliance with the Laws and the requirements of the technical standards stated in the Contract. Sub-Clause 4.8 explicitly obliges the Contractor to comply with the applicable health and safety (**HS**) regulations and Laws, and all HS obligations specified in the Contract. Pursuant to Sub-Clause 4.8, the Contractor shall maintain records and make reports as specified in the Specification/Employer's Requirements in compliance with the applicable health and safety regulations and Laws. Sub-Clause 4.21 allocates to the Contractor the responsibility for the security of the Site. Sub-Clause 6.4 requires compliance with all labour Laws, including those related to HS and welfare. Sub-Clause 4.18 allocates to the Contractor the responsibility to ensure that the emissions, surface discharges, effluent, and other pollutants from the Contractor's activities shall not exceed the values indicated in the Specification/Employer's Requirements and the applicable Laws. The HSSE provisions in the Red/Yellow/Emerald/Silver Books are further discussed in the following.

A UK study compared HS contractual provisions in four standard forms with key features of the UK *Construction (Design and Management) Regulations 2007* (***CDM***).[1] It concluded that though FIDIC contracts are unlikely to fully embrace secondary legislation of a particular country, they are carefully thought out and suitable for construction works in the UK.[2] For example, the *CDM* requirements have been successfully integrated in FIDIC contracts for the construction of the Hinkley Point C Nuclear Power Station in England.[3]

15.2.2 Case studies – additional requirements of international financial institutions (IFIs)

Financing institutions often develop specific environmental, social, health, and safety (**ESHS**) requirements.

The *Agence Française de Développement* (**AFD**) has included in its Standard Procurement Documents[4] specific ESHS criteria for selection of contractors for FIDIC contracts. AFD requires the Standard Bidding Documents for Works to be adjusted considering the available

1 <https://www.legislation.gov.uk/uksi/2007/320/contents/made> accessed on 9 December 2022

2 Dingayo Mzyece and others, 'Contractual Provisions for Health and Safety: Standard Form Contracts in the UK Construction Industry' (CIB W099 International Conference on Modelling and Building Health and Safety 2012) 428–437 <https://www.irbnet.de/daten/iconda/CIB_DC25838.pdf> accessed 9 December 2022

3 Reported by EDF Senior Legal Counsel Paul Merrett on the FIDIC International Contract User's Conference 2022

4 AFD, *Standard Bidding Documents for Procurement of Works* (AFD 2019) <https://www.afd.fr> accessed 17 August 2022

ESHS documents (ESHS impact study, environmental and social management plan (**ESMP**), Resettlement Action Plan (**RAP**), Environmental and Social Commitment Plan (**ESCP**), etc.) and the regulations applicable to the Employer. The suggested ESHS qualification criteria include ESHS certification, ESHS management in-house policies and procedures acceptable to the Employer, relevant experience, specific ESHS knowledge transfer experience, and available in-house personnel dedicated to ESHS issues. The evaluation criteria include specific experience for the Key Experts. Bidding forms include ESHS Cost Schedules. The Bidder will need to produce a detailed ESHS Methodology in the Technical Bid, explaining how the ESHS requirements will be achieved. Appendix B requires suppliers, contractors, subcontractors, consultants, or sub-consultants to undertake in the Statement of Integrity to:

- comply with and ensure that all their subcontractors or sub-consultants comply with international environmental and labour standards, consistent with applicable law and regulations in the country of implementation of the contract, including the fundamental conventions of the International Labour Organisation (**ILO**) and international environmental treaties;
- implement environmental and social risks mitigation measures when specified in the ESMP provided by the Contracting Authority.

The World Bank has developed Particular Conditions for the FIDIC Red Book. Special Provisions amend the FIDIC GCs, adding additional requirements related to HSSE Management,[5] for example:

- Particular Conditions Part D – Environmental and Social (**ES**) Metrics for Progress Reports and Part E – Sexual Exploitation and Abuse (**SEA**) and/or Sexual Harassment Performance Declaration for Subcontractors;
- Sub-Clause 4.8 is amended adding additional HS obligations for the Contractor;
- Sub-Clause 4.21 is amended extending the Contractor's security obligations, including submission for Review of a Security Management Plan that sets out the security arrangements for the Site.

15.2.3 Requirements of the performing Consultant/Contractor

Most of the consultants/contractors using FIDIC contracts are ISO certified to demonstrate social, HS, and environmental commitments. Usually, the companies develop an integrated management system, complying with ISO 9001[6] for quality management, ISO 14001[7] for environmental management, and ISO 45001[8] for occupational health and safety

5 The World Bank, *Standard Procurement Documents, Request for Bid Works (Without Prequalification)* (2021)

6 *ISO 9001:2015 Quality Management Systems* <https://www.iso.org/iso-9001-quality-management.html> accessed 28 December 2022

7 *ISO 14001:2015 Environmental Management Systems* <https://www.iso.org/standard/60857.html> accessed 28 December 2022

8 *ISO 45001:2018 Occupational Health and Safety Management Systems* <https://www.iso.org/standard/63787.html> accessed 28 December 2022

management, some adding ISO 26000[9] for social responsibility. Digital platforms are available to support integrated management, providing online access to all documents, reviews, audits, and records. Built-in version controls and automated task management functionality should ensure that the right documents are used and the ownership responsibilities are clearly defined.[10]

15.3 Health, safety, and security (HSS) provisions in FIDIC Contracts

15.3.1 HSS provisions in the FIDIC Agreements for Services

Pursuant to Sub-Clause 3.3 of the White Book, the Consultant shall perform the Services exercising reasonable skill, care, and diligence to be expected from an experienced consultant, and the compliance with the HSS requirements shall always be a priority.

According to the FIDIC Code of Ethics, the consulting engineer shall maintain knowledge and skills at levels consistent with development in technology, legislation, and management, and perform services only when competent to perform them.[11] If these principles are followed, the risk of disasters like Grenfell Tower[12] will be minimised.

Depending on the scope of the Services, the Consultant has different HSS responsibilities:

- as a designer or advisor, the Consultant has to secure the HSS of its personnel in the office and during site visits;
- as a designer or technical control, the Consultant has to secure compliance of design with the HSS requirements;
- as a Red/Yellow/Emerald Book FIDIC Engineer, the Consultant shall monitor the Contractor's HSS compliance;
- if appointed as a coordinator for safety and health matters, besides monitoring the Contractor's compliance, the Consultant has additional duties under the local Laws.[13]

Sub-Clause 3.8 of the White Book entitles the Consultant to suspend all or part of the Services and remove its personnel from the Country if in the reasonable opinion of the Consultant, the HSS of those personnel is compromised by an Exceptional Event.

9 *ISO 26000:2010 Guidance on Social Responsibility* <https://www.iso.org/standard/42546.html> accessed 28 December 2022

10 QMS International, 'What Is an Integrated Management System?' (2019) <https://www.qmsuk.com/news/what-is-an-integrated-management-system> accessed 8 December 2022

11 <https://fidic.org/fidic-code-ethics> accessed 8 December 2022

12 Robert Booth, '"Every death was avoidable": Grenfell Tower inquiry closes after 400 days' (*The Guardian* 2022): 'switching from zinc cladding to combustible plastic-filled aluminium panels with "significantly worse" fire performance resulted in 72 deaths' <https://www.theguardian.com/uk-news/2022/nov/10/every-death-was-avoidable-grenfell-tower-inquiry-closes-after-400-days> accessed 28 December 2022

13 For example, the Agreement for Services for FIDIC Engineer and the Particular Conditions amending Sub-Clause 6.7 of the FIDIC 1999 *Red/Yellow Book* GCs in the mandatory contract documents for construction of water projects financed under the EU Operational Programme 'Environment 2014-2020' in Bulgaria assign to the FIDIC Engineer the duties and responsibilities of the HS Coordinator in compliance with the relevant *Ordinance 2 dated 22 March 2004 on minimum requirements for healthy and safe working conditions in construction*, implementing the *EU Directive 92/57/EEC of 24 June 1992* <https://www2.aop.bg/obrazci-i-spisyci/standartizirani-dokumenti/> accessed 28 December 2022

In 2021, FIDIC issued the Guidance Memorandum – COVID-19 Impact to the Consultant's Services under the FIDIC Suite of Contracts & Agreements to help Clients and Consultants understand how the White Book GCs and the Model Joint Venture (Consortium) Agreement will respond to a range of scenarios arising during the pandemic.[14]

15.3.2 HSS provisions in the FIDIC Works Contracts during construction

The Red/Yellow/Silver/Emerald Books impose extensive HSS duties on the Contractor during construction. The Contractor is responsible for:

- the adequacy, stability, and safety of its operations, activities, and methods of construction (Sub-Clause 4.1);
- packing, loading, transporting, unloading, storing, and protecting all Goods and other things required for the Works (Sub-Clause 4.16);
- using properly equipped facilities and non-hazardous Materials, except as otherwise specified in the Contract (Sub-Clause 7.1);
- the care of the Works, Goods, and Contractor's Documents until the issuance of the TOC, retaining responsibility for outstanding work (Sub-Clause 17.1);
- compliance with the HS regulations, Laws, and obligations specified in the Contract (Sub-Clauses 4.8, 6.4);
- the appointment of a HS officer under Sub-Clause 6.7 and compliance with his/her directives (Sub-Clause 4.8);
- the preparation of a project-specific HS Manual within 21 days of the Commencement Date, before commencing any construction on the Site, in addition to any other similar mandatory documents under the local Laws, with a continuing obligation to revise the HS manual as necessary (Sub-Clause 4.8);
- reporting HS statistics in the Sub-Clause 4.20 progress reports and details of any accident (Sub-Clause 4.8);
- maintaining records and making reports in compliance with the local Laws concerning the HS of persons/ damage to property (Sub-Clause 4.8);
- keeping unauthorised persons off the Site, limiting authorised persons to the Contractor's and Employer's Personnel, except as notified (Sub-Clause 4.21);
- the HS of all persons **entitled to be** on the Site and other places where the Works are being executed (Sub-Clause 4.8);
- taking all necessary precautions to maintain the HS of the Contractor's Personnel, ensuring in collaboration with the local health authorities availability of medical services at the Site and at any accommodation of the Contractor's and Employer's Personnel, and suitable arrangements for all welfare and hygiene requirements and for the prevention of epidemics (Sub-Clause 6.7);
- keeping the Site and Works (and other places where the Works are being executed) clear of unnecessary obstruction to avoid danger to persons (Sub-Clause 4.8);

14 https://fidic.org/sites/default/files/FIDIC%20COVID%2019%20Guidance%20Memorandum%20for%20Consultant_Services_270821_2.pdf> accessed 28 December 2022

- providing fencing, lighting, safe access, guarding and watching of the Works until taking over, a continuing obligation for outstanding works or remedying defects during the DNP (Sub-Clause 4.8);
- providing Temporary Works (including roadways, footways, guards, and fences) necessary for the use and protection of the public and of owners and occupiers of adjacent land and property (Sub-Clause 4.8);
- providing safe access, facilities, permissions, and safety equipment to the Employer's Personnel to examine, inspect, measure, and test the Materials, Plant, and workmanship (Sub-Clause 7.3);
- carrying out any remedial work that is urgently required for the safety of the Works, whether because of an accident, unforeseeable event, or otherwise (Sub-Clause 7.6).

The GCs provide the following remedies for non-compliance with HSS requirements:

- removal of Contractor's Personnel employed on the Site or Works, including the Contractor's Representative and Key Personnel, persisting in conduct prejudicial to HS (Sub-Clause 6.9(d));
- suspension of the Works (Sub-Clause 8.9);
- withholding of amounts for unfulfilled HS obligations, e.g. construction of suitable welfare facilities, until compliance is achieved (Sub-Clause 14.6.2);
- Notice to Correct (Sub-Clause 15.1);
- Claim for damages for breach of contract subject to Sub-Clause 20.2;
- termination for non-compliance with a Notice to Correct when the failure constitutes a material breach of the Contractor's HS obligations under the Contract (Sub-Clause 15.2.1 (a)).

Pursuant to Sub-Clause 2.3, the Employer's Personnel and other contractors have to comply with the same HS obligations as the Contractor under Sub-Clause 4.8 related to:

- compliance with the HS regulations, Laws, and obligations specified in the Contract, and the directives issued by the Contractor's HS officer;
- the HS of all persons **entitled to be** on the Site and other places where the Works are being executed;
- keeping the Site and Works (and other places where the Works are being executed) clear of unnecessary obstruction to avoid danger to persons entitled to be there.

As discussed in Chapter 13, Sub-Clauses 17.4–17.6 stipulate obligations for both Parties to indemnify and hold harmless each other against and from third-party claims in respect of bodily injury and damage to property, often attributable to negligence or non-compliance with the HSSE obligations.

'Safety first' is an important consideration, leading to some exceptions in the FIDIC procedures:

- work necessary for the protection of life or property or for the safety of the Works may be carried out on the Site during official holidays or outside the normal working hours stated in the Contract Data (Sub-Clause 6.5);
- the Contractor may object to the execution of a Variation instructed under Sub-Clause 13.3.1, giving promptly a Notice that it will adversely affect the Contractor's ability to comply with its HS obligations under Sub-Clause 4.8 (Sub-Clause 13.1);
- when the CA instructs suspension, the Contractor shall perform work necessary to protect and secure the suspended Works against deterioration, loss, and damage (Sub-Clause 8.9);
- after termination, the Contractor shall leave the Site, but before that shall comply immediately with notified instructions for the protection of life or property or for the safety of the Works (Sub-Clauses 15.2.3(a)(ii), 16.3(a));
- the GPPC suggests an amendment to Sub-Clause 3.5 (Silver Book: 3.4) where the applicable law prevents the Contractor from complying with an instruction that may adversely affect the HS of the Contractor's Personnel.[15]

15.3.3 HSS provisions in the FIDIC Works Contracts for the completed Works

Pursuant to Sub-Clause 4.1 of the Yellow/Emerald/Silver Books, the Contractor is responsible for the fitness of the Works for the intended purposes as defined in the Employer's Requirements (or for their ordinary purposes, if a definition/description is missing). The Contractor is responsible for executing **any work necessary for stability or safe and proper operation of the Works,** no matter whether it is expressly stated or implied by the Contract. These requirements are not included in the Red Book, where the design risk is allocated to the Employer. However, the Red Book Sub-Clause 10.1, as with the other books, allows taking over of the Works with minor outstanding work/defects only if they will **not affect substantially** the safe use of the Works or Section for the intended purposes until their completion/remedy during the DNP.

The Contractor has an obligation to notify the CA if an instruction under Sub-Clause 3.5 (Silver Book: 3.4) does not comply with applicable Laws or will reduce the safety of the Works. The Contractor may object to the execution of a Variation instructed under Sub-Clause 13.3.1, giving promptly a Notice that it may adversely affect the Contractor's obligation under Sub-Clause 4.1 to complete the Works so that they shall be fit for the intended purposes (Sub-Clause 13.1).

As discussed in Chapter 10, mandatory requirements from the local/international regulations may differ from those in the Specification/Employer's Requirements, and the higher (more conservative) requirements become the law of the contract. The *EU Regulation No 305/2011*[16] includes requirements for construction products, mandatory for the completed Works in EU countries.

15 FIDIC, *Yellow Book* (2022) GPPC 21

16 *Regulation (EU) No 305/2011 of the European Parliament and of the Council of 9 March 2011 laying down harmonised conditions for the marketing of construction products* <https://eur-lex.europa.eu/legal-content/EN/TXT/?uri=CELEX:32011R0305> accessed 14 September 2022

> Construction works as a whole and in their separate parts must be fit for their intended use taking into account in particular the health and safety of persons involved throughout the life cycle of the works. Subject to normal maintenance, construction works must satisfy these basic requirements for construction works for an economically reasonable working life.

All of the following seven requirements listed in Annex 1 of *Regulation No 305/2011* are related to HSSE (the first five concern HS, and the third and the last two concern sustainability):

1. Mechanical resistance and stability
2. Safety in case of fire
3. Hygiene, health, and the environment
4. Safety and accessibility in use
5. Protection against noise
6. Energy economy and heat retention
7. Sustainable use of natural resources

Interpreting a fitness for purpose clause in a contract loosely based on FIDIC, the UK Supreme Court held recently that where there are apparently inconsistent provisions or standards, rather than concluding they are inconsistent, the proper interpretation is that the higher standard must prevail, and the less rigorous standard will be treated as a minimum requirement. In this case, the Contractor designed and installed foundations for offshore wind farms, certified as compliant with the specified standard J101. Several months later, an error was found in J101, putting at risk the stability and safety of the Works. The Contractor was held responsible where the Works needed to be designed to any additional or more rigorous requirements or parameters in order to secure the intended fitness for purpose for 20 years' design life without major retrofits or refurbishments.[17]

15.4 Environmental management provisions in FIDIC Contracts

15.4.1 FIDIC and sustainability

The construction industry has a huge impact on the sustainability agenda. Aside from the potential for building over wild habitats, construction is one of the most resource-consuming sectors. Its transformation from the big polluter and resource consumer towards sustainability enabler is happening, but rather slowly.[18] The UN Sustainable Development Goals (SDGs),[19] the Paris Agreement[20] on climate change, and the European Green Deal[21] committing to zero net emissions of greenhouse gases in 2050 reinforce the status of sustainable development as a public- and private-sector imperative. Several of the

17 *MT Hojgaard A/S v E.On Climate and Renewables UK Robin Rigg East Ltd* [2017] UKSC 59

18 Adriana Spassova, 'Construction Industry and Sustainability' Chapter 15 in Sara Valaguzza and Mark Alan Hughes (eds), *Interdisciplinary Approaches to Climate Change for Sustainable Growth* (Springer 2022) 284

19 UN Department of Economic and Social Affairs, *Transforming Our World: the 2030 Agenda for Sustainable Development* (UN 2015) <https://sdgs.un.org/publications/transforming-our-world-2030-agenda-sustainable-development-17981> accessed 17 August 2022

20 UN, *Paris Agreement* (UN 2015) <https://unfccc.int/sites/default/files/english_paris_agreement.pdf> accessed 17 August 2022

21 <https://ec.europa.eu/info/strategy/priorities-2019-2024/european-green-deal_en#Highlights> accessed 17 August 2022

SDG goals are of particular relevance for the construction sector, with their emphasis on access to housing and infrastructure and improvements in energy efficiency. The Glasgow Climate Pact[22] drives further action on long-term strategies for phasing down coal power, halting and reversing deforestation, speeding up the switch to electric vehicles, and reducing methane emissions.

FIDIC advocates that life-cycle thinking is very important from the earliest stage of a project. The FIDIC State of the World Reports 2021[23] focuses on strategic planning to enable better allocation of resources, selecting the right project and moving the focus from delivery to whole-life sustainability. Sustainability must be considered from the very earliest stages, as the potential environmental impacts are very significant. Once it has been decided to build a new building, as opposed to rehabilitating/refurbishing an existing building, a huge commitment to consume resources has already been made. Designers and contractors may be able to help limit that consumption, but they cannot change the overall commitment.[24] The life-cycle approach is essential to reduce long-term costs for projects. Sustainable choices can be made along the way: selection of materials, prefabrication, etc. From a holistic point of view, to drive the sustainability agenda, consultants should advise the clients upfront, from the idea, to develop and compare alternatives using the available sustainability tools.

Sustainability has been one of FIDIC's core principles since 1913. In 2021, FIDIC issued the Climate Change Charter[25] to encourage engineers to take the lead in planning, designing, and developing projects in line with the UN SDGs. The charter recognises that project procurement and risk allocation in contracts need significant change and innovation. Carbon reduction objectives and metrics are required at project level.

FIDIC adopts the net zero definitions provided by the Intergovernmental Panel on Climate Change (**IPCC**): net zero carbon dioxide (**CO_2**) emissions are achieved when anthropogenic CO_2 emissions are balanced globally by anthropogenic CO_2 removals over a specified period. Net zero CO_2 emissions are also referred to as carbon neutrality.[26]

The UN SDGs are due to be achieved by 2030. Net zero is widely considered to be achievable by 2050. Having considered the net zero carbon target and the increasing efforts to integrate circular economy thinking into activities, FIDIC proposed in the 2021 State of the World Report 'Net Zero – What Next?' that the net zero target should be moved forward and a net resource target should be achieved in 2060 (see Figure 15.1). Resources should be used and continue to be used in such a way that the use, creation, and/or need for materials, products, and services are designed to eliminate waste in the first place and generate minimal detrimental effect on the need for additional resources in the future, thereby ensuring sustainability for future generations with no overall waste due to the underutilisation.

The FIDIC Contracts Committee suggested how to incorporate net zero and net resource requirements for the design/construction/operation of the Works in tender evaluation

---

22 UN, *COP 26: The Glasgow Climate Pact* (UN 2021) https://ukcop26.org/wp-content/uploads/2021/11/COP26-Presidency-Outcomes-The-Climate-Pact.pdf> accessed 17 August 2022

23 FIDIC, <https://fidic.org/node/31087> accessed 17 August 2022

24 Surrey Hills Wiki, 'Sustainability in Building Design and Construction' (2022) <https://www.designingbuildings.co.uk/wiki/Sustainability_in_building_design_and_construction> accessed 17 August 2022

25 FIDIC, 'FIDIC Climate Change Charter' (FIDIC 2021) <https://issuu.com/fidic/docs/climate_charter_with_foreword/s/13931502> accessed 17 August 2022

26 FIDIC, State of the World Report '*Net Zero – What Next?*' (FIDIC 2021) 4 <https://fidic.org/sites/default/files/SOTW_Oct_2021_Net_Zero%20FINAL%20-%202.pdf> accessed 17 August 2022

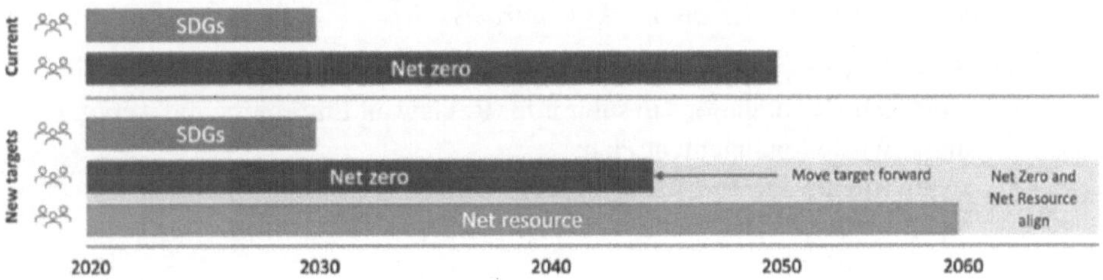

Figure 15.1 SDG Targets suggested by FIDIC

(From FIDIC State of the World Report 2021 '*Net Zero – what next?*' 32)

criteria and FIDIC contracts on the webinar 'Greening FIDIC's rainbow contract suite – let's do it!' held on 10 May 2022.[27] FIDIC formed TG23 to draft net zero guidance for FIDIC users.

15.4.2 Net zero requirements for the scope of consulting services

Consulting engineers can provide the skills, expertise, and services necessary for achieving sustainable buildings, plants, and infrastructure as:

- providers of technical assistance during the project preparation;
- designers;
- supervisors, project managers (FIDIC Engineer/Employer's representative);
- Consultants during operation.

Besides monitoring and managing the risks, costs, time, and stakeholders, the consultants can monitor and improve the sustainability during the whole project life cycle.

Consultants can influence the choice of the clients to:

- renovate, not demolish;
- create nature-based infrastructure for waste water treatment and flood protection;
- cut the losses of water and electricity with smart management of infrastructure;
- develop bankable sustainable projects with clear green targets;
- implement green procurement;
- base project decisions on carbon neutrality, not just profit.

Such requirements may be included in Appendix 1 of the White Book/Sub-Consultancy Agreement 2017.

The World Bank Request for Proposals for Consulting Services now includes ESHS considerations to apply to the supervision of civil works.[28]

27 FIDIC, <https://www.youtube.com/watch?v=COlzXPuEqIA&t=955s> accessed 17 August 2022

28 The World Bank, *Summary of Environmental, Social, Health and Safety (ESHS) Enhancements* (IBRD & IDA 2017) <https://thedocs.worldbank.org/en/doc/324811490634997634-0290022017/render/SummaryofkeyESHSenhancementsforprocurementMarch102017.pdf> accessed 17 August 2022

15.4.3 Net zero requirements for the Works Contracts

- The **Specification/Employer's Requirements** may include:
    - Contractor's obligations to submit for Review an Environmental Action Plan and a Waste Management Plan;
    - requirements for the use of green suppliers; reusable materials; electric equipment; off-Site manufacturing; local personnel; local materials; local subcontractors;
    - Net Zero Targets for construction activities (Scope 1: direct emissions; Scope 2: emissions from purchased energy; Scope 3: supply chain of a product upstream and downstream);
    - measurements and reporting against the Net Zero Targets.[29]
- The **Schedules** may include Net Zero Performance Guarantees for the completed Works linked with Net Zero Performance Damages and Net Zero Performance Bonuses.
- The **GCs** include the following provisions relevant to net zero obligations (examples from the Yellow Book):
    - pursuant to Sub-Clause 2.3, the Employer's Personnel and other contractors shall comply with the obligations under Sub-Clause 4.18 [Protection of the Environment];
    - under Sub-Clause 2.5 [Site Data and Items of Reference], the Employer has a continuous obligation to make available to the Contractor data in its possession on subsurface, hydrological, climatic, and environmental conditions at the Site;
    - under Sub-Clause 4.14 [Avoidance of Interference], the Contractor has a duty to minimise the use and occupation of roads and footpaths;
    - under Sub-Clause 4.18 [Protection of the Environment], the Contractor shall ensure that emissions, surface discharges, effluent, and any other pollutants from the Contractor's activities shall not exceed the values indicated in the Employer's Requirements or prescribed by applicable Laws;
    - Sub-Clause 4.20 [Progress Reports] in (g) obliges the Contractor to report activities relating to environmental aspects;
    - Sub-Clause 4.22 [Contractor's Operations on Site] obliges the Contractor to keep Contractor's Equipment and Contractor's Personnel off adjacent land; to keep the Site free from all unnecessary obstruction; and to clear away and remove any wreckage, rubbish, hazardous waste, and Temporary Works that are no longer required;
    - Sub-Clause 5.4 [Technical Standards and Regulations] refers to environmental Laws and other standards **specified in the Employer's Requirements**;
    - Sub-Clause 6.9 [Contractor's Personnel] enables any Contractor's person persisting in environmental misconduct to be removed;
    - DAAB Procedural Rule 3 enables online meetings;
- The **Particular Conditions** may include the following Net Zero provisions (reference to the Yellow Book Clauses):

29 Standards for reporting and external verification are available to be integrated at <https://ghgprotocol.org/standards> accessed 17 August 2022

- new definitions amending Sub-Clause 1.1 [Definitions] for Net Zero Targets and Environmental Impact Damages (during construction), Net Zero Incentives, Net Zero Performance Guarantees, Net Zero Performance Damages;
- hard copies may be abolished under Sub-Clauses 1.3 [Notices and Other Communications] and 1.8 [Care and Supply of Documents];
- a Sustainability Monitoring Assistant may be introduced in the Engineer's team under Sub-Clause 3.4 and an Environmental Management Officer for the Contractor under Sub-Clause 6.12 [Key Personnel];
- quarterly sustainability meetings with the Employer, main Subcontractors, and suppliers may be added under Sub-Clause 3.8 [Meetings];
- compliance with Net Zero Targets may be expressly added under Sub-Clause 4.1 [Contractor's General Obligations];
- requirements for green suppliers and compliance with the Net Zero Targets under Sub-Clause 4.4 [Subcontractors];
- Sub-Clause 4.10 [Use of Site Data] may state that when examining the Site and interpreting the data submitted by the Employer under Sub-Clause 2.5 [Site Data and Items of Reference], the Contactor shall plan measures to minimise the carbon footprint related to access, accommodation, facilities, personnel, power, transport, water, and any other utilities or services;
- the Contractor may be obliged to use the shortest routes under Sub-Clause 4.15 [Access Route];
- obligations for smart logistics for minimum vehicles and minimum carbon footprint, using sea, rail, electric vehicles, securing a charging point on the Site under Sub-Clause 4.16 [Transport of Goods];
- preference for non-polluting (electrical) Contractor's Equipment and rented from the region to reduce transportation may be added to Sub-Clause 4.17 [Contractor's Equipment];
- requirements for a Sustainability Plan (including a Waste Management Plan) and Environmental Impact Damages (for generating more waste/pollution than specified in the Contract on a quarterly basis) may be added to Sub-Clause 4.18 [Protection of the Environment];
- requirements for use of existing utilities only after checking their status and rehabilitation in case of losses/ no diesel generators but renewable energy sources (**RES**) in case of non-availability of infrastructure/ charging points for electric vehicles may be added in Sub-Clause 4.19 [Temporary Utilities];
- Carbon reporting on a quarterly basis, including comparison of actual and planned quantities (linked to Net Zero Targets/ Net Zero Incentives/ Environmental Impact Damages) may be added to Sub-Clause 4.20 [Progress Reports];
- Sub-Clause 4.22 [Contractor's Operations on Site] may include requirements for minimum handling and transportation of Goods (e.g. 'just in time' supplies, mounted from the vehicle);
- requirements for sustainable design, using local recyclable materials with lowest embodied carbon, blending traditions with innovations, application of prefabricated modular structures, Net Zero Incentives, depending on the project

(e.g. percentage self-generation of renewable energy, Zero Carbon Housing, minimum use of energy and water with smart infrastructure), digital twin, and BIM may be added under Sub-Clause 5.1 [General Design Obligations];

- Greenhouse Gas (**GHG**) standards may be added under Sub-Clause 5.3 [Contractor's Undertaking];
- a requirement for engagement of a certain minimum percentage of Contractor's Personnel from the region (Country) of the Site may be added under Sub-Clause 6.1 [Engagement of Staff and Labour];
- a requirement to maintain a detailed GHG protocol or ISO 14064[30]/14067[31] compliant records of each major construction or O&M activity to be added under Sub-Clause 6.10 [Contractor's Records]; results from such documentation could be used to fully offset lifetime emissions attributable to the project;
- requirements for durable, reusable materials; adherence to circular manufacturing requirements; in the case of demolition, most of the materials, with a minimum percentage stated, to be reused or recycled under Sub-Clause 7.1 [Manner of Execution];
- Net Zero Value Engineering (VE), including incentives for Green modifications, to be added to Sub-Clause 13.2 [Value Engineering], also adding sub-paragraph (e) as a requirement for any VE proposal to have a positive environmental impact;
- amounts for Net Zero Incentives for innovations reducing carbon footprint, such as drones, AI, digital twin; bonuses for exceeding the targets; amounts for green VE, to be allocated under Sub-Clause 13.4 [Provisional Sums];
- Performance Damages for failure to meet Net Zero Performance Guarantees may be added, related to Sub-Clauses 9.4 [Failure to Pass Tests on Completion], 11.4 [Failure to Remedy Defects], and 12.4 [Failure to Pass Tests after Completion];
- failures to comply with specified Net Zero Targets/ Net Zero Performance Guarantees, constituting material breach of the Contractor's obligations, may be added to the grounds for termination under Sub-Clause 15.2 [Termination for Contractor's Default]; and
- Clause 21 [Disputes and Arbitration] may include additional arrangements for virtual adjudication and arbitration.

15.5 AI tools for HSSE management

Using video analytics for workplace safety reduces manual inspections and accidents. Smart safety applications identify objects in a 3D environment. Workers who haven't worn proper safety equipment, like helmets and high-visibility jackets, will be detected. AI may reduce the injury rate in construction areas by using computer vision.[32] AI has

30 *ISO 14064-1:2018 Greenhouse Gases – Part 1: Specification with guidance at the organization level for quantification and reporting of greenhouse gas emissions and removals*

31 *ISO 14067: Greenhouse Gases – Carbon footprint of products – Requirements and guidelines for quantification*

32 viAct, 'Return on Investment (ROI) to Use Computer Vision in Construction' (2021) <https://www.viact.ai/post/return-on-investment-roi-to-use-computer-vision-in-construction> accessed 17 August 2022

been trained to identify indicators of risk in photos, video, and other project data, and even to predict accidents before they happen. Danger zone alert and safety measures detection can be performed by AI to reduce accidents.[33] Monitoring of cranes, work in confined spaces/at high level, detection of fire hazards, missing barricades, and non-compliance behaviour automates the HS management. For example, viMOV is capable of supporting multiple AI modules simultaneously, enabling control on safety aspects in confined spaces. AI modules like PPE Detection, Atmospheric & Dusty Condition Monitoring, Danger Zone Intrusion, Flowing Solid & Liquid Detection, Counting Workers on Entry Site, and the like enable detection of any HSSE non-compliances.[34]

AI may detect whether any dump truck is uncovered or overweight before it leaves the site, truck wheel cleanliness, and site entrance hygiene status. AI can also issue on-site instant audio-visual alerts in the case of any waste being dumped into the wrong container and send a violation record to a dashboard for management to follow up and take action. AI can record air quality and environmental data in real time and send instant alerts for gas leakage and air hazards. An online dashboard can keep a record of all the non-compliances taking place on the Site and display a trend report.[35]

15.6 Example Notice

Appendix 1 includes an example for Engineer's Notice under Sub-Clause 6.9 (d) of the Yellow Book for removal and replacement of Contractor's Personnel due to their persistent non-compliance with the provision of Sub-Clauses 6.7 and 7.3.

33 viAct, 'Artificial Intelligence: How Will It Affect Construction Industry' (2021) <https://www.viact.ai/post/artificial-intelligence-how-will-it-affect-construction-industry> accessed 17 August 2022

34 viAct, 'Conserving the Confined: AI to Ensure Better Safety Compliances for Confined Space Works' <https://www.viact.ai/post/conserving-the-confined-ai-to-ensure-better-safety-compliances-for-confined-space-works> accessed 17 August 2022

35 viAct, 'Artificial Intelligence Brings New Possibilities for Construction & Demolition Waste Management' <https://www.viact.ai/post/artificial-intelligence-brings-new-possibilities-for-construction-demolition-waste-management > accesses 17 August 2022

CHAPTER 16

Project Claims Management with FIDIC Contracts

16.1 Project Claims Management processes

The PMBOK® Guide does not include a section dedicated to Project Claims Management, but the Construction Extension includes Annex A1 'Managing Claims in Construction'. It defines a claim as a demand for something due or believed to be due. Claims management describes the processes required to prevent claims, to mitigate the effects of those that do occur, and to manage claims quickly and effectively.[1]

Early recognition of potential problems, and open communications regarding possible alternatives or changes to the plan, create a collaborative environment in which claims are less likely to occur.[2] In the absence of an agreement, the claim may proceed to formal negotiation or dispute resolution before it is resolved.

Poorly drafted or incomplete and unsubstantiated claims became the top cause of construction disputes for 2021, according to Arcadis. Errors and/or omissions in the contract documents also appeared on their 2021 list as a top cause.[3] Neither of these appeared in the Arcadis top three for 2020, when the main reasons for construction disputes were failure to understand and/or comply with the contractual obligations, scope changes, and

1 PMI Inc., *Construction Extension to the PMBOK® Guide* (PMI 2016) 167

2 PMI Inc., *Construction Extension to the PMBOK® Guide* (PMI 2016) 168

3 Arcadis, 2022 Global Construction Disputes Report 'Successfully Navigating Through Turbulent Times' 11 <https://images.connect.arcadis.com/Web/Arcadis/%7Bcb063f2c-be31-410c-9807-d7a9bf16f666%7D_2022_Global_Construction_Disputes_Report_-_Successfully_navigating_through_turbulent_times.pdf> accessed 24 January 2023

DOI: 10.4324/9781003272892-16

third-party or force majeure events,[4] which are the top-ranking causes in the last 5 years until the end of July 2022 according to another survey.[5] The research shows the importance of claims management, underpinned by efficient contract management, facilitated by collaboration. Obviously, a focus on procurement management, scope management, and risk management with fair risk allocation may minimise disputes.

16.2 Claims Management planning

According to the Construction Extension,

> the goal for claims management is to prevent claims entirely and, if not, to resolve them at the earliest opportunity for the least cost and least disruption to the project.
>
> A well-written contract with fair contract terms provides a basis for minimising claims when it comes to scope changes and unknown site conditions, *force majeure* delays and fortuitous events, and timely reporting and fair-notice provisions.[6]

The GCs of the FIDIC Contracts include detailed provisions for fair risk allocation in case of changed circumstances and provide remedies in case of poor performance and breach of contract.

16.3 Claims Management in the FIDIC Suite for Services

Disagreements may arise under the White Book in relation to the entitlement to an EOT, payment of additional costs, or damages for breach of the agreement.

Sub-Clause 5.1.3 sets out the claims management procedure related to Variations to the Services. Disagreements may escalate into disputes when:

1. The Client issues an instruction, or other circumstances result in change to the Services (e.g. changes in legislation under Sub-Clause 1.5).
2. The Consultant gives a Notice to the Client as soon as reasonably practicable that the above constitutes a Variation to the Services, including details of the estimated impact upon the Programme and cost of the Services.
3. Within 14 days, the Client gives a Notice that the above does not constitute a Variation to the Services.
4. The Consultant refers the disagreement as a dispute under Clause 10 within 7 days of receipt of the Notice.

Chapter 8 comments on the entitlement to an EOT under Sub-Clause 4.4. Chapter 9 comments on the entitlement to additional fees under Clauses 5 and 7.

Pursuant to Sub-Clause 7.2.3, the Client shall not withhold payment of any part of an invoice for any amount properly due to the Consultant under the Agreement by reason of claims or alleged claims against the Consultant unless the amount to be held has been:

- agreed with the Consultant to be due to the Client; or
- awarded by an adjudicator or an arbitrator to the Client pursuant to a referral under Clause 10.

4 Arcadis, 2021 Global Construction Disputes Report 'The Road to Early Resolution' 5

5 HKA, Fifth Annual CRUX Report 'Battling the Headwinds: a Regional Analysis of Claims and Dispute Causation' (2022) 2 <https://www.hka.com/crux-insight-fifth-edition-battling-the-headwinds/> accessed 24 January 2023

6 PMI Inc., *Construction Extension to the PMBOK® Guide* (PMI 2016) 169

Sub-Clause 6.5.2(c) stipulates an exception in the case of termination for Consultant's default: payments may be withheld until all reasonably foreseeable costs incurred by the Client have been established and all deliverables have been received.

Sub-Clause 7.5 [Disputed Invoices] sets out the procedure to be followed when the Client considers any part of an invoice as not properly due:

- the Client shall give a Notice of its intention to withhold payment of any item in an invoice with reasons within 7 days of the issue of the invoice, paying the non-contested remainder of the invoice;
- the Consultant may disagree and refer the dispute under Clause 10;
- the Consultant shall be entitled to financing charges under Sub-Clause 7.2.2 for all contested amounts finally determined to have been due.

Clause 8 sets out the provisions for the Parties' claims for breach of the Agreement:

- damages shall be limited to the amount of reasonably foreseeable loss and damage suffered as a direct result of such breach;
- the proportion of damages payable by a Party jointly liable with third parties shall be limited to the proportion of liability that is attributable to its breach;
- the maximum amount shall not exceed the amount stated in the Particular Conditions;
- the period for an admissible claim shall be stated in the Particular Conditions;
- the limitations shall not apply to claims arising out of deliberate, manifest, and reckless default, fraud, fraudulent misrepresentation, or reckless misconduct.

16.4 Claims Management in the FIDIC Contracts for Works

16.4.1 Scope of FIDIC Works Contracts Claims Management procedures

The 2017 Edition of the Red/Yellow/Silver Book includes the first definitions of 'Claim' and 'Dispute'. The 2022 Reprint of the 2017 Edition introduces amendments clarifying the claims management procedure. The 2022 Sub-Clause 1.6 (Red)/1.5 (Yellow)/1.3 (Silver Book) distinguishes between a 'Claim' and the new category **'matter to be agreed or determined'**.

'Claim' is defined as:

- **a request or assertion by one Party to the other Party for an entitlement or relief** under any Clause of these Conditions or otherwise in connection with, or arising out of, the Contract or the execution of the Works,
- excluding a matter to be agreed or determined under Sub-Clause 3.7 (Silver Book: 3.5) (a).

Both categories are discussed in relation to the claims management, since they:

- are related to entitlements/reliefs;
- are governed by the Sub-Clause 3.7 (Silver Book: 3.5) agreement/determination procedure led by the CA (discussed in Chapter 6);
- may lead to disagreements that escalate into Disputes when poorly managed.

Sub-Clause 3.7 (Silver Book: 3.5) of the Reprint 2022 includes fixed lists of 'matters to be agreed or determined' expressly provided in 13 Sub-Clauses in the Red Book, 12 in the Yellow Book, and 9 in the Silver Book, shown in Table 16.3. Each of the listed Sub-Clauses specifies what the CA has to agree/determine and when the 42 days for agreement commence.

Clause 20 [Employer's and Contractor's Claims] includes detailed procedure setting out how the Claim is to be notified and dealt with, and within what timescales. It does not apply to the category 'matter to be agreed or determined', or to any claim with a small 'c' (like claims that may be made under the Performance Security/ a guarantee/ an insurance policy or third-party claims).

For proper claims management planning, FIDIC users must carefully read several provisions, starting with the definitions of 'Claim' and 'Dispute', and then the procedures under:

- Sub-Clause 3.7 (Silver Book: 3.5) for agreement/determination (presented in Figures 6.2 and 6.3);
- Clause 20 for Claims (see Figures 16.1 and 16.2);
- additional requirements of relevant Sub-Clauses (see Figures 12.1 and 16.3);
- Clause 21 for Dispute avoidance and resolution (see Figures 17.3 and 17.4).

16.4.2 Claims procedure in Clause 20

16.4.2.1 Claim categories

The entitlement or relief that is requested or asserted by way of a Claim falls into three categories under Sub-Clause 20.1 (see Figure 16.1):

- **time**: extension of the DNP for the Employer/ Time for Completion for the Contractor;
- **money**: additional payment (or reduction of the Contract Price) for the Employer/ additional payment for the Contractor;
- **another entitlement or relief.**

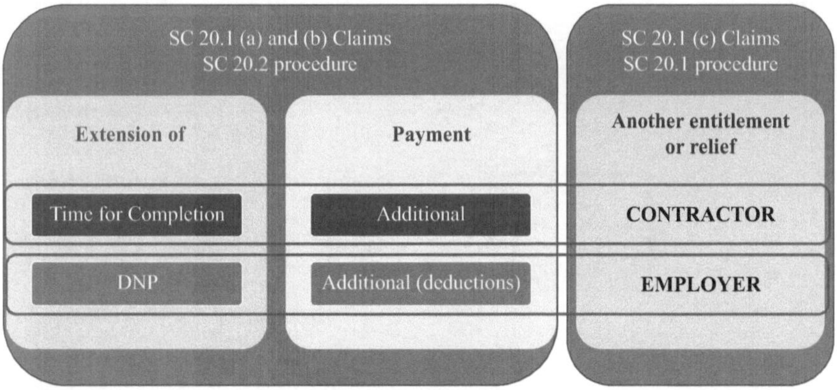

Figure 16.1 Sub-Clause 20.1 Claim categories

The third category can be for anything that one of the Parties has promised to the other Party under the Contract. It may be related to any certificate, instruction, Notice, opinion, valuation, etc.

16.4.2.2 Management of other Claims not for time and money

The Sub-Clause 20.1 procedure for management of the other Claims is simplified:

1. Either Party requests an entitlement/relief;
2. The other Party/CA disagrees (or does not respond within a reasonable time);
3. The claiming Party may, by giving a Notice, refer the Claim to the CA for agreement/determination under Sub-Clause 3.7 (Silver Book: 3.5). The Notice shall:
   a. be given as soon as practicable after the claiming Party becomes aware of the (deemed) disagreement; and
   b. include details of the claiming Party's case and the disagreement.

16.4.2.3 Management of Claims for time and money

The time and money Claims follow the Sub-Clause 20.2 detailed procedure summarised in Figure 16.2:

- **Sub-Clause 20.2.1**: the claiming Party shall **give a Notice of Claim as soon as practicable, but no later than 28 days after it became aware, or should have become aware,** of the event/circumstance giving rise to the cost, loss, delay, or extension of DNP;
- **Sub-Clause 20.2.2**: the following scenarios may happen in the next 2 weeks:
  a. if the CA considers that the Notice of Claim was late:
     i. the CA shall give a Notice to the claiming Party with reasons;
     ii. if the claiming Party disagrees or considers there are circumstances justifying late submission, it shall include details in its fully detailed Claim under Sub-Clause 20.2.4;
  b. if the CA does not give such a Notice within 14 days:
     i. the Notice of Claim shall be deemed to be valid;

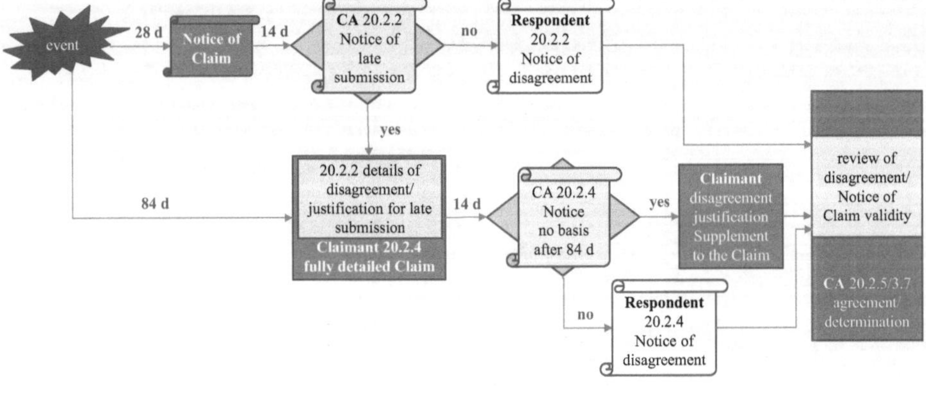

Figure 16.2 Sub-Clause 20.2 Claims procedure

   ii. if the other Party disagrees with the validity, it shall give a Notice to the CA with details of the disagreement, to be reviewed by the CA during the Sub-Clause 20.2.5 agreement/determination of the Claim.

- **Sub-Clause 20.2.3**: the claiming Party shall keep contemporary records to substantiate the Claim; the CA may monitor them/ instruct the Contractor to keep additional records, without admitting the Employer's liability;
- **Sub-Clause 20.2.4**: the claiming Party shall submit a **fully detailed Claim within 84 days** after it became aware, or should have become aware, of the event/ circumstance giving rise to the Claim or other agreed period; if the claiming Party fails to submit a statement of the contractual/other legal basis of the Claim within the time limit, the following scenarios are possible in the next 2 weeks:
  a. the CA gives a Notice to the claiming Party that the Notice of Claim shall be deemed to have lapsed:
     i. the Notice of Claim shall no longer be valid;
     ii. if the claiming Party disagrees or considers there are circumstances justifying late submission, it shall supplement its fully detailed Claim with details;
  b. the CA does not give such a Notice within 14 days:
     i. the Notice of Claim shall be deemed to be valid;
     ii. if the other Party disagrees with the validity, it shall give a Notice to the CA with details of the disagreement, to be reviewed by the CA during the Sub-Clause 20.2.5 agreement/determination of the Claim;

- **Sub-Clause 20.2.5:** the CA shall proceed to agree/determine the entitlement within the time limits under Sub-Clause 3.7.3 (Silver Book: 3.5.3):
  i. the CA shall promptly give a Notice to the claiming Party if additional particulars are necessary;
  ii. the CA shall nevertheless give a Notice to the claiming Party within 42 days for agreement/rejection of the contractual or other legal basis of the Claim;
  iii. the claiming Party shall submit the additional particulars as soon as practicable;
  iv. the CA shall proceed to agree/determine the entitlement (the 42 days for agreement commence on the date of receipt of the additional particulars).

Sub-Clause 20.2.6 sets out the procedure for Claims of continuing effect. In this case:

a) the fully detailed Claim shall be considered as interim;
b) the CA shall give response on the basis of the Claim;
c) the claiming Party shall submit further interim Claims at monthly intervals giving accumulated amounts of the claimed time/money;
d) within 28 days after the end of the effects, the claiming Party shall submit a final fully detailed Claim giving the total amounts claimed.

After receiving the Notice of Claim, the Engineer shall include in each IPC (the Employer in each payment under Sub-Clause 14.7 of the Silver Book) any amounts reasonably substantiated as due.

Table 16.1 Time limit and contents of the Notice of Claim and fully detailed Claim

	Notice of Claim	*Fully detailed Claim*
Time	As soon as practicable and within **28 days** after becoming aware (or should have become aware) of the event or circumstance	Within **84 days** after becoming aware (or should have become aware) of the event or circumstance, or as agreed by the other Party
Contents	Written description of the event or circumstance giving rise to the cost, loss, delay, or extension of DNP expressly identified as a Notice of Claim	a. A detailed description of the event/circumstance b. A statement of the contractual and/or other legal basis of the Claim c. Contemporary records d. Detailed supporting particulars
Notice of Claim to be valid	To be submitted within 28 daysTo comply with Sub-Clause 1.3	The statement of the basis of the Claim to be submitted within 84 days (or as agreed by the other Party)

The Employer shall only be entitled to claim any payment from the Contractor and/or to extend the DNP, or set off against or make any deduction from any amount due to the Contractor, by complying with the Sub-Clause 20.2 procedure.

16.4.2.4 Requirements for Claims for time and money

Table 16.1 compares the Sub-Clause 20.2 requirements for the Notice of Claim and the fully detailed Claim.

The Claim shall be time barred if the claiming Party fails to give a Notice of Claim within 28 days and/or a statement of the basis of the Claim within 84 days. The other Party shall be discharged from any liability in connection with the event/circumstance giving rise to the Claim. However, enforcement may be denied in some cases in some jurisdictions.

The Sub-Clause 20.2 procedure gives an opportunity to both Parties to justify their position in relation to a late submission. In all cases, the CA shall complete the agreement/determination procedure, including review of the relevant disagreements and the validity of the Notice of Claim. The circumstances that may be taken into account may include:

- the starting date of awareness as the date when the delay becomes on the critical path and/or additional Costs start to be incurred (which may be much later than the impact of the event/circumstance is felt due to successful risk mitigation);[7]

7 *OHL SA v Her Majesty's Attorney General for Gibraltar* – [2014] EWHC 1028 (TCC)

J. Akenhead held: 'The "event or circumstance" … in the appropriate context can mean either the incident (variation, exceptional weather or one of the other specified grounds for extension) or the delay which results or will inevitably result from the incident in question' [312]

- whether or to what extent the other Party would be prejudiced by acceptance of the late submission;
- any evidence of the other Party's prior knowledge of the event/circumstance giving rise to the Claim for a late Notice of Claim or the basis of the Claim in the case of a delayed statement.

However, according to Sub-Clause 4.20, nothing stated in a progress report shall constitute a Notice. If an event or circumstance discussed on a meeting or in a report is a risk, it may be mitigated by revising the Programme under Sub-Clause 8.3 or implementing a Variation after an advance warning under Sub-Clause 8.4. The fact that a risk event causes delay, disruption, and/or additional Costs does not mean that the Contractor is entitled to an EOT and/or compensation: the delay may be mitigated or may be not on the critical path at all; there might be no entitlement to Costs under the Contract. That is why the Notice of Claim is an important contractual provision, and the failure to comply with the time limit may have severe consequences in most jurisdictions.[8,9]

The requirements of Sub-Clause 20.2 are in addition to those of any other Sub-Clauses applying to the Claim. Figure 16.3 gives an example for an Exceptional Event with continuing effect. If the claiming Party fails to comply with any of those requirements, the entitlement to time and money shall take account of the extent to which the failure has prevented or prejudiced proper investigation of the Claim by the CA (Sub-Clause 20.2.7).

The Parties must carefully read the relevant provision on which they rely for an entitlement or relief to identify which procedure to follow, what is the possible entitlement, and what are the conditions to be met. For example:

- the Contractor may be entitled only to an EOT (but no compensation) and only under the Red/Yellow/Emerald Books in the case of exceptionally adverse climatic conditions following the Sub-Clause 20.2 Claims procedure, in so far as that the delay is on the critical path (stated as 'to the extent that completion for the purposes of Sub-Clause 10.1 [Taking Over the Works and Sections] is or will be delayed');
- there is no requirement for the Contractor to comply with the Sub-Clause 20.2 Claims procedure in respect of any Variation, as stated in Sub-Clauses 8.5(a) and 13.3.1: the CA follows the agreement/determination procedure under Sub-Clause 3.7 (Silver Book: 3.5); at the same time, where the Contractor has prepared and submitted a proposal at the request of the CA under Sub-Clause 13.3.2, but the CA does not give consent to it, the Contractor has to follow the Sub-Clause 20.2 Claim procedure to recover the incurred Cost for the preparation of the proposal.

8 Polish *Supreme Court V CSK 449/16*: 'strict contractual time limits that lead to forfeiture of rights if not observed cannot be contested on the grounds of the principle of freedom of contract', Sylwia Moreu-Zak, 'Strict Time Limits in Construction Work Contracts' (*WSG* 2019) <https://www.worldservicesgroup.com/publications.asp?action=article&artid=11251> accessed 5 March 2023

9 Nicolas Bouchardie and Dimitar Kondev, 'Time Bars in the FIDIC Conditions of Contract' (presentation for BSCL 2023) 'The tribunal in ICC Case 24652/MHM/HBH dismissed the great majority of the Contractor's claims as time-barred without considering their merits.' <https://bscl.bg/wp-content/uploads/2022/10/Time-Bars-in-the-FIDIC-Conditions-of-Contract1.pdf> accessed 9 April 2023

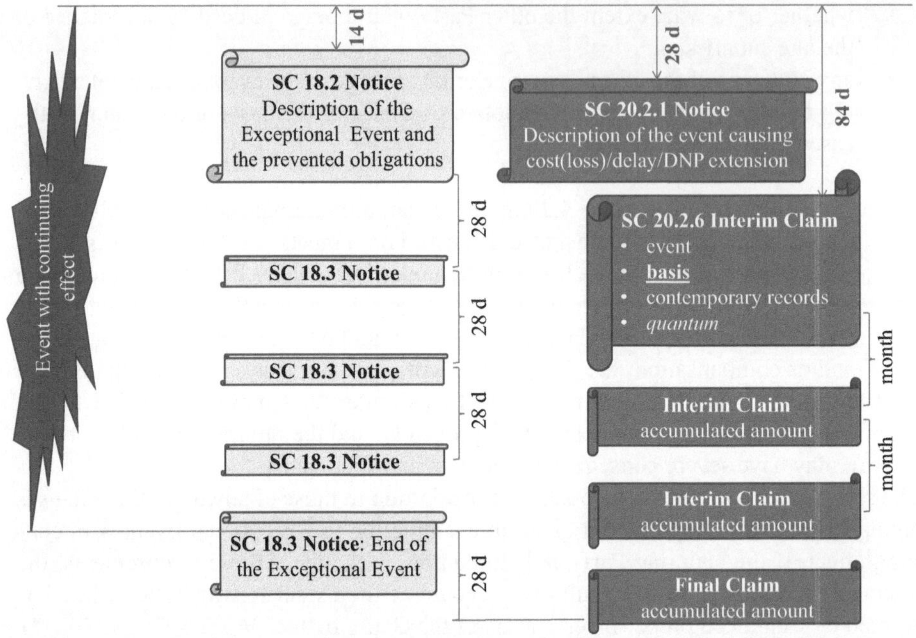

Figure 16.3 Requirements for Claims for Exceptional Events with continuing effect

Chapter 8 discusses the conditions to be met where the Contractor is entitled to an EOT and the Employer to Delay Damages. Table 8.1 lists the Sub-Clauses entitling the Contractor to an EOT. Chapter 9 lists the provisions entitling the Employer to a reduction of the Contract Price and the Contractor to additional payment.

16.5 Project Claims Management monitoring and control

According to the Project Management Institute (PMI),

> Being proactive, such as keeping project documents in order, traceable, and retrievable in a timely manner, is a sound approach toward claims prevention and is the least invasive and least costly route in resolving potential claims and contract disputes.[10]

Besides the mandatory list of Claim Notices in the monthly reports (Sub-Clause 4.20 (f)), the CA is advised to keep a Claims Register, following up in separate spreadsheets the status of Claims and Sub-Clause 3.7 (Silver Book: 3.5)(a) 'matters to be agreed or determined'.

The FIDIC 2017 Contracts Guide recommends a reasonable and proactive approach for effective and efficient claims management and dispute avoidance:

> A Claim is not a Dispute, although the subject of a Claim may become a Dispute under the Contract ... Claims should not be regarded as either inevitable or unpalatable, and making a Claim should not be regarded as being an aggressive act. Construction projects normally give

10 PMI Inc., *Construction Extension to the PMBOK® Guide* (PMI 2016) 168

> rise to major risks, which have to be dealt with if they occur. While one or both of the Parties might prefer the price accepted just after the tender stage to remain unchanged, it will not be helpful to the project if the Parties instinctively seek to attribute blame where it transpires (as is so often the case in practice) that circumstances arise or events occur which give rise to an adjustment to the Contract Price. In these events, the procedures for making and dealing with Claims – as set out in Clause 20 – are specified so as to provide the degree of structure and formality considered necessary for the proper contract administration of an engineering or construction project. By complying with these procedures, and by maintaining a co-operative approach to the determination of all adjustments, the Parties should enhance the likelihood of achieving a successful project.[11]

The CA shall proactively manage all risks before they result in any issues, then find solutions in collaboration with the Parties, using the Variation procedure to minimise Claims. Upon the receipt of a Notice of Claim, the CA shall:

- visit the Site to investigate the facts when applicable and give instructions for contemporary records under Sub-Clause 20.2.3;
- check the validity of the Notice of Claim;
- collaborate with the Contractor to mitigate the risks;
- register the Notice of Claim in the Claims Register (Table 16.2 presents an example).

The following Claim prevention actions are recommended:

- the Parties to give Notices of Claim as soon as possible as early warning, to be registered/followed up in the Claims Register and listed in the Contractor's and CA's Progress Reports;
- the Parties/CA to follow strictly the timelines of the Claim procedure and the other requirements of the relevant Clauses;
- the CA to organise regular risk management meetings for mitigation of risks and solving of issues using the Sub-Clause 13.3 Variation procedure;
- early recognition of potential risk events and open communication regarding possible alternatives or changes to the programme or design, in a collaborative environment;
- prompt certification and payments for instructed and executed, but not agreed Variations, with provisional quantities/rates and prices until agreed/determined;

Table 16.2 Claims Register

No	*Notice of Claim*	*Admissible?*	*Fully detailed Claim*	*CA's response*	*Agreement*	*CA's determination*	*NOD?*
1	Date, title		Date, ID	Date, ID	Date, ID	Date, ID	Date, ID
2	Date, title		Date, ID	Date, ID	Date, ID	Date, ID	Date, ID

11 FIDIC, *The FIDIC 2017 Contracts Guide* (FIDIC 2022) 7

- withholding amounts for non-fulfilled obligations/defects under SC 14.6.2 until the default is remedied, proceeding the payment even under the minimum percentage to secure the Contractor's cash flow, while incentivising timely compliance;
- regular checking and agreement on the Contractor's contemporary records related to the Claim event;
- use of BIM as a transparent source of agreed data;
- the Contractor to present the impact of the Claim events on the critical path of the revised Programme and include mitigation measures in the supporting report (as stipulated under Sub-Clause 8.3 (k)(v));
- the Contractor to link the resources to the time schedule in the Programme under SC 8.3 (not a FIDIC requirement, but necessary for efficient claims management);
- the CA and the Contractor to agree which method for delay analysis is to be used and which baseline time schedule.

The author's experience shows that creating a collaborative no-blame environment and applying these measures may reduce the Claims to 0–15 per cent of the submitted Notices of Claims. Since prevention is the best cure, this approach saves time and money, and improves relationships, enabling the Parties and CA to focus on the project's success.

When a Claim is prepared, it has to be properly drafted, structured, and substantiated to demonstrate the causation leading to the claimed entitlement. If properly done, the CA's determination follows its structure in order to address each statement. According to Andy Hewitt,[12] the essential elements to include in any Claim or determination are:

- the **cause**: the event that has given rise to the claim, to be established and substantiated in the claim;
- the **effect**: a demonstration of how the event affected the claimant by linking the cause with the effect; this could be in terms of time or money and must be substantiated in the Claim;
- **entitlement**: the claimant's right under the Contract or at law, which must also be established and substantiated in the claim;
- **substantiation**: contemporary records or other supporting evidence.

Making the document concise, clear, stand-alone, and logically reaching the conclusions (about the Claim basis, causation and entitlement/relief) is important to persuade the reviewers and prevent any Disputes.

16.6 Escalation of disagreements into Disputes

Figure 16.4 summarises the claims management process from escalation of issues into disagreements, and then into Disputes (Red/Yellow/Emerald Book Sub-Clauses):

12 Andy Hewitt, *Construction Claims and Responses: Effective Writing and Presentation* (John Wiley & Sons 2011) 75

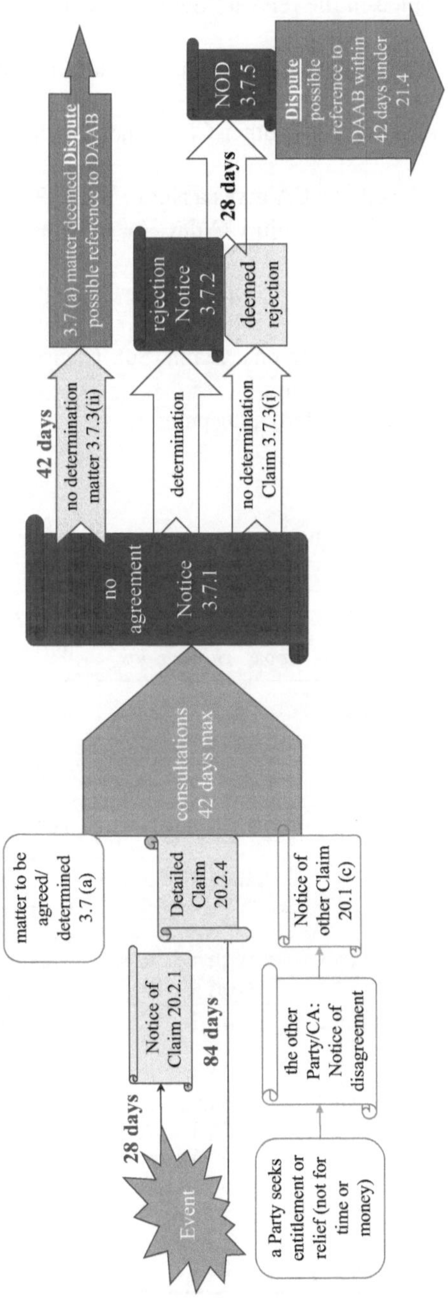

Figure 16.4 Escalation of disagreements into Disputes

(The references are to the *Red/Yellow/Emerald Book* provisions)

1. One Party has made a Claim, or there has been a matter to be agreed or determined under Sub-Clause 3.7 (a);
2. The CA consults with the Parties and proactively leads the discussions to reach an agreement under Sub-Clause 3.7.1 within 42 days (or other time limit proposed by the CA and agreed by both Parties) commencing as presented in Table 16.3:
   a. on the date specified in the relevant Sub-Clause for a matter to be agreed/ determined;
   b. after receiving a fully detailed/interim/final fully detailed Claim for time and/or money;
   c. on the date of referral of any other Claim (not for time and money) to the CA.
3. If no agreement is achieved, the CA gives a Notice to the Parties and proceeds to determine the entitlement/relief within 42 days (or other agreed time limit).
4. If the CA:
   a. gives a Notice of determination, rejecting the Claim (the Party's assertion in respect of the matter); or
   b. does not give a determination on a Claim (Sub-Clause 3.7 (a) matter) within the Sub-Clause 3.7.3 time limit:
5. The disagreement may escalate into a Dispute:

Table 16.3 Commencement of the time limit for agreement under Sub-Clause 3.7.3 (Silver Book: 3.5.3)

3.7 *(Silver Book: 3.5)* Category	*Sub-Clauses*			*Start of time limit for agreement*
	Red Book	*Yellow Book*	*Silver Book*	
(a) matter to be agreed/ determined		1.9		Contractor's Notice of error
	4.7.3	4.7.3		Contractor's 4.7.2 Notice of error
	10.2	10.2		TOC issued/ Contractor's Notice
	11.2	11.2	11.2	Contractor's Notice
	12.1			Contractor's Notice
	12.3			Contractor's Notice
	13.3.1	13.3.1	13.3.1	Contractor's submission
	13.5	13.5	13.5	completed varied work
	14.4	14.4	14.4	CA finds difference of progress
	14.5	14.5	14.5	conditions are fulfilled
	14.6.3	14.6.3	14.6.3	Contractor's Notice
	15.3	15.3	15.3	date of termination
	15.6	15.6	15.6	CA receives particulars
	18.5	18.5	18.5	CA receives particulars
Claim for time and money	20.2.4	20.2.4	20.2.4	submission of detailed Claim
other Claim	20.1(c)	20.1(c)	20.1(c)	Notice of other Claim

a. when either Party has given a Notice of Dissatisfaction (NOD) with the CA's determination to the other Party with a copy to the CA within 28 days; if the Dispute is not referred to the DAAB within 42 days under Sub-Clause 21.4, the NOD shall no longer be valid, and there is no Dispute;
b. under Sub-Clause 3.7.3 (ii), where no determination is given on a Sub-Clause 3.7 (a) matter within 42 days; the Dispute may be referred directly to the DAAB without the need for a NOD.

16.7 Innovative Project Claims Management

The best Claim presentation may be with a short video, which may be an extract of web camera/drone footage archive or 4D animations, extracted from BIM.

BIM and AI support the creation and archiving of contemporary records. Data management software enables the extraction of relevant evidence from the model or video footage.

An iContract could run scenarios for any situation, from legislative changes in requirements impacting on construction methodology or materials, to any number of events that arise on a construction site daily. By having an intuitive and sophisticated digital contract engine, a user could run a scenario, either as a simulated possibility or based on a real event, with the iContract informing the user what the contract consequence will be. As the flow of data from external sources such as BIM models, schedules, and cost control software becomes more reliable, the iContract can begin to run autonomously and begin to incorporate AI in contract clash detection and forecasting dispute avoidance as trends are recognised.[13]

16.8 Example Notice

Appendix 1 includes an example Contractor's Notice of Claim under Sub-Clause 20.2.1 of the Yellow Book.

13 Alan McNamara, 'Automating the Chaos: Intelligent Construction Contracts' Chapter in *Smart Cities and Construction Technologies* (IntechOpen 2020) 11

CHAPTER 17

Project Dispute Management with FIDIC Contracts

17.1 Dispute management under the FIDIC Suite for Services

Clause 10 [Disputes and Arbitration] sets out a multi-tier dispute resolution procedure summarised in Figure 17.1:

- **Sub-Clause 10.1.1:** if any dispute arises, the authorised senior representatives of the Parties shall meet within 28 days of a written request from one Party to the other to attempt to resolve the dispute amicably;
- **Sub-Clause 10.1.2:** if the dispute is not resolved within 56 days of receipt of the request, either Party may refer it to adjudication, even if the representatives have not met;
- **Sub-Clause 10.2.1:** the adjudicator shall be agreed between the Parties (failing agreement, he/she shall be appointed in accordance with the Rules for Adjudication in Appendix 5);
- **Rule 23:** the adjudicator shall give a decision within 56 days after the reference or, if later, the day on which the Adjudicator's Agreement came into effect;
- **Sub-Clause 10.2.6:** the decision becomes binding on both Parties upon its release;
- **Sub-Clauses 10.2.3–10.2.4:** if either Party is dissatisfied with the adjudicator's decision, or if the adjudicator fails to give it within the time limit, either Party may give a Notice of Dissatisfaction (NOD) to the other Party with a copy to the adjudicator within 28 days;
- **Sub-Clause 10.2.7:** if no NOD is given within 28 days, the decision becomes final and binding on both Parties;

DOI: 10.4324/9781003272892-17

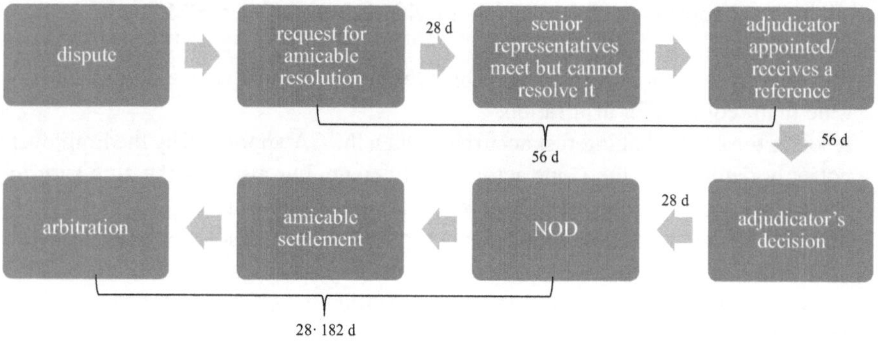

Figure 17.1 Dispute resolution under the White Book

- **Sub-Clause 10.2.5:** neither Party shall be entitled to commence arbitration unless a NOD is given, except to refer the failure of a Party to comply with a decision of the adjudicator; if a NOD is given but neither Party commences arbitration within 182 days of giving or receiving the NOD, the NOD shall be deemed to have lapsed and no longer be valid;
- **Sub-Clause 10.3.1:** where a NOD has been given, both Parties shall attempt to settle the dispute amicably before the commencement of arbitration;
- **Sub-Clauses 10.3.1–10.4.1:** arbitration may be commenced 28 days after the date when the NOD is given, even if no attempt at amicable settlement has been made, to finally settle any dispute under the Rules of Arbitration of the International Chamber of Commerce (**ICC**);
- **Sub-Clause 10.4.2:** one or three arbitrators shall be appointed, authorised to open up, review and revise any ruling or decision of the adjudicator.

17.2 Dispute management in FIDIC Works Contracts

17.2.1 The dispute avoidance role of the DAAB

17.2.1.1 Benefits of dispute boards for efficient PM

The first mandatory tier of dispute resolution in the FIDIC 2017 Edition and the Emerald Book is the DAAB: a standing dispute board of a sole member or three members appointed from the beginning to support the project team in dispute avoidance.

The Sub-Clause 21.4 provision stipulating that disputes shall be decided by a dispute board before referring them to arbitration has been confirmed by case law in relation to the identical requirement in Sub-Clause 20.4 of the 1999 Edition.[1,2] GP5 asserts the role of DAAB:

> Unless there is a conflict with the governing law of the Contract, all formal disputes must be referred to a Dispute Avoidance/Adjudication Board (or a Dispute Adjudication Board, if applicable) for a provisionally binding decision as a condition precedent to arbitration.

1 Swiss Supreme Court, *case no. 4A_124/2014 – Yellow Book* 1999

2 *Peterborough City Council v Enterprise Managed Services Ltd* [2014] EWHC 3193 (TCC) – *Silver Book* 1999

FIDIC explains the importance of DAAB as a mechanism for efficient PM that:[3]

- provides a procedure for resolving Disputes at much lower cost and in much less time than required for arbitration;
- resolves the conflict of interest occurring when the CA appointed by the Employer not only determines the Contractor's entitlements but also has the authority to issue a decision as to whether its determination was correct;
- provides for an independent third party promoting early resolution of disputes to enable the project to proceed without unnecessary disruption;
- assists in maintaining appropriate communication between the Parties;
- assists in dispute avoidance;
- gives decisions providing provisional relief.

Recent data[4] demonstrates the dispute avoidance role of proactive standing dispute boards:

- only 0.53 per cent of the issues resolved by them have been referred to arbitration, against 14.05 per cent for disputes resolved by *ad hoc* boards appointed after a dispute arises;
- they settle 50 per cent of the issues with informal opinions; only the remaining 50 per cent are referred to them as disputes for a formal decision.

The data summarise the benefits of standing and *ad hoc* boards appointed on 231 projects:

- Sixty-six per cent of the dispute board decisions are immediately complied with, with NODs in 51 per cent;
- only 6 per cent are referred to arbitration, and only 22 per cent of those are overturned.

The 2017 edition of the FIDIC books for major works strengthens the role of the dispute board as an independent member of the PM team, supporting timely and efficient dispute avoidance:

- the 1999 *ad hoc* Dispute Adjudication Board (DAB) appointed only after a dispute arises under the Yellow and Silver Books is replaced with the standing **DAAB** appointed before the Commencement Date, adding in the name 'avoidance' to emphasise its preventive function;
- the Clause 20 Claims procedure is identical for both Parties, and a new Clause 21 is created for Disputes, demonstrating that the Notice of Claim is not an adversarial act leading to disputes but an early warning important for the risk management (in the 1999 Edition, Clause 20 governs the procedures for Contractor's claims, disputes, and arbitration, while the Employer's claims are in Clause 2 with no time bar);

3 FIDIC, *The FIDIC Golden Principles* (2019) 12
4 DRBF, reported on the Conference in Tokyo (2018)

- Claims and Disputes are defined terms with detailed procedures and strict time limits;
- when the Parties fail to appoint the DAAB, and the appointing official does so under Sub-Clause 21.2, the Parties and the members shall be deemed to have signed and be bound by the DAAB Agreement, with fees as stated in the terms of the appointment;
- arbitrators may take account of the extent to which a Party has failed to co-operate in constituting a DAAB in any award dealing with costs of the arbitration (Sub-Clause 21.6);
- Sub-Clause 21.3 [Avoidance of Disputes] is dedicated to dispute prevention;
- the Parties shall be entitled to object to a DAAB Member under Clause 11 of the GCs of DAAB Agreement (**DAABA**) for an alleged lack of independence/impartiality or otherwise; the challenge procedure is to be decided by the ICC and administered by the ICC International Centre for Alternative Dispute Resolution (**ADR**);
- if the Dispute is not referred to the DAAB within 42 days of a NOD, such NOD shall be deemed to have lapsed and be no longer valid (Sub-Clause 21.4.1);
- if a Party is dissatisfied with only part of the DAAB's decision, this part shall be identified in the NOD, and the remainder of the decision shall become final and binding (Sub-Clause 21.4.4);
- an additional ground for termination is introduced in Sub-Clauses 15.2.1(a)(iii) and 16.2.1(d)(ii) (Silver Book 16.2.1(c)(ii)) for the failure of a Party to comply with a DAAB decision when constituting a material breach of its obligations under the Contract;
- the failure to comply with the DAAB decision may be referred under Sub-Clause 21.7 to arbitration for enforcement by way of summary or other expedited procedure as an interim/provisional measure or award, which may include an order or award of damages or other relief.

17.2.1.2 Selection of adjudicators

Pursuant to Sub-Clause 21.1 (Red/Yellow/Silver/Emerald Books)/14.1.1 (Green Book), each Party nominates at least three adjudicators in the Contract Data. Under the Green Book Sub-Clause 14.1.1, the Parties shall jointly appoint one Adjudicator from those listed. The Employer shall state in the Contract Data under Sub-Clause 21.1 of the Red/Yellow/Silver/Emerald Books whether the DAAB shall comprise one member; otherwise, the default is three members, which may be too expensive for a project with a value below $50 million. If the project is complex with major risks, the Employer may prefer to have a three-member DAAB for a project with lower value, selecting professionals with different expertise (e.g. a geotechnical expert, a quantity surveyor, and an expert with legal qualifications). The nominated persons shall be 'suitably qualified' (Sub-Clause 21.1). Before nominating them, the nominating Party should ask whether they are available and request a warranty required under Sub-Clause 3.3 of the GCs of DAABA that the professional is:

(a) experienced and/or knowledgeable in the type of work that the Contractor is to carry out under the Contract;

(b) experienced in the interpretation of construction and/or engineering contract documentation; and
(c) fluent in the language for communications stated in the Contract Data (or agreed between the Parties and the DAAB).

The candidate has to warrant that he/she is and will remain impartial and independent of the Employer, the Contractor, the Employer's Personnel, and the Contractor's Personnel (being free of conflict of interest as defined under Sub-Clause 4.1 of the GCs of DAABA).[5]

The Employer/tenderers may nominate members of the FIDIC President's List of Approved Adjudicators,[6] or from national lists, or from among other professionals complying with the above requirements. After the selection of the Contractor, the Parties jointly agree upon the preferred nominations and request confirmation of availability and no conflict of interest.

17.2.1.3 Appointment of the DAAB/Green Book Adjudicator

The Parties shall jointly appoint the member(s) of the DAAB (Green Book Adjudicator) within the time stated in the Contract Data. For the purposes of dispute avoidance, it is important to ensure that the appointment is before the Commencement Date. If no time limit is stated, the default is:

- Silver Book: 28 days after the Contract is signed;
- Red/Yellow/Emerald Books: 28 days after the Letter of Acceptance, while the Contract shall be signed within 35 days after the Letter of Acceptance;
- Green Book: 28 days after the date when the Contract comes into effect.

If the DAAB is to comprise three members, each Party shall select one member for the agreement of the other Party. The Parties shall consult both these members and shall agree the third member, who shall be appointed to act as chairperson. The DAAB shall be deemed to be constituted on the date that the Parties and the sole member or the three members of the DAAB have all signed a DAABA. The terms of the remuneration of either the sole member or each of the three members, including the remuneration of any expert whom the DAAB consults, shall be mutually agreed by the Parties when agreeing the terms of the DAABA. Each Party shall be responsible for paying one-half of this remuneration.

The conditions of the DAABA for the Red/Yellow/Silver/Emerald Books comprise:

- Clause 21 and any other provisions of the Contract applicable to the DAAB's activities;
- the GCs of DAABA appended to the GCs of the relevant book;
- the DAAB Procedural Rules annexed to the GCs of DAABA;
- any protocols that may be proposed by the DAAB and agreed by the Parties (Figure 17.2).

5 *ICC Case 19581 Final Award* (August 2014, ICC Dispute Resolution Bulletin 2015 Issue 1): the arbitral tribunal ruled that the Contractor was justified in referring the dispute directly to arbitration, as the DAB must be considered inexistent due to the fact that the person appointed as its sole member lacked the required independence and impartiality.

6 <https://fidic.org/president-list> accessed 19 January 2023

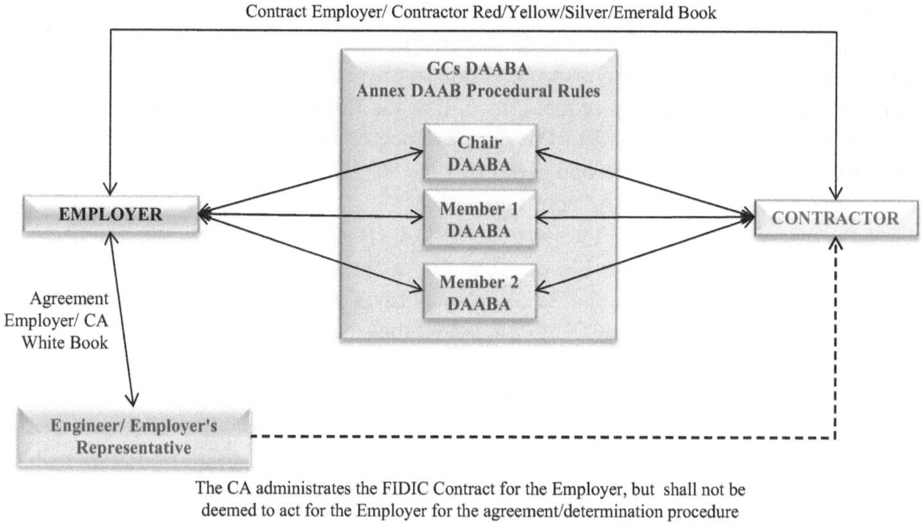

Figure 17.2 FIDIC Contract with DAAB: Contractual relationships

Under Sub-Clause 21.1, the appointment of any member may be terminated by mutual agreement of both Parties, but not by the Employer or the Contractor acting alone. Unless otherwise agreed by both Parties, the term of the DAAB (including the appointment of each member) shall expire either on the date the discharge under Sub-Clause 14.12 (confirming the Final Statement) shall have become (or is deemed to have become) effective, or 28 days after the DAAB has given its decision on all Disputes referred to it under Sub-Clause 21.4, whichever is later. However, if the Contract is terminated, the term of the DAAB (including the appointment of each member) shall expire 28 days after the DAAB has given its decisions on all Disputes (referred to it within 224 days after the date of termination) or the date that the Parties reach a final agreement on all matters (including payment) in connection with the termination, whichever is earlier.

Sub-Clause 21.2 gives a solution where the Parties are unable to reach agreement on the appointment of the DAAB or replacement of a member. Either or both Parties may apply to the president of FIDIC to appoint the members after due consultation with both Parties. The appointment shall be final and conclusive.

17.2.1.4 Avoidance of disputes

The DAAB members cannot give advice to the Parties (GCs of DAABA Sub-Clause 5.1(b)), and the Parties shall not request advice or consultation concerning the conduct of the Contract (GCs of DAABA Sub-Clause 6.4), except as required for the DAAB members to carry out the DAAB's activities. At the same time, the DAAB members shall be available to give informal assistance when requested jointly by the Parties (GCs of DAABA Sub-Clause 5.1(e)).

In order to prevent Disputes, the DAAB members have an obligation to become and remain knowledgeable about the Contract and informed about the Parties' performance of the Contract, the Site and its surroundings, and the progress of the Works. To achieve this, they shall regularly visit the Site, meet with the Parties, and study and archive all documents received (GCs of DAABA Sub-Clause 5.1(d)).

Under Rule 4.3, the Parties have an obligation to provide the DAAB with a copy of all documents that the DAAB may request, including the documents forming the Contract and any other document relevant to the performance of the Contract, such as progress reports, the Programme, instructions given by the CA, Variations, Statements submitted by the Contractor, certificates, and communications.

After the initial meeting with the Parties, the DAAB shall hold meetings, and/or visit the Site, at regular intervals (70–140 days) and/or at the written request of either Party. Either Party may request additional Site visits/meetings at times of critical construction events and in the case of urgency. Rule 3.1 points that the purpose of meetings and Site visits is not only to enable the DAAB to become and remain informed about the project development, but to become aware of any issue or disagreement between the Parties, in order to assist them in dispute prevention. The Parties and the CA shall attend each meeting. The Contractor in co-operation with the Employer and the CA shall be responsible for the logistics, including health, safety, and security (HSS) management, but the DAAB shall manage the meetings, including:

- time management (setting the time and agenda in consultation with the Parties);
- stakeholder management (restricting the number of participants, seeking balance between the Parties and efficiency);
- scope management (sticking to the agreed timing and agenda, unless informal assistance is needed);
- communications management (leading the meetings in a polite but confident manner, giving equal opportunity to the Parties to report/state their position, preparing a report on the activities during the meeting/Site visit for the Parties);
- cost and resource management (using virtual meetings and videos to reduce travel costs and CO_2 emissions);
- risk management (asking questions without giving advice for the project execution, to focus the attention of the Parties on issues spotted on the Site);
- claims management (requesting information for pending payments, changes, non-conformities, compliance with the Variation and Claims procedures, establishing the facts, and inviting the Parties to hold meetings in an attempt to reach an agreement on the issues or to use the procedure for informal assistance).

The Sub-Clause 21.3 (Red/Yellow/Silver/Emerald Book)/14.1.2 (Green Book) sets out the informal assistance procedure, which may commence with:

- a joint request by the Parties for assistance/informal discussion of any issue or disagreement; or
- an invitation by a proactive DAAB (Green Book Adjudicator) for such a request in the case of disagreement.

The advice given during informal meetings or Site visits is not binding on the Parties/DAAB (Green Book Adjudicator).

17.2.2 The dispute resolution procedure

17.2.2.1 Multi-tier dispute resolution

Figure 17.3 summarises the dispute resolution procedure under the FIDIC Red/Yellow/Silver/Emerald Books:

1. A Dispute may be referred to the DAAB under Sub-Clause 21.4.1:
   a. within 42 days of giving/receiving a NOD with the CA's determination;
   b. when no determination is given within the time limit on a Sub-Clause 3.7 (a)[7] matter within 42 days;
   c. in case of failure to certify/pay, referred to under Sub-Clause 16.2.1;
   d. in case of non-payment of financing charges due under Sub-Clause 14.8, within 28 days after the request for payment;
   e. in case of disagreement with a Notice of intention to terminate the Contract under Sub-Clauses 15.2.1/16.2.1, or a Notice of termination under Sub-Clauses 15.2.2/16.2.2/18.5/18.6.
2. The DAAB shall give its decision under Sub-Clause 21.4.3 within 84 days after receiving the reference, or such period as proposed by the DAAB and agreed by the Parties (with no obligation to give the decision until any outstanding invoices due are paid in full to the DAAB). The decision shall be binding on both Parties, which shall promptly comply with it, whether or not a Party gives a NOD.
3. Under Sub-Clause 21.4.4, if either Party is dissatisfied with the DAAB's decision, it may give within 28 days a NOD to the other Party with a copy to the DAAB and the CA setting out the matter in Dispute and the reasons for dissatisfaction. Also, either Party may give a NOD if the DAAB fails to give its decision within

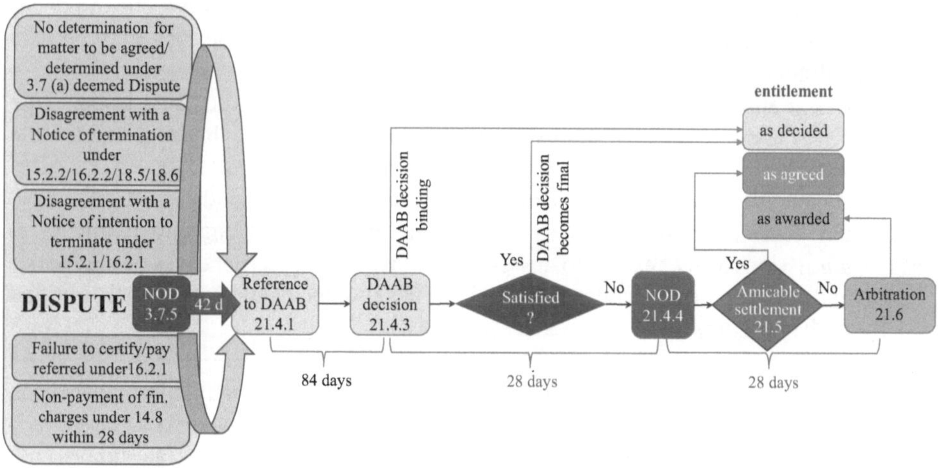

Figure 17.3 Dispute resolution procedure

7 Silver Book Sub-Clause 3.5 (a)

the Sub-Clause 21.4.3 period. If no NOD is given within 28 days after receiving the DAAB decision, the decision shall become final and binding on both Parties.

4. Both Parties shall attempt to settle the Dispute amicably under Sub-Clause 21.5 before commencing arbitration.
5. Unless the Parties agree otherwise, arbitration may be commenced under Sub-Clause 21.6 on or after the 28th day after the day on which the NOD is given, even if no attempt at amicable settlement has been made.
6. Arbitration shall finally settle any Dispute in respect of which the DAAB's decision has not become final and the Parties have not settled it amicably.

17.2.2.2 DAAB as the first mandatory step

As soon as any Dispute is referred to the DAAB under Sub-Clause 21.4.1, the DAAB starts to manage the dispute resolution process. The CA has a reactive role to support the DAAB's activities as Employer's consultant. Rule 5.1(k) empowers the DAAB to open up, review, and revise any certificate, decision, determination, instruction, opinion, or valuation of the Engineer (Silver Book: Employer) that is relevant to the Dispute. The DAAB applies most of the PM Knowledge Areas:

- time management (establishing the procedure under Rule 5.1(b); deciding on the duration of any hearing under Rule 7.1(b));
- communications management (requesting documents/clarifications to establish the facts under Rule 5.1(g); written documentation and arguments from the Parties under Rule 7.1(c) prior to any hearing; documents/oral submissions under Rule 7.1(e) at any hearing);
- scope management (deciding on its own jurisdiction, and the scope of the referred Dispute under Rule 5.1(c));
- cost and resource management (deciding under Rule 5.1(e) to work only on documents where possible; if facts have to be established, the hearing may be held online);
- risk management (taking the initiative in ascertaining the facts under Rule 5.1(g) and adopting an inquisitorial procedure during any hearing under Rule 7.1(d));
- stakeholder management (requesting under Rule 7.1(f) attendance of persons at any hearing that may assist in ascertaining the facts; refusing under Rule 7.1(g) admission to any hearing to any persons other than representatives of the Parties and CA; proceeding under Rule 7.1(h) in the absence of any party timely notified of the hearing).

The DAAB members shall use all reasonable endeavours to reach a unanimous decision (Rule 8.2(b)). If this is not possible, the decision shall be made by a majority of the DAAB members, who may require the other DAAB member to prepare a separate report with reasons and supporting particulars to be issued to the Parties (Rule 8.2(c)).

Rule 8 enables a typographical, clerical, or arithmetical error found by the DAAB or either Party to be corrected by an addendum to the original decision issued to the Parties. The period for NOD shall be calculated from the date the Parties receive the addendum.

17.2.2.3 Amicable settlement as the next step

After a NOD with the DAAB's decision has been given, Sub-Clause 21.5 obliges the Parties to attempt to settle the dispute amicably before the commencement of arbitration. However, sometimes the dispute escalation leads to adversarial behaviour, and the Parties cannot even talk with each other. If direct negotiations by senior executives are no longer possible, other ADR methods may be used (less formal, time-consuming, and costly than arbitration). FIDIC recommends both Parties to engage actively to settle the Dispute amicably, accepting consensually the most suitable procedure. FIDIC proposes for the Parties to consider the ICC Expert Rules if expert determination is the preferred method, and the ICC Mediation Rules if mediation is preferred. The GPPC includes alternative Mediation Rules.[8]

17.2.2.4 Arbitration as the final step

FIDIC recommends that arbitration under Sub-Clause 21.6 should not be replaced with litigation, since:[9]

- in international contracts, international commercial arbitration has numerous advantages over litigation in national courts and is likely to be more acceptable to the Parties;
- compared with litigation, arbitration is a private procedure in which a sole arbitrator or a panel of three arbitrators will decide the dispute and render an award, which should be enforceable by due legal process;
- arbitration is particularly suited to the resolution of Disputes that arise under engineering and construction contracts, because the Parties themselves can choose and agree the arbitrators, who can be qualified and experienced technical professionals, not just legal professionals;
- in many countries, arbitration is enshrined in the legislation, and the 1958 New York 'Convention on the Recognition and Enforcement of Foreign Arbitral Awards' makes enforcement of the award much easier and more straightforward than enforcing a court judgement in a different jurisdiction, since most countries have ratified this Convention.[10]

FIDIC has chosen for the Red/Yellow/Silver/Emerald/Green Books the ICC as the arbitral institution whose arbitration rules shall be used and followed during the arbitration.[11] The justification is that ICC's arbitration rules are well suited to international engineering and construction contracts, since they provide for appropriate case management and for the scrutiny of any award issued by the arbitral tribunal.[12] However, before entering into a Contract, it is recommended that the Parties agree the number of arbitrators and the language of arbitration, while the ICC Court can be relied upon to choose a suitable place

8 FIDIC, *Yellow Book* (FIDIC 2022) GPPC 53

9 FIDIC, *The FIDIC 2017 Contracts Guide* (FIDIC 2022) 544

10 New York Arbitration Convention, List of Contracting States <http://www.newyorkconvention.org/list+of+contracting+states> accessed 23 January 2023

11 ICC Rules of Arbitration (2021) <https://iccwbo.org/dispute-resolution-services/arbitration/rules-of-arbitration/> accessed 23 January 2023

12 FIDIC, *The FIDIC 2017 Contracts Guide* (FIDIC 2022) 545

of arbitration. It is important to choose a language for the arbitration that is fair to both Parties.

The Parties may replace ICC with other institutions, ensuring that the choice that is made will guarantee independence and fairness, and that the institution:[13]

- is indeed a well-recognised international arbitral institution;
- is willing to administer arbitrations under the Contract;
- is knowledgeable in the experience and competence that are required of arbitrators under FIDIC Contracts;
- has the capacity and is, in fact, able to make an independent and appropriate appointment of arbitrators;
- has international arbitration rules compatible with the provisions of Clause 21 (and with other relevant matters set out in the Contract Data).

For major projects tendered internationally, FIDIC recommends that the place of arbitration be situated in a country other than that of the Parties. This country should have a modern and liberal arbitration law and should have ratified a bilateral or multilateral convention (such as The 1958 New York Convention), or both, that would facilitate the enforcement of an arbitral award in the states of the Parties.[14]

Sub-Clause 21.6 empowers the arbitrators to open up, review, and revise any certificate, determination (other than a final and binding determination), instruction, opinion, or valuation of the Engineer (Silver Book: Employer) and any decision of the DAAB (other than a final and binding decision) relevant to the Dispute. Nothing shall disqualify the CA from being called as a witness and giving evidence before the arbitrator(s) on any matter whatsoever relevant to the Dispute. At the same time, the DAAB members shall not be called as witnesses to give evidence concerning any Dispute in any arbitration under the Contract (GCs of DAABA Sub-Clause 8.1). Any decision of the DAAB shall be admissible in evidence in the arbitration. Neither Party shall be limited in the arbitration proceedings to the evidence or arguments previously put before the DAAB to obtain its decision, or to the reasons for dissatisfaction given in the Party's NOD under Sub-Clause 21.4.

Figure 17.4 presents the routes to arbitration:

- the main route is, following Sub-Clauses 21.4 and 21.5, to commence arbitration only after the DAAB and amicable settlement procedures are exhausted, on or after the 28th day after the NOD with the DAAB's decision was given;
- Sub-Clause 21.8 sets out an alternative direct route when no DAAB is in place, which should not be misused to avoid the constitution of the DAAB, as demonstrated by the case law;[1]
- Sub-Clause 21.7 sets out a direct route to arbitration for the failure to comply with an agreement of the Parties/ a final and binding determination under Sub-Clause 3.7 (Silver Book: 3.5)/ a binding DAAB decision (final or not). The arbitral tribunal may enforce the agreement/determination/decision by an interim or provisional measure or an award, using summary or other expedited procedure.

13 FIDIC, *The FIDIC 2017 Contracts Guide* (FIDIC 2022) 545
14 FIDIC, *Yellow Book* (FIDIC 2022) GPPC 55

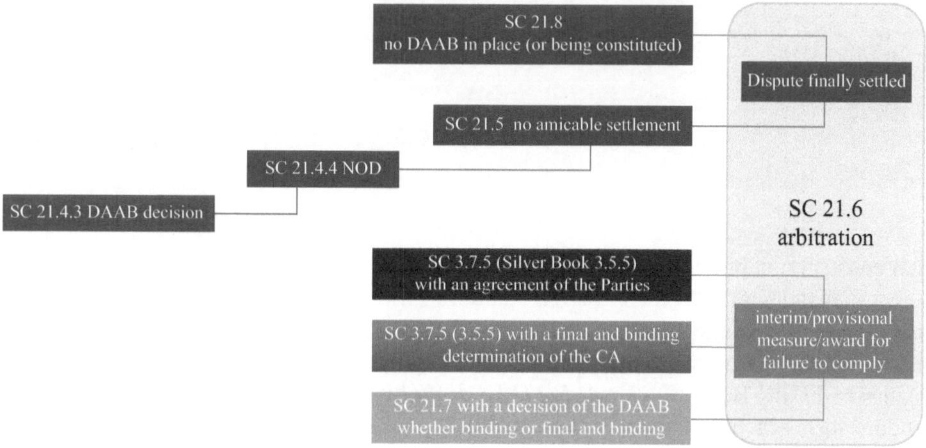

Figure 17.4 Routes to arbitration

Arbitration may be commenced before or after completion of the Works. The obligations of the Parties, the CA, and the DAAB shall not be altered by reason of any arbitration.

17.3 Innovative dispute management

A DAAB appointed at the start of the Contract plays a major dispute avoidance role as a member of the PM team by following the compliance with the FIDIC procedures and giving informal advice.

In 2012, JICA recognised the standing dispute boards in FIDIC contracts 'as a mechanism for sound contract management rather than conflict settlement procedures'.[15] Ten years later, its example is followed by the most important MDBs.

Virtual DAAB meetings and Site visits using real-time video or drone footage save travel time and costs. Online platforms enabled DAAB activities and virtual arbitration during the COVID-19 restrictions.

The use of technologies like blockchain to track and record contractual events will generate a more robust and reliable audit trail of past events if and when a Dispute arises. The design of the blockchain should help Parties agree on a 'single source of truth' and therefore to understand better where a problem may have arisen and where responsibility for that issue lies. In turn, this will help to narrow the scope of Disputes or avoid Disputes altogether, because Parties will have better-quality data available to them at the outset. It will also improve the efficiency of the dispute resolution process. In the near future, Parties will be able to automate the occurrence of digital actions in the event of a breach or failed performance.[16]

15 JICA, *Dispute Board Manual* (2012) iv

16 Charlie Morgan and Rebecca Reed, 'Dispute Resolution in the Era of Big Data and AI' (Herbert Smith Freehills Insight 2019) <https://www.herbertsmithfreehills.com/insight/dispute-resolution-in-the-era-of-big-data-and-ai> accessed 24 January 2023

CHAPTER 18

Conclusion

Each construction project is unique, but familiarity with best practice enables successful project management (PM). The FIDIC standard forms of contract are recognised as the most used for international projects. Besides the numerous contracts in the FIDIC Suites for Services and Works, FIDIC publishes many guidance documents, all of them comprising the FIDIC Body of Knowledge (**FBOK**). This book demonstrates that the FIDIC standard forms include PM tools in each PM Knowledge Area identified in the globally used PMBOK® Guide and its Construction Extension.

The FIDIC family of contracts enable an integrated PM approach aligned in the Suites for Services and Works. This book links the PMBOK with the provisions of the White Book for the appointment of designers and contract administrators, and the Red/Yellow/Emerald/Silver/Green Books for the appointment of contractors. In some cases, it also refers to the provisions of the Gold Book and the supply-chain forms (the Sub-Consultancy Agreement, the Subcontract for Construction, the Subcontract for Plant and Design-Build) and other FBOK documents. The book also outlines the innovative features of the 2021 Green Book, enhancing the PM efficiency.

Each new edition of the FIDIC Contracts upgrades the PM role of the Contract Administrator (CA), named Engineer in the Red/Yellow/Blue-Green/Emerald/Green Book or Employer's Representative in the Gold/Silver Books. The 2022 Reprint with Amendments of the Red/Yellow/Silver/Books 2017 edition further clarifies the PM procedures and strengthens the CA's agreement/determination role. The FIDIC 2017 Contracts Guide published in 2022 is an important component of the FBOK, explaining with practical examples the interpretation of those forms. It is not easy to use these contracts for major works, navigating between more than 100 pages of interacting detailed provisions. That is why the role of the captain (the FIDIC CA acting as an overall project manager) is essential, especially in stormy waters or shallow reefs. For projects without major risks, those who want simplified PM procedures would enjoy the 2021 Green Book General Conditions (GCs) in 26 pages, presenting a user-friendly alternative to the Red/Yellow Books.

Training, and reading case law and books, may minimise disagreements resulting from deficiencies in the contract documents, poorly drafted claims, and failure to understand the contractual obligations, identified among the top causes of construction disputes. It is hoped that the practical tips, case studies, and flow charts presented in this book will help the readers to properly apply the FIDIC PM procedures. The accent is on project procurement management, where the potential for whole-life benefits may be unlocked with fair risk allocation, focus on long term sustainability outcomes, and incentives for proactive risk management and continuing improvements. The FBOK's constantly developing guidance supports FIDIC users in each Knowledge Area (such as the FIDIC Procurement Procedures Guide, the FIDIC Definition of Services Guidelines, Infrastructure Construction/Civil Works, and the FIDIC Guidelines for the Selection of Consultants in the Project Procurement Management).

DOI: 10.4324/9781003272892-18

The old triangle of quality, time, and money no longer leads PM. Proactive risk management, starting with fair risk allocation, is the recipe for project success in a constantly changing environment. Safety and sustainability are important considerations for improved value to the community. Innovations are disrupting all industries. Construction was lagging behind, but is quickly catching up in recent years. Uncertainty accompanying each project may be better mitigated using the new technologies. Building Information Modelling (BIM) and artificial intelligence (AI) enable transparent data sharing and facilitate PM in all Knowledge Areas. The industrialisation of construction by digitisation and standardisation underpins innovative PM. Though the FIDIC GCs present the best practice developed after more than 60 years of use, they cannot easily integrate innovations, due to the long conservative process of their amendments and the stability and certainty expected from the market. At the same time, the FIDIC forms may be tailored for each project. The Specification/Employer's Requirements/Schedules, and Special Provisions amending the GCs, may mandate PM innovations and integrate incentives for improved value. While at present we are far away from smart contracts, distributed ledger and blockchain may be used for immediate payment to the supply chain.

Collaboration is essential for timely risk management. FIDIC is working on a collaborative form for complex projects, but meanwhile, all FIDIC forms may be adapted to create a collaborative environment for the use of BIM, following the Advisory Notes (the latest version reflecting recent developments in digital technologies is included in the 2022 Reprint with amendments of the 2017 Edition of the Red/Yellow/Silver Book). In the case of a framework or many linked contracts in a complex project, the FAC-1 multi-party standard form may integrate all FIDIC Contracts for works and services, enabling early supply-chain involvement and incentives for whole-life benefits and innovations. Even if the Employers are not ready to go so far, this book recommends the use of well-known tools like risk registers, risk management meetings, resource-driven planning, the Last Planner® System (LPS), and platforms for data sharing, plus modern tools developed by start-ups for efficient PM.

The proactive and collaborative solving of issues enables focusing on project success, saving time and money that would otherwise be spent on claims management. The CA in FIDIC Contracts has a leading role in dispute avoidance, supported by the independent Dispute Avoidance/Adjudication Board (DAAB), constituted before the commencement of the Works as part of the PM team. The innovative PM in construction uses a hybrid methodology, introducing agile and lean elements and new PM methods partially disrupting the traditional methodology, while keeping the predictive elements necessary to satisfy numerous stakeholders and regulatory regimes. The FIDIC Contracts have fair risk allocation and detailed PM procedures. They may accommodate any innovations during the preparation of the contract documents, or later using Variations initiated by the Employer, the CA, or the Contractor.

In a nutshell, in order to innovate the PM with FIDIC Contracts, the drafters of the tender documents may:

- mandate the use of BIM following the FIDIC Advisory Notes;
- incorporate lean planning, amending Sub-Clause 8.3;
- integrate bonuses for Sub-Clause 13.2 Value Engineering (VE) incentives in the Schedules (including incentives for early completion, net zero, etc.);

- integrate net zero provisions linked with net zero damages in case of failure to meet the targets;
- provide separate contingencies for known risks and unknown risks using the Sub-Clause 13.4 mechanism of Provisional Sums;
- integrate pain/gain sharing mechanisms in the adjustment of the Contract Price under Sub-Clause 14.1;
- amend the Sub-Clause 14.1 payment mechanism enabling open book pricing for innovations; Cost Plus Profit may be combined with guaranteed maximum price (GMP) and pain/gain share;
- plug in the Particular Conditions innovative provisions from new FIDIC forms like:
  - Risk Register, Risk Management Plan, Contract Risk Management meetings, Milestones, Schedules of Baselines from the Emerald Book 2019;
  - tabulated entitlements/requirements, liquidated damages for prolongation costs and for termination, insurance certificates from the Green Book 2021.

Contractors may further innovate the PM, introducing in their proposals digital tools enhancing safety, transparency, and efficiency (the winning proposal to be integrated into the Contract).

Consultants have a key role in all stages of the project life cycle. The early involvement of the consultant allows focusing on whole-life outcomes, where innovations play a major role. Digitalisation secures instant access to correct data, easy comparison of scenarios, and collaboration of virtual teams. The proactive CA may use innovative PM methods, monitor the Parties' compliance with the requirements, and enable transparent, efficient communications and timely risk management. The 'honeymoon' period after the commencement of the works contracts is the best time for the CA to agree with the Parties additional PM tools upgrading the contractual requirements and streamlining the existing procedures, taking into account the Contractor's quality management (QM) and Compliance Verification Systems.

The PMBOK and FBOK are constantly developing. The FIDIC Committees are elaborating new guidance to support the Contract users to implement sustainable and digital solutions, and enhance project resilience using the contractual mechanisms. The technologies are here, and FIDIC Contracts can be tailored to accommodate them. 'Disrupt or be disrupted' is not only a slogan but a trend, which should incentivise FIDIC users to overcome the fear of innovations and to strive for whole-life improved value supported by the well-known standard forms.

Appendix 1

Scenario with Examples of Notices and Logs

A.1 The story

Dinlandia is a small country with beautiful landscapes and cultural heritage. It is most famous for the discovery of a dinosaur in the 1800s, where its name comes from. Unfortunately, the dinosaur remains are exhibited in a European museum, and Dinlandia cannot attract as many tourists as the neighbouring Marinaland, which boasts sea beaches and marinas. The new government elaborated a 30-year strategy with goals including sustainable development, targeting twice the number of tourists. The parliament approved the strategy and mandated the government to sign credit agreements and use budget funds for the development of sustainable infrastructure, enabling good connectivity.

The Ministry of Infrastructure (**MoI**) commissioned a concept design for a high-speed railway, connecting all major cities with the border of Marinaland.

MoI started negotiations for a credit agreement with the International Bank for Innovations and Sustainability (**IBIS**) for the design and construction of a high-speed railway passing through three major cities and reaching the border of Marinaland. IBIS gave MoI a small grant for technical assistance services. The consultant appointed by the MoI elaborated a feasibility study for the high-speed railway with a cost-benefit analysis (**CBA**). The CBA demonstrated that the project would be viable only if it connected the capital of Dinlandia with the capital of Marinaland and the biggest port of Marinaland.

A.2 Project procurement management

IBIS would only finance the Dinlandia project if the following conditions were fulfilled:

1) Dinlandia to sign a bilateral contract with Marinaland, including obligations of both countries to execute the high-speed railway in the same timeline with compatible installations. Henceforth, this will be called the '**DIMA Corridor**'.
2) The Dinlandia project to achieve the IBIS innovations and sustainability indicators.
3) The Dinlandia project to include energy efficiency renovations of the existing railway stations that will service the high-speed railway.
4) The IBIS Standard Tender Documents for Works and Services to be used, including FIDIC Contracts.

This was a good opportunity for Marinaland to sign a credit agreement for its infrastructure as well, since it needed financing for the modernisation of the main port. At the same time, in order to have a better bargaining position in the negotiations with its neighbour, Marinaland asserted that it did not need the connection to Dinlandia. The government wanted to gain additional benefits for the local business and labour force. Being a richer country with a modernised railway, they told the Dinlandian government that they were ready to help them come out of the dinosaur era, but not for free.

The prime minister and the minister of infrastructure of Dinlandia started the negotiations with the Marinaland government. The negotiations were successful only after the Dinlandia government agreed the following benefits for the Marinaland business:

- signalling, communications, and automation for the high-speed railway to be executed by **RailSignal Ltd**, which had performed such works for the Marinaland railway sections reaching the port and would be executing similar works for the new Marinish sections reaching the border;
- the Dinlandian tender documents for all works contracts financed by IBIS for the DIMA Corridor to include a requirement for hiring at least 35 per cent of personnel from Marinaland.

IBIS did not object to the inclusion of RailSignal Ltd as a nominated Subcontractor and the condition for employment of 35 per cent personnel from Marinaland, only requesting that those personnel should have evidence of former experience in similar works. IBIS took into account the lack of expertise in Dinlandia and the good quality of the infrastructure in Marinaland. IBIS expected those arrangements to facilitate the transfer of knowledge, improve the relationship between the neighbours, and thus facilitate the trade, tourism, and industrial development of the region.

After the bilateral agreement was signed, Dinlandia signed the credit agreement with IBIS for the DIMA Corridor Dinlandia project. The project had to be ready for operation within 5 years. IBIS had signed a licence agreement with FIDIC for the use of the FIDIC 2017 White Book for services, and the Blue-Green Book 2016, Red, Yellow, and Silver Books 2017 reprinted 2022 with amendments, Emerald Book 2019, and Green Book 2021 for works. The IBIS Standard Tender Documents for Works and Services stipulated mandatory use of those FIDIC forms.

Marinaland applied to IBIS for a loan for the financing of the port and the railway section from its capital to the Dinlandian border. Six months later, the Marinaland government signed the credit agreement with IBIS with four and a half years' time for completion of the railway up to the Dinlandian border (DIMA Corridor Marinaland project) and the reconstruction of the port.

MoI signed a contract for services with **PM Kings JV** for the preparation of tender strategy, elaboration of tender documents, permit management, support to MoI during the tender procedures for selection of a designer for Dinia Station, and contractors for three stations and high-speed railway sections. According to the Scope of Services (Appendix 1 to the White Book 2017 Model Agreement), after the selection of contractors, PM Kings JV had to support MoI until the signing of the contracts and then become the FIDIC Engineer/Employer's Representative. Several consulting companies had signed the FIDIC Model Joint Venture Agreement to form the Consulting JV. One of the JV partners, Ferrovia Disegno srl, specialised in railway design. The other JV partner, InfraConsulting Ltd, had experience in FIDIC Contract administration of infrastructure projects. The JV appointed the boutique consulting company Heritage Consult Ltd, with certified experts in the restoration of cultural monuments, under a FIDIC Sub-Consultancy Agreement for the drafting of the technical tender documents for selection of the designer and contractor, and the supervision of the design and construction of Dinia Central Station.

After the inception phase, PM Kings JV organised surveys, elaborated reports for the site investigations, and prepared a tender strategy. Due to the high cost, the criticality of the completion of the railway, and the differences in the terrain, the Consultant proposed that the railway should be executed in two Lots by different contractors:

- **Lot 1: High-speed Railway Section Dinia (Dinlandian capital) Central Station – Mountgate Station,** passing through a valley, with the reconstruction of the existing one-track railway line into a double-track high-speed railway line. The Consultant recommended the Yellow Book 2017 for the appointment of a specialised contractor to design and build the railway, after prequalification of companies experienced in similar projects. The justification was to transfer the design risk to the contractor in order to save time and achieve best value;
- **Lot 2: High-speed Railway Section Mountgate Station – Marinish Border,** new route, passing through the mountains. Lot 2 included several tunnels, the longest being 6 kilometres. PM Kings JV proposed the FIDIC Emerald Book 2019 as the most suitable for underground construction.

The Employer's Requirements included green targets and net zero incentives.

Besides the high-speed railway, the tender strategy envisaged the contracts for the three stations being executed by different contractors in parallel:

- **Lot 3: Marinish Border Station** – a new station to accommodate, besides the Railway Infrastructure administration, also the border administration of the Customs Directory at the Ministry of Finance, the Border Police (part of the Ministry of Interior), the Border Veterinary Control (part of the Ministry of Agriculture), and the Road Taxes administration. PM Kings JV drafted the

Employer's Requirements for the FIDIC Silver Book, transferring most of the risks to the contractor to be appointed after a restricted procedure. The prequalified companies had 4 months to prepare their tenders. The evaluation of the technical proposals had to consider the architecture, net zero performance, and other innovative features.

- **Lot 4: Mountgate Station** – a small existing station in relatively good condition, but with low energy efficiency. In order to save funds, the Employer insisted that an Energy Performance Contract should be signed. The Consultant proposed that the FIDIC Green Book 2021 should be used.
- **Lot 5: Dinia Central Station** – a cultural heritage with beautiful architecture to be renovated. The consultant recommended that the FIDIC Red Book should be used with the Employer's design. First, a designer experienced in cultural heritage projects had to be selected and appointed. While preserving the architectural elements, net zero performance of the renovated building was mandatory for the project to be financed by IBIS.

Italiano Vero Srl signed a contract with the FIDIC Yellow Book GCs for the design and construction of the railway Lot 1.

Tunnelling Ltd signed a contract with the FIDIC Emerald Book GCs for the design and construction of the railway Lot 2.

DIMA United JV signed an engineering, procurement, construction (EPC)/Turnkey contract with the MoI with the FIDIC Silver Book GCs for the new Marinish Border Railway Station.

MoI selected **Baurin GmbH** for the energy efficiency renovation of the existing Mountgate Station. The parties signed an Energy Performance Contract with the FIDIC Green Book GCs.

MoI selected **Artefact Ltd** as a designer, and after the design was approved, **DinArt Ltd** as a Contractor for the renovation of Dinia Central Station. The design agreement incorporated the White Book form, while the works contract included the FIDIC Red Book GCs.

The Engineer PM Kings JV, being a joint venture of legal entities, gave Notices to the Parties for each Lot pursuant to Sub-Clause 3.1. The Notice for Lot 1 authorising the natural person to act on behalf of the Engineer follows.

Project:	Lot 1 DIMA Corridor: High-speed Railway Section Dinia Central Station – Mountgate Station
To:	Mr. Giovanni Salente Contractor's Representative Italiano Vero Srl 16 Stratos St. Dinia 1100
	Mr. Nobi Dini Minister of Infrastructure 6 Dino Blvd. Dinia 1000

Subject: **Notice of the natural person** authorised to act on behalf of the Engineer pursuant to Sub-Clause 3.1 [*The Engineer*] of the General Conditions

Date: 20 February 2024

Dear Sirs,

The Engineer PM Kings JV appoints and authorises Mr. Marin Yordan to act on behalf of the Engineer under the Contract for design and construction of Lot 1 DIMA Corridor: High-speed Railway Section Dinia Central Station – Mountgate Station.

Mr. Yordan is a professional engineer with suitable qualifications, experience, and competence and is fluent in English.

Attachment: CV of Mr. Marin Yordan

Best regards
Mr. Fulvio Conti
PM Kings JV Representative

A.3 Project integration management

The Engineer PM Kings JV, represented by the natural person Mr. Yordan, gave Notice of the Commencement Date for each Lot. The Notice for Lot 1 follows.

Contract: Lot 1 DIMA Corridor: High-speed Railway Section Dinia Central Station – Mountgate Station

To: Mr. Giovanni Salente
Contractor's Representative
Italiano Vero Srl
16 Stratos St. Dinia 1100

Copy: Mr. Nobi Dini
Minister of Infrastructure
6 Dino Blvd. Dinia 1000

Subject: **Notice of the Commencement Date** pursuant to Sub-Clause 8.1 [*Commencement of Works*] of the General Conditions

Date: 1 March 2024

Dear Sirs,

We write to notify you that **15 March 2024** will be the Commencement Date for the Works. The Contractor is instructed to commence the execution of the Works on, or as soon as is reasonably practicable after, the Commencement Date.

Best regards
For the Engineer PM Kings JV
Mr. Marin Yordan

The Engineer gave Notices to the Parties for each Lot pursuant to Sub-Clause 3.4 for delegation of authority. The Notice for Lot 1 follows.

Contract:	Lot 1 DIMA Corridor: High-speed Railway Section Dinia Central Station – Mountgate Station
To:	Mr. Giovanni Salente Contractor's Representative Italiano Vero Srl 16 Stratos St. Dinia 1100 Mr. Nobi Dini Minister of Infrastructure 6 Dino Blvd. Dinia 1000
Subject:	**Notice of delegation of authority to assistants** pursuant to Sub-Clause 3.4 [*Delegation by the Engineer*] of the General Conditions.
Date:	3 March 2024

The Engineer hereby delegates authority to the Engineer's Representative Mr. Hurry Boy to act on behalf of the Engineer at the Site, excluding powers to replace the Engineer's Representative, to act under Sub-Clause 3.7 [*Agreement or Determination*] and Sub-Clause 15.1 [*Notice to Correct*]. Mr. Boy is a natural person with suitable qualifications, experience, and competence to carry out the duties and exercise the authority of the Engineer's Representative, fluent in English. A copy of his curriculum vitae is attached to this Notice.

The Engineer hereby delegates authority and assigns duties to Ms. Fani Rosh to act on behalf of the Engineer at the Site as the Resident Engineer for Lot 1, excluding powers to act under Sub-Clause 3.7 [*Agreement or Determination*] and Sub-Clause 15.1 [*Notice to Correct*]. Ms. Rosh is an engineer with suitable qualifications, experience, and competence to carry out the duties and exercise the authority of the Resident Engineer for Lot 1, fluent in English and Dinlandian. A copy of her curriculum vitae is attached to this Notice.

The Engineer hereby delegates authority and assigns duties to assistants to perform Review of the Contractor's Documents under Sub-Clause 5.2, inspect the Works under Sub-Clause 7.3, and attend the tests under Sub-Clause 7.4 as follows:

1. Mr. Elis Samish for structural works;
2. Ms. Poli Giovanni for electrical works;
3. Mr. Valeri Petrov for mechanical works;
4. Ms. Nadia Petkova for fire safety;
5. Mr. Radi Veri for railway;
6. Mr. Kori Dunisi for signalling and telecommunications.

The Engineer confirms that the assistants have suitable qualifications, experience, and competence and are fluent in the language for communications, being Dinlandian.

Attachment: CVs of the above Engineer's assistants

Best regards
For the Engineer PM Kings JV
Mr. Marin Yordan

A.4 Project scope management

During the design Review, the Employer requested the Engineer to instruct the Contractor to reconstruct 360 m of railway connecting the Lot 1 route to Lake Station. The Engineer gave the following Notice.

Contract:	Lot 1 DIMA Corridor: High-speed Railway Section Dinia Central Station – Mountgate Station
To:	Mr. Giovanni Salente Contractor's Representative Italiano Vero Srl 16 Stratos St. Dinia 1100
Copy:	Mr. Nobi Dini Minister of Infrastructure 6 Dino Blvd. Dinia 1000
Subject:	**Notice of Variation** pursuant to Sub-Clause 13.3.1 [*Variation by Instruction*] of the General Conditions
Date:	1 September 2024

Dear Sir,

The Engineer instructs the Contractor to design and execute reconstruction of the 360-m railway connection to Lake Station pursuant to Sub-Clause 3.5 [*Engineer's Instructions*]. The reconstruction shall follow the existing route and align with the new level of the high-speed railway at the turnover at the exit of Mountgate Station. The additional works shall include all necessary retaining walls, drainage, etc.

The Contractor is instructed to proceed with the execution of the Variation and to submit within 28 days of receiving this Notice of Variation to the Engineer detailed particulars including:

(a) a description of the varied work, including design for Engineer's Review and details of the resources and methods to be adopted by the Contractor;
(b) any necessary modifications to the Programme according to Sub-Clause 8.3 [*Programme*] and to the Time for Completion; and
(c) the Contractor's proposal for adjustment to the Contract Price, with supporting particulars.

Best regards
For the Engineer PM Kings JV
Mr. Marin Yordan

The Contractor objected to the Variation, promptly giving a Notice to the Engineer:

Contract:	Lot 1 DIMA Corridor: High-speed Railway Section Dinia Central Station – Mountgate Station
To:	Mr. Marin Yordan PM Kings JV 8 Dino Blvd. Dinia 1000
Copy:	Mr. Nobi Dini Minister of Infrastructure 6 Dino Blvd. Dinia 1000
Subject:	**Notice of objection to the Engineer's Notice of Variation dated 1 September 2022** pursuant to Sub-Clause 13.1 (c) of the General Conditions
Date:	3 September 2024

Dear Mr. Yordan,

The Contractor refers to the Engineer's Notice of Variation dated 1 September 2024 pursuant to Sub-Clause 13.3.1 [*Variation by Instruction*] of the General Conditions. The Contractor requests the Variation to be cancelled, since the Variation will adversely affect the Contractor's ability to comply with Sub-Clause 4.18 [*Protection of the Environment*]. The railway connection to Lake Station crosses a protected area, and the reconstruction activities will be dangerous for the natural habitat.

In case you decide to confirm the instruction for Variation, we cannot take responsibility for the environmental procedures under Sub-Clause 1.13 and will be able to give you within 28 days only basic design with description of the varied work in the scope necessary for the environmental procedure. We can give you detailed design for Review, details about the impact on the Programme, and Contract Price after the completion of the environmental procedure.

Best regards
Mr. Giovanni Salente
Contractor's Representative
Italiano Vero Srl

APPENDIX 1

The Engineer confirmed the instruction for Variation:

Contract:	Lot 1 DIMA Corridor: High-speed Railway Section Dinia Central Station – Mountgate Station
To:	Mr. Giovanni Salente Contractor's Representative Italiano Vero Srl 16 Stratos St. Dinia 1100
Copy:	Mr. Nobi Dini Minister of Infrastructure 6 Dino Blvd. Dinia 1000
Subject:	**Notice of confirmation of the Engineer's Notice of Variation dated 1 September 2022** pursuant to Sub-Clause 13.1 of the General Conditions
Date:	9 September 2024

Dear Mr. Salente,

The Engineer refers to the Contractor's Sub-Clause 13.1 Notice dated 3 September 2024 in which it requested the Engineer's Sub-Clause 13.3.1 Notice of Variation dated 1 September 2024 to be cancelled.

The Engineer has reviewed the Contractor's request, consulted with the Employer, and confirms the instruction, because the railway connection enables environmentally friendly transportation of the tourists and it has to be reconstructed to align with the new level of the high-speed railway.

We hereby instruct you to proceed with the basic design and the preparation of the information in the minimum scope required under the Biodiversity Protection Act. Different methods of execution may be presented with minimum interference in the existing infrastructure and proposal for compensation measures in case of limited deforestation or other negative impact. The method statement must include environmentally friendly methods.

We hereby confirm that the Employer shall be responsible for the environmental procedure under Sub-Clause 1.13, but the Contractor, being responsible for the design, must do its best to minimise the impact of the reconstruction on the biodiversity and after the completion of the environmental procedure, to comply with the mitigation and compensatory measures. We agree that you may submit the proposal for the modification of the Programme/adjustment of the Contract Price after the completion of the environmental procedure.

Looking forward to the submission of the basic design for Review under Sub-Clause 5.2.

Best regards
For the Engineer PM Kings JV
Mr. Marin Yordan

The Lot 1 Contractor made the following Value Engineering (VE) proposal:

Contract: Lot 1 DIMA Corridor: High-speed Railway Section Dinia Central Station – Mountgate Station

To: Mr. Marin Yordan
PM Kings JV
8 Dino Blvd. Dinia 1000

Copy: Mr. Nobi Dini
Minister of Infrastructure
6 Dino Blvd. Dinia 1000

Subject: **Value engineering proposal** pursuant to Sub-Clause 13.2 of the General Conditions

Date: 30 September 2024

Dear Mr. Yordan,

The Contractor proposes a pedestrian overpass at km +56+800 to replace the pedestrian underpass. The cost for the construction will be the same, but the cost for the maintenance will be less (no need for pumping installation for drainage of the elevators and underpass), and it will reduce the consumption of energy for lighting. Additionally, flowers may be planted and easily maintained. The visibility of the structure will reduce the crime threats. We are sure that this alternative will be preferred by the community.

The proposal shall have no impact on the Time for Completion, if we receive your consent within the next two months.

Best regards
Mr. Giovanni Salente
Contractor's Representative
Italiano Vero Srl

The Engineer gave consent to the proposal and instructed the Variation:

Contract: Lot 1 DIMA Corridor: High-speed Railway Section Dinia Central Station – Mountgate Station

To: Mr. Giovanni Salente
Contractor's Representative
Italiano Vero Srl
16 Stratos St. Dinia 1100

Copy: Mr. Nobi Dini
Minister of Infrastructure
6 Dino Blvd. Dinia 1000

Subject: **Notice of consent to the Contractor's VE proposal dated 30 September 2024 and Notice of Variation** pursuant to Sub-Clause 13.2 of the General Conditions

Date: 9 October 2024

Dear Mr. Salente,

The Engineer refers to the Contractor's Sub-Clause 13.2 Notice dated 30 September 2024 including a VE proposal for a pedestrian overpass replacing the underpass at km +56+800.

The Engineer has reviewed the Contractor's request, consulted with the Employer and the Chief Engineer of the Municipality, and gives consent to the proposal.

We hereby instruct you to proceed with the Variation and submit design for Review under Sub-Clause 5.2.

Best regards
For the Engineer PM Kings JV
Mr. Marin Yordan

Italiano Vero Srl filled in the following Issue Log and Variation Log.

Issue Log

ID	*Who raised it*	*Date*	*Description*	*Responsible*	*Target resolution date*	*Status*	*Final solution*
I1	Railway Infrastructure Agency	20 March 2024	Additional turnout at Dinia Central Station	Contractor's design manager	6 September 2024	instructed varied works	Closed with Variation 01
I2	Contractor	25 March 2024	Existing building in conflict with the new alignment	Employer	30 June 2024	pending decision for demolition	
I3	Mayor Hatony	20 June 2024	New pedestrian overpass km 30+120	Contractor's design manager	30 August 2024	pending proposal	
I4	Employer	1 September 2024	Connection to Lake Station	Engineer	9 September 2024	instructed varied works	Closed with Variation 02

APPENDIX 1

Variation Log

Issue ID	*Variation Sub-Clause*	*Description*	*Proposed Cost Plus Profit [Din]*	*Funds*	*Proposed EOT*	*Engineer comment*	*date*	*Status*
I1	13.3.1	Additional turnout at Dinia Central Station	50,000	Provisional Sums	NA	50,000	20 September 2024	Variation 01
I2	13.3.2	Existing building in conflict with the new alignment	10,000		0	pending decision for demolition		pending
I3	13.3.2	New pedestrian overpass km 30+120				pending proposal		
I4	13.3.1	Connection to Lake Station	3,360,000	State budget	30 days	2,680,000 no EOT	9 September 2024	Variation 02 Claim 06
NA	13.2	A pedestrian overpass at km +56+800 to replace the pedestrian underpass	0	NA	NA	consent	9 October 2024	Variation 03
	Total		**3,420,000**		**30 days**	**2,730,000**		

EOT: extension of time.

The Fire Act amendment issued after the Commencement Date of Lot 5 imposed higher requirements for the fire extinguishing systems. The Employer's agreement with the designer had expired. The Employer preferred to use the contingency in the Red Book Contract for the design of the varied works. The Engineer gave a Notice to the Contractor and requested a proposal for design and execution of the varied works necessary for compliance with the amendment to the Fire Act.

Contract:	Lot 5 DIMA Corridor: Dinia Central Station
To:	Mr. Ron Albani Contractor's Representative DinArt Ltd. 26 Victory Blvd. Dinia 1100
Copy:	Mr. Nobi Dini Minister of Infrastructure 6 Dino Blvd. Dinia 1000
Subject:	**Notice of adjustment to the execution of the Works as a result of a change in the Laws** pursuant to Sub-Clause 13.6 and **request for proposal** pursuant to Sub-Clause 13.3.2 of the General Conditions

Date: 1 December 2024

Dear Mr. Albani,

The Engineer gives Notice that an amendment to the Fire Act in force from 1 January 2025 requires a water tank for fire needs to be designed and constructed.

The Engineer requests the Contractor to prepare a proposal pursuant to Sub-Clause 13.3.2 [*Variation by Request for Proposal*] together with supporting documentation.

The proposal should estimate the volume and location of the tank in compliance with the amended Fire Act.

Following the receipt of this Notice, the Contractor should, as soon as practicable, submit to the Engineer for consideration detailed particulars including:

(a) a description of the varied work to be performed, including details of the resources and methods adopted or to be adopted by the Contractor;
(b) any necessary modifications to the Programme according to Sub-Clause 8.3 [*Programme*] and to the Time for Completion; and
(c) the Contractor's proposal for adjustment to the Contract Price, with supporting particulars.

Best regards
For the Engineer PM Kings JV
Mr. Marin Yordan

Contract: Lot 5 DIMA Corridor: Dinia Central Station

To: Mr. Marin Yordan
PM Kings JV
8 Dino Blvd. Dinia 1000

Copy: Mr. Nobi Dini
Minister of Infrastructure
6 Dino Blvd. Dinia 1000

Subject: **Notice of objection to the Engineer's request for proposal** dated 1 December 2024 pursuant to Sub-Clause 13.3.2 (b) by reference to Sub-Clause 13.1 (a) of the General Conditions

Date: 3 December 2024

Dear Mr. Yordan,

The Contractor refers to the Engineer's Notice dated 1 December 2024 requesting proposal for additional works necessary because of changes in the Laws. The Contractor is entitled to object under Sub-Clause 13.3.2 (b) by reference to Sub-Clause 13.1 (a) of the General Conditions and insists the Engineer to exclude from the varied works the design, since it is Unforeseeable having regard to the scope and nature of the Works described in the Specification. The Contractor cannot comply pursuant to Sub-Clause 13.3.2 (b) since the FIDIC Red Book Contract is for construction only and the design is Employer's responsibility. The Contractor does not have design expertise and asserts that any attempt by the Engineer to transfer design responsibility to the Contractor will be an infringement of the FIDIC Golden Principles.

The Contractor is ready to submit a proposal for the execution of the additional construction works within 21 days after the Engineer submits the design of the water tank agreed by the Fire Authority, and approved by the Local Authority with an amendment of the Construction Permit. The design agreements and construction permits are Employer's responsibility under Sub-Clause 1.13.

Best regards
Mr. Ron Albani
Contractor's Representative
DinArt Ltd.

The Engineer confirmed the request for proposal for Variation:

Contract:	Lot 5 DIMA Corridor: Dinia Central Station
To:	Mr. Ron Albani Contractor's Representative DinArt Ltd. 26 Victory Blvd. Dinia 1100
Copy:	Mr. Nobi Dini Minister of Infrastructure 6 Dino Blvd. Dinia 1000
Subject:	**Notice of confirmation of the Engineer's request for proposal** pursuant to Sub-Clause 13.1 and **instruction** under Sub-Clause 13.4 (a) of the General Conditions
Date:	9 December 2024

Dear Mr. Albani,

The Engineer refers to the Contractor's Sub-Clause 13.3.2 (b) Notice dated 3 December 2024 in which it requested that the Engineer's Sub-Clause 13.3.2 Request for proposal dated 1 December 2024 be varied excluding the design and permitting.

The Engineer has reviewed the Contractor's request, consulted with the Employer, and confirms the request for proposal for the following reasons:

1. The water tank is a minor part of the Works, and our request for the Contractor to design it under the Red Book General Conditions complies with the Golden Principles (see item

4.3 in the FIDIC Golden Principles published by FIDIC in 2019, available on the FIDIC website).

2. The Employer has difficulties in assigning the design, since the agreement with Artefact Ltd. has expired, and a new tender procedure will take much time and effort.
3. The Employer has difficulties in arranging financing for the design services, while under the Lot 5 Contract, the design can be paid from the Provisional Sum for Unforeseeable works.

We hereby instruct you under Sub-Clause 13.4 [*Provisional Sums*] to employ as a Subcontractor Artefact Ltd for the design of the water tank. The design and the as-built records for the water tank under Sub-Clause 4.4.2 shall be in the varied scope of the Contractor's Documents under Sub-Clause 1.1.15, to be submitted to the Engineer for Review under Sub-Clause 4.4.1 by Notice stating that the Contractor's Documents are ready for Review and comply with the Contract.

We hereby confirm that the Employer shall be responsible for the design agreements and construction permit under Sub-Clause 1.13, which will be done within the design Review procedure under Sub-Clause 4.4.1.

We hereby inform you that the Employer has already contacted Artefact Ltd, and they are ready to sign an agreement with you for the design of the water tank. They already have information for the Site and are ready to give you their best quotation and proceed with the design.

We are looking forward to your proposal under Sub-Clause 13.3.2 as soon as possible after the design is ready for Review.

Best regards
For the Engineer PM Kings JV
Mr. Marin Yordan

A.5 Project schedule management

After the Review of the initial Programme, the Engineer gave the following Notice to the Contractor.

Contract:	Lot 1 DIMA Corridor: High-speed Railway Section Dinia Central Station – Mountgate Station
To:	Mr. Giovanni Salente Contractor's Representative Italiano Vero Srl 16 Stratos St. Dinia 1100
Copy:	Mr. Nobi Dini Minister of Infrastructure 6 Dino Blvd. Dinia 1000

Subject:	**Notice of non-compliance of the initial programme** pursuant to Sub-Clause 8.3 of the General Conditions
Date:	15 April 2024

Dear Sir,

We refer to our Notice dated 1 March 2024 setting the Commencement Date as 15 March 2024 and to the Contractor's initial programme for the execution of the Works submitted on 28 March 2024.

The Engineer has reviewed the initial programme and considers that it does not comply with the Contract. It does not include the Review periods for the design and the Tests on Completion. The supporting report has lists of the Contractor's Personnel and Contractor's Equipment, but they are not allocated to the major stages, and the designers are not listed.

The Contractor is required to submit a revised programme within 14 days of the date of receiving this Notice.

Best regards
For the Engineer PM Kings JV
Mr. Marin Yordan

The Engineer PM Kings JV gave the following Notice to the Contractor.

Contract:	Lot 1 DIMA Corridor: High-speed Railway Section Dinia Central Station – Mountgate Station
To:	Mr. Giovanni Salente Contractor's Representative Italiano Vero Srl 16 Stratos St. Dinia 1100
Copy:	Mr. Nobi Dini Minister of Infrastructure 6 Dino Blvd. Dinia 1000
Subject:	**Notice for slow progress pursuant to Sub-Clause 8.7** [*Rate of Progress*] of the General Conditions
Date:	15 September 2024

Dear Sir,

We hereby notify you that the Engineer's tracking of the Contractor's Programme shows that the progress has fallen behind the Programme by 1 month.

The Engineer instructs the Contractor to submit, under Sub-Clause 8.3 [*Programme*], a revised programme describing the revised methods which the Contractor proposes to adopt in order to expedite progress and complete the Works within the Time for Completion. We remind you that if these revised methods cause the Employer to incur additional costs, the Employer shall be entitled subject to Sub-Clause 20.2 [*Claims For Payment and/or EOT*] to payment of these costs by the Contractor, in addition to Delay Damages (if any).

Best regards
For the Engineer PM Kings JV
Mr. Marin Yordan

A.6 Project financial management

The Lot 1 Contractor gave the following Notice.

Contract:	Lot 1 DIMA Corridor: High-speed Railway Section Dinia Central Station – Mountgate Station
To:	Mr. Nobi Dini Minister of Infrastructure 6 Dino Blvd. Dinia 1000
Copy:	Mr. Marin Yordan PM Kings JV 8 Dino Blvd. Dinia 1000
Subject:	**Request for reasonable evidence for financial arrangements** pursuant to Sub-Clause 2.4 of the General Conditions
Date:	18 June 2025

Dear Mr. Dini,

After the Engineer's instruction to execute a Variation dated 6 June 2025, the accumulated total of Variations exceeds 35 per cent of the ACA (Accepted Contract Amount). Please be so kind to provide reasonable evidence that financial arrangements have been made and are being maintained which will enable the Employer to pay the remaining part of the Contract Price.

We already have difficulties with the cash flow due to the Engineer's abuse of the provision under Sub-Clause 14.6.2 (c) used improperly to withhold amounts in IPCs for insignificant discrepancies in the supporting documents. If we do not receive the requested reasonable evidence within 28 days as stipulated under Sub-Clause 2.4, we shall follow the procedure under Sub-Clause 16.1 to suspend work until the evidence is presented.

Best regards
Mr. Giovanni Salente
Contractor's Representative
Italiano Vero Srl

A.7 Project cost management

The Engineer PM Kings JV gave the following Notices.

Contract:	Lot 5 DIMA Corridor: Dinia Central Station
To:	Mr. Ron Albani Contractor's Representative DinArt Ltd. 26 Victory Blvd. Dinia 1100
Copy:	Mr. Nobi Dini Minister of Infrastructure 6 Dino Blvd. Dinia 1000
Subject:	**Notice of the date on which and place where measurement will be made on Site** pursuant to Sub-Clause 12.1 [*Works to be Measured*] of the General Conditions
Date:	1 March 2025

Dear Mr. Albani,

We hereby notify you that the measurement of the flooring will take place on 8 March starting at the first floor at 9 AM.

The Contractor has to supply the following particulars:

- Conformity Declarations and
- Certificates of Origin for the Materials.

If the Contractor fails to attend or send a representative at the time or place specified above (or otherwise agreed with the Contractor), the measurement made by the Resident Engineer shall be deemed to have been made in the Contractor's presence, and the Contractor shall be deemed to have accepted the measurement as accurate.

Best regards
For the Engineer PM Kings JV
Mr. Marin Yordan

Contract:	Lot 5 DIMA Corridor: Dinia Central Station
To:	Mr. Ron Albani Contractor's Representative DinArt Ltd. 26 Victory Blvd. Dinia 1100

Copy:	Mr. Nobi Dini Minister of Infrastructure 6 Dino Blvd. Dinia 1000
Subject:	**Notice of the date on which and place where measurement records may be agreed** pursuant to Sub-Clause 12.1 [*Works to be Measured*] of the General Conditions
Date:	1 March 2025

Dear Mr. Albani,

We hereby notify you that the Engineer has prepared the records for measurement of the wall finishing works. The Engineer invites the Contractor's Representative to examine and agree the records with the Resident Engineer on 10 March 2023 at 10 AM in the Engineer's office.

If the Contractor fails to attend or send a representative at the time or place specified above (or otherwise agreed with the Contractor), the Contractor shall be deemed to have accepted the records as accurate.

Best regards
For the Engineer PM Kings JV
Mr. Marin Yordan

A.8 Project quality management

The Lot 1 Contractor sent to the Engineer the following Notice:

Contract:	Lot 1 DIMA Corridor: High-speed Railway Section Dinia Central Station – Mountgate Station
To:	Mr. Marin Yordan PM Kings JV 8 Dino Blvd. Dinia 1000
Copy:	Mr. Nobi Dini Minister of Infrastructure 6 Dino Blvd. Dinia 1000

Subject: **Notice that reinforcement is ready for inspection** pursuant to Sub-Clause 7.3 of the General Conditions

Date: 14 March 2025

Dear Mr. Yordan,

We write to notify you that the reinforcement of the east foundation of the overpass at km +56+800 is ready for inspection. The Employer's Personnel are invited to inspect it on 15 March at 10 AM. Please find attached the Request for Inspection with the drawing, as-built record, and Conformity Declaration for the reinforcement.

We have planned to pour the concrete on 15 March at 2 PM if you do not delay the inspection.

ATTACHMENTS: per text

Best regards
Mr. Giovanni Salente
Contractor's Representative
Italiano Vero Srl

Promptly on receipt of the Contractor's Notice, the Engineer PM Kings JV gave the following Notice to the Contractor.

Contract:	Lot 1 DIMA Corridor: High-speed Railway Section Dinia Central Station – Mountgate Station
To:	Mr. Giovanni Salente Contractor's Representative Italiano Vero Srl 16 Stratos St. Dinia 1100
Copy:	Mr. Nobi Dini Minister of Infrastructure 6 Dino Blvd. Dinia 1000
Subject:	**Notice that inspection is not required** pursuant to Sub-Clause 7.3 of the General Conditions
Date:	14 March 2025

Dear Sir,

The Engineer refers to your Notice submitted today in which you informed us that the east foundation of the overpass at km +56+800 is ready for inspection by the Employer's Personnel and request the inspection to be done on 15 March at 10 AM.

Please note that the Engineer has received two Notices requesting inspections of structural works at other locations at the same time. Our authorised inspector already examined the reinforcement today, and the Engineer does not require an inspection on 15 March, and so the Contractor can proceed to cover up the reinforcement. We would like to remind you to secure the necessary vibrators and follow the submitted method statement, executing the works with minimum interference with the traffic as planned.

Best regards
For the Engineer PM Kings JV
Mr. Marin Yordan

A.9 Project resource management

The Lot 1 Contractor gave a Notice to the Engineer for the delivery to the Site of major items of Contractor's Equipment:

Contract:	Lot 1 DIMA Corridor: High-speed Railway Section Dinia Central Station – Mountgate Station
To:	Mr. Marin Yordan PM Kings JV 8 Dino Blvd. Dinia 1000
Copy:	Mr. Nobi Dini Minister of Infrastructure 6 Dino Blvd. Dinia 1000
Subject:	**Notice of the delivery of Contractor's Equipment** pursuant to Sub-Clause 4.17 of the General Conditions
Date:	16 November 2024

Dear Sir,

We write to notify you that on **12 November 2024** a 1.5-t compactor, a two-way excavator, a portal crane, a ballast cleaning machine, and a track motor vehicle were delivered to the Site. All of these items are owned by the Contractor.

Best regards
Mr. Giovanni Salente
Contractor's Representative
Italiano Vero Srl

A.10 Project communications management

The Lot 1 Contractor gave the following Notices to the Engineer.

Contract:	Lot 1 DIMA Corridor: High-speed Railway Section Dinia Central Station – Mountgate Station
To:	Mr. Marin Yordan PM Kings JV 8 Dino Blvd. Dinia 1000
Copy:	Mr. Nobi Dini Minister of Infrastructure 6 Dino Blvd. Dinia 1000

Subject: **Notice of errors identified in the Employer's Requirements** pursuant to Sub-Clause 1.9 of the General Conditions

Date: 3 April 2024

Dear Sir,

The Contractor identified errors in the Employer's Requirements during the initial period of review, which followed notification of the Commencement Date, pursuant to Sub-Clause 8.1 [*Commencement of Works*] of the General Conditions.

The Contractor found that 35 of the 60 water culverts to be rehabilitated following the scope of works identified under the Employer's Requirements no longer exist, and new ones have to be constructed. The finding could not have been reasonably possible during the Site examination in the tender period. If an experienced contractor exercising due care could have discovered the error before submitting the Tender, the related questions and answers would have been distributed to all tenderers.

The Contractor invites the Engineer to participate in a joint inspection of the relevant locations specified in the Employer's Requirements and proceed as specified under Sub-Clause 3.7 [*Agreement or Determination*].

Best regards
Mr. Giovanni Salente
Contractor's Representative
Italiano Vero Srl

Contract: Lot 1 DIMA Corridor: High-speed Railway Section Dinia Central Station – Mountgate Station

To: Mr. Marin Yordan
PM Kings JV
8 Dino Blvd. Dinia 1000

Copy: Mr. Nobi Dini
Minister of Infrastructure
6 Dino Blvd. Dinia 1000

Subject: **Notice of error/fault/defect identified in the Employer's Requirements** pursuant to Sub-Clause 1.9 of the General Conditions

Date: 3 April 2025

Dear Sir,

The Contractor has identified an error in the Employer's Requirements outside of the initial period of review, which followed notification of the Commencement Date, pursuant to Sub-Clause 8.1

[*Commencement of Works*] of the General Conditions. The railway track does not align with the new design for the Dinia Central Station. The error could not have been found earlier, since the Engineer submitted the design to us on 16 March 2025.

The Contractor requests the Engineer to review the Employer's Requirements and proceed as specified under Sub-Clause 3.7 [*Agreement or Determination*].

Best regards
Mr. Giovanni Salente
Contractor's Representative
Italiano Vero Srl

A.11 Project risk management

The Lot 1 Contractor gave the following Notices to the Engineer PM Kings JV:

Contract: Lot 1 DIMA Corridor: High-speed Railway Section Dinia Central Station – Mountgate Station

To: Mr. Marin Yordan
PM Kings JV
8 Dino Blvd. Dinia 1000

Copy: Mr. Nobi Dini
Minister of Infrastructure
6 Dino Blvd. Dinia 1000

Subject: **Notice of an Exceptional Event** preventing performance of obligations under the Contract pursuant to Sub-Clause 18.2 of the General Conditions

Date: 13 May 2025

Dear Sir,

The Contractor gives Notice that an Exceptional Event has occurred as defined in Sub-Clause 18.1 [*Exceptional Events*]: specifically, a tornado occurred in the vicinity of the Site on 11 May 2025. It was followed by heavy rains and large hail, and the Site was flooded. The flood triggered landslides and debris flow. The Exceptional Event prevented and is continuing to prevent the Contractor from proceeding with the Works with due expedition and without delay.

Best regards
Mr. Giovanni Salente
Contractor's Representative
Italiano Vero Srl

Contract:	Lot 1 DIMA Corridor: High-speed Railway Section Dinia Central Station – Mountgate Station
To:	Mr. Marin Yordan PM Kings JV 8 Dino Blvd. Dinia 1000
Copy:	Mr. Nobi Dini Minister of Infrastructure 6 Dino Blvd. Dinia 1000
Subject:	**Notice of damage to the Works, Goods, or Contractor's Documents** pursuant to Sub-Clause 17.2 [*Liability for Care of the Works*] of the General Conditions
Date:	16 May 2025

Dear Sir,

The Contractor gives Notice that the Works, the approved design stamped by the authorities, and some Materials stored on the Site have been damaged as a result of the tornado, hail, and flood notified to you on 13 May 2025.

The Contractor requests the Engineer to issue instructions for the rectification of the losses and damages.

Best regards
Mr. Giovanni Salente
Contractor's Representative
Italiano Vero Srl

A.12 Project stakeholder management

The Lot 1 Contractor gave the following Notice to the Engineer:

Contract:	Lot 1 DIMA Corridor: High-speed Railway Section Dinia Central Station – Mountgate Station
To:	Mr. Marin Yordan PM Kings JV 8 Dino Blvd. Dinia 1000
Copy:	Mr. Nobi Dini Minister of Infrastructure 6 Dino Blvd. Dinia 1000

Subject:	**Notice that work is required on Site during the locally recognised days of rest** pursuant to Sub-Clause 6.5 [*Working hours*] of the General Conditions
Date:	19 May 2025

Dear Sir,

The Contractor gives Notice that it is necessary to work during the locally recognised days of rest due to the instructed rectification of damages. The work is necessary to prevent further damages to the Works from the landslides following the flood.

If the Engineer gives consent to work during the weekend, the Contractor will comply with the provisions of Sub-Clause 4.14 to restrict interference with the convenience of the public. We do not expect any tension with the community, who are also repairing the damages from the flood to their property.

Best regards
Mr. Giovanni Salente
Contractor's Representative
Italiano Vero Srl

A.13 Project health, safety, security, and environmental (HSSE) management

The Engineer gave the following Notice.

Contract: **Lot 1 DIMA Corridor: High-speed Railway Section Dinia Central Station – Mountgate Station**

Contract:	Lot 1 DIMA Corridor: High-speed Railway Section Dinia Central Station – Mountgate Station
To:	Mr. Giovanni Salente Contractor's Representative Italiano Vero Srl 16 Stratos St. Dinia 1100
Copy:	Mr. Nobi Dini Minister of Infrastructure 6 Dino Blvd. Dinia 1000
Subject:	**Notice for removal and replacement of Contractor's Personnel** pursuant to Sub-Clause 6.7 [*Health and Safety of Personnel*] and 6.9 [*Contractor's Personnel*] of the General Conditions
Date:	1 October 2024

Dear Sir,

The Engineer requests the Contractor to remove the health and safety officer Mr. Dan Hushi and the Site Manager Mr. Trace Muri in compliance with Sub-Clause 6.9 (d), due to their persistent non-compliance with the provision of Sub-Clauses 6.7 and 7.3 as follows:

- No welfare facilities with toilets and running water are available for the Contractor's Personnel working in the area far from Dinia and Mountgate Stations;
- Workers are often without personal protection equipment, including helmets;
- The Contractor did not provide safe access and safety equipment for the Site inspections of the Employer's Personnel on more than ten occasions, resulting in delays and two incidents, as you have been notified on 7 September and 25 September 2024.

The Contractor shall promptly appoint suitable replacements, following the procedure under Sub-Clause 6.12.

The replacement of the above Key Personnel would not prejudice the Engineer to withhold the value of the unfulfilled obligations in the next IPC under Sub-Clause 14.6.2, and the Employer to claim damages for breach of contract under Sub-Clause 20.2 seeking compensation for the additional costs incurred due to the delayed inspections, numerous health and safety meetings, and correspondence.

Looking forward to receiving as soon as possible CVs for suitable replacements under Sub-Clause 6.12 and a report on the measures to cure the non-compliance under Sub-Clauses 4.8/6.7.

Best regards
For the Engineer PM Kings JV
Mr. Marin Yordan

A.14 Project claims management

The Lot 1 Contractor gave the following Notice.

Contract:	Lot 1 DIMA Corridor: High-speed Railway Section Dinia Central Station – Mountgate Station
To:	Mr. Marin Yordan PM Kings JV 8 Dino Blvd. Dinia 1000
Copy:	Mr. Nobi Dini Minister of Infrastructure 6 Dino Blvd. Dinia 1000
Subject:	**Notice of Claim** pursuant to Sub-Clause 20.2.1 of the General Conditions related to the Exceptional Event notified on 13 May 2025
Date:	20 May 2025

Dear Sir,

The Contractor gives Notice to the Engineer of a Claim for delay suffered due to the Exceptional Event duly notified on 13 May 2025, which was a tornado occurring in the vicinity of the Site on 11 May 2025. It was followed by heavy rains and large hail, and the Site was flooded. The flood triggered landslides and debris flow. The Exceptional Event and its effect prevented the Contractor from proceeding with the Works with due expedition and without delay. The Contractor shall be entitled to EOT under Sub-Clause 18.4 [*Consequences of an Exceptional Event*].

The Contractor is preparing the fully detailed Claim to be submitted following the Sub-Clause 20.2 procedure. Alternatively, in the Variation by instruction under Sub-Clauses 13.3.1 and 17.2, the Engineer may add the incurred 10 days' delay during the period of prevention commencing on 11 May 2023 and the next delay incurred for the instructed rectification of the losses and damages to the Works and Goods, resulting from the same Exceptional Event.

Best regards
Mr. Giovanni Salente
Contractor's Representative
Italiano Vero Srl

GLOSSARY

Term	*Meaning*	*Defined in Section*
5G	fifth generation mobile network with increased availability and higher data speed	3.4.2.1
ACA	Accepted Contract Amount. FIDIC definition (Red/Yellow/Gold/Emerald Books)	FIDIC GCs
ACAD	AutoCAD: a commercial computer-aided design and drafting software application developed by Autodesk	13.8.4
adaptive approach	a development approach in which the requirements are subject to a high level of uncertainty and volatility and are likely to change throughout the project	3.2.2
ADB	Asian Development Bank	1.2
ADR	alternative dispute resolution	17.2.1.1
Advance Payment Certificate	FIDIC definition (Red/Yellow/Emerald Books)	FIDIC GCs
Advance Payment Guarantee	FIDIC definition (Red/Yellow/Silver/Emerald Books)	FIDIC GCs
AFD	*Agence Française de Développement*	1.2
AfDB	African Development Bank	1.2
agile	a term used to describe a mindset of values and principles as set forth in the Agile Manifesto	3.3.2
AI	artificial intelligence	3.4.2.1
AIIB	Asian Infrastructure Investment Bank	1.2
AIM	asset information model	5.12.4
Base Date	FIDIC definition (Red/Yellow/Silver/Gold/Emerald/Green Books)	FIDIC GCs
BIM	Building Information Modelling	3.4.2.5

Term	*Meaning*	*Defined in Section*
Blue-Green Book	FIDIC Form of Contract for Dredging and Reclamation Works	1.2
BOK	body of knowledge	1.1
BOT	build-operate-transfer	5.7.1
BQs	bill(s) of quantities	4.1.4
Bronze Book	FIDIC Operate-Design-Build-Operate Contract for brownfield projects	4.1.15
CA	Contract Administrator	1.3
CAD	computer-aided design	3.4.2.4
CBA	cost-benefit analysis	5.6
CDB	Caribbean Development Bank	1.2
CDE	common data environment	5.12.2
CDM	UK Construction (Design and Management) Regulations 2007	15.2.1
CEO	Chief Executive Officer	4.3.4
CHEK	EU Change toolkit for digital building permit	12.4
civil law	a system of law in which the laws are codified in a large number of general rules and principles	5.8.4
Claim	FIDIC definition (Red/Yellow/Silver/Emerald Books)	FIDIC GCs
Clause (Sub-Clause)	provision in the FIDIC standard forms	1.2
Client	FIDIC definition (White Book)	FIDIC GCs
CNC	computer numerical control	3.4.2.7
CO_2	carbon dioxide	15.4.1
Commencement Date	FIDIC definition (White/Red/Yellow/Silver/Gold/Emerald/Green Books)	FIDIC GCs
common law	the body of law developed by judges from around the 11th century in England, and later exported to its various colonies such as Canada, the USA, Australia, Fiji, Hong Kong, India, Malaysia, and Singapore	4.1.1
Completion Schedule	FIDIC definition (Emerald Book)	FIDIC GCs
Compliance Verification System	FIDIC definition (Red/Yellow/Emerald/Silver Books)	FIDIC GCs
Concurrent engineering	collaborative sharing of information in the process of executing different tasks in a project with the aim of producing a functional design, a product with good quality, and a productive process	3.3.1
Conditions of Contract	FIDIC definition (Red/Yellow/Silver/Emerald/Green Books)	FIDIC GCs
Construction Extension	Construction Extension to the PMBOK® Guide published by PMI	1.1
Consultant	FIDIC definition (White Book)	FIDIC GCs
Contract	FIDIC definition (Red/Yellow/Silver/Gold/Emerald/Green Books)	FIDIC GCs
Contract Agreement	FIDIC definition (Red/Yellow/Silver/Gold/Emerald/Green Books)	FIDIC GCs
Contract Data	FIDIC definition (Red/Yellow/Silver/Gold/Emerald/Green Books)	FIDIC GCs

Term	*Meaning*	*Defined in Section*
Contract Price	FIDIC definition (Red/Yellow/Silver/Gold/ Emerald/Green Books)	FIDIC GCs
Contract Risk Management Plan	FIDIC definition (Emerald Book)	FIDIC GCs
Contract Risk Register	FIDIC definition (Emerald Book)	FIDIC GCs
Contractor	FIDIC definition (Red/Yellow/Silver/Gold/ Emerald/Green Books)	FIDIC GCs
Contractor's Documents	FIDIC definition (Red/Yellow/Silver/Gold/ Emerald/Green Books)	FIDIC GCs
Contractor's Equipment	FIDIC definition (Red/Yellow/Silver/Gold/ Emerald/Green Books)	FIDIC GCs
Contractor's Key Equipment	FIDIC definition (Emerald Book)	FIDIC GCs
Contractor's Key Personnel	FIDIC definition (Emerald Book)	FIDIC GCs
Contractor's Personnel	FIDIC definition (Red/Yellow/Silver/Gold/ Emerald Books)	FIDIC GCs
Contractor's Proposal	FIDIC definition (Yellow/Gold/Emerald Books)	FIDIC GCs
Contractor's Representative	FIDIC definition (Red/Yellow/Silver/Gold/ Emerald Books)	FIDIC GCs
Contracts Committee	FIDIC Contracts Committee: a working committee of FIDIC appointed by the FIDIC Board, responsible for developing and updating the FIDIC suite of contracts and related documents	1.2
Cost	FIDIC definition (Red/Yellow/Silver/Gold/ Emerald/Green Books)	FIDIC GCs
Cost Plus Profit	FIDIC definition (Red/Yellow/Silver/Gold/ Emerald/Green Books)	FIDIC GCs
Cost-reimbursable contracts	the Employer takes the risk and pays the Contractor the incurred costs plus a percentage for overheads and profit	5.7.5
Country	FIDIC definition (White/Red/Yellow/Gold/Em erald/Silver/Green Books)	FIDIC GCs
CPM	critical path method	8.3
D	dimension	3.4.2.4
DAAB	Dispute Avoidance/Adjudication Board: FIDIC definition (Red/Yellow/Emerald/Silver Books)	FIDIC GCs
DAAB Agreement	FIDIC definition (Red/Yellow/Emerald/Silver Books)	FIDIC GCs
DAABA	Dispute Avoidance/Adjudication Board Agreement	17.1.1
DAB	Dispute Adjudication Board: FIDIC definition (1999 Red/Yellow/Silver/Gold Books)	FIDIC GCs
Date of Completion	FIDIC definition (Red/Yellow/Silver/Emerald Books)	FIDIC GCs
day	FIDIC definition (White/Red/Yellow/Gold/Em erald/Silver/Green Books)	FIDIC GCs

Term	*Meaning*	*Defined in Section*
Daywork Schedule	FIDIC definition (Red/Yellow/Emerald/Silver Books)	FIDIC GCs
decennial liability	Statutory provisions imposing 'strict liability', arising without the claimant having to prove fault or causation. The Contractor and the supervising Consultant are jointly and severally liable to the Employer for a period of 10 years from the date of delivery of the Works in respect of serious structural defects or collapse	5.10
Delay Damages	FIDIC definition (Red/Yellow/ Silver/Emerald Books)	FIDIC GCs
Delivery Performance Domain	addresses activities and functions associated with delivering the scope and quality that the project was undertaken to achieve	2.1
Development Approach and Life-Cycle Performance Domain	addresses activities and functions associated with the development approach, cadence, and life-cycle phases of the project	2.1
DfD/A	design for disassembly and adaptability	3.4.3.2
Dispute	FIDIC definition (Red/Yellow/Silver/Gold/Emerald Books)	FIDIC GCs
disruption	disturbance, hindrance, or interruption to a Contractor's normal working methods, resulting in lower productivity or efficiency in the execution of particular work activities	9.2.8
DLT	distributed ledger technology	3.4.2.12
DNP	Defects Notification Period. FIDIC definition (Red/Yellow/Silver/Emerald/Green Books)	FIDIC GCs
Drawings	FIDIC definition (Red/Green Books)	FIDIC GCs
EBRD	European Bank for Reconstruction and Development	1.2
EC	European Commission	5.12.6
EFTA	European Free Trade Association	10.3.2
EHS	environmental, health, and safety	6.3.2
EIB	European Investment Bank	1.2
EIR	exchange information requirements	5.12.4.2
Emerald Book	FIDIC Conditions of Contract for Underground Works	1.2
Employer	FIDIC definition (Red/Yellow/Gold/Emerald/Silver/Green Books)	FIDIC GCs
Employer-Supplied Materials	FIDIC definition (Red/Yellow/Silver/Emerald Books)	FIDIC GCs
Employer's Equipment	FIDIC definition (Red/Yellow/Gold/Emerald/Silver Books)	FIDIC GCs
Employer's Personnel	FIDIC definition (Red/Yellow/Gold/Emerald/Silver Books)	FIDIC GCs
Employer's Representative	FIDIC definition (Silver/Gold Books)	FIDIC GCs

Term	*Meaning*	*Defined in Section*
Employer's Requirements	FIDIC definition (Yellow/Gold/Emerald/Silver Books)	FIDIC GCs
Engineer	FIDIC definition (Red/Yellow/Emerald/Green Books)	FIDIC GCs
Engineer's Representative	FIDIC definition (Red/Yellow/Emerald Books)	FIDIC GCs
EOT	extension of time. FIDIC definition (Red/Yellow/Emerald/Silver Books)	FIDIC GCs
EPC	engineering, procurement, construction	1.2
EPCM	engineering, procurement, and construction management	4.2
ES	environmental and social	15.2.2.1
ESCP	Environmental and Social Commitment Plan	15.2.2.1
ESHS	environmental, social, health, and safety	15.2.2
ESMP	environmental and social management plan	15.2.2.1
EU	European Union	5.8
Eurocodes	a series of ten European Standards, providing a common approach for the design of buildings and other civil engineering works and construction products	10.3.2
EVM	earned value management: a method integrating project scope, actual cost, budget, and schedule for assessment of progress and performance	9.1.3
Excavation	FIDIC definition (Emerald Book)	FIDIC GCs
Exceptional Costs	FIDIC definition (White Book)	FIDIC GCs
Exceptional Event	FIDIC definition (Red/Yellow/Gold/Emerald/Silver Books)	FIDIC GCs
F4E	Fusion for Energy	7.6.3.2
FAC-1	Framework Alliance Contract FAC-1	5.12.10
FBiH	Federation of Bosnia and Herzegovina	9.1.2.4
FBOK	FIDIC Body of Knowledge: includes all FIDIC Contracts, management systems, manuals, guidance documents, and training materials	1.2
FIDIC	International Federation of Consulting Engineers (*Fédération Internationale des Ingénieurs-Conseils*)	1.2
Final Payment Certificate or FPC	FIDIC definition (Red/Yellow/Gold/Emerald Books)	FIDIC GCs
Final Statement	FIDIC definition (Red/Yellow/Gold/Emerald/Silver Books)	FIDIC GCs
float	the amount of time that the activity may be delayed beyond its early start/finish dates without delaying the Time for Completion	8.3
FPC	Final Payment Certificate	9.2.14
FS	Final Statement	9.2.14
GBR	Geotechnical Baseline Report. FIDIC definition (Emerald Book)	FIDIC GCs
GCs	FIDIC General Conditions	1.2

Term	*Meaning*	*Defined in Section*
GDR	Geotechnical Data Report. FIDIC definition (Emerald Book).	FIDIC GCs
GHG	greenhouse gas	15.4.3
GMP	guaranteed maximum price	5.7.5
Gold Book	FIDIC Conditions of Contract for Design, Build and Operate Projects	1.2
Goods	FIDIC definition (Red/Yellow/Gold/Emerald/Silver Books)	FIDIC GCs
GPPC	Guidance for the Preparation of Particular Conditions	1.3
GPS	Global Positioning System	11.2.3
GPs	FIDIC Golden Principles: the essential features of a FIDIC Contract that make the risk/reward allocation fair and balanced	1.3
Green Book	FIDIC Short Form of Contract	1.2
HS	health and safety	15.2.1
HSS	health, safety, and security	15.3
HSSE	health, safety, security, and environmental	2.2
IADB	Inter-American Development Bank	1.2
ICC	International Chamber of Commerce	16.3
ICE	Institution of Civil Engineers	4.1.1
iContracts	intelligent contracts (smart contracts)	3.4.2.12
IFAD	International Fund for Agricultural Development	1.2
IFIs	international financial institutions	1.2
ILO	International Labour Organisation	15.2.2.1
incremental approach	an adaptive development approach in which the deliverable is produced successively, adding functionality until the deliverable contains the necessary and sufficient capability to be considered complete	3.2.3
IoT	Internet of Things	3.4.2.1
IPC	Interim Payment Certificate. FIDIC definition (Red/Yellow/Gold/Emerald Books)	FIDIC GCs
IPCC	Intergovernmental Panel on Climate Change	15.4.1
IPD	integrated project delivery	3.3.1
ISO	International Organisation for Standardisation	2.2
ISPA	Instrument for Structural Policies for Pre-accession	9.1.2.4
IT	information technology	2.1
ITER	International Thermonuclear Experimental Reactor	1.3
iteration	a timeboxed cycle of development on a product or deliverable in which all of the work that is needed to deliver value is performed	3.2.3
iterative approach	a development approach that focuses on an initial, simplified implementation and then progressively elaborates, adding to the feature set until the final deliverable is complete	3.2.3
JICA	Japan International Cooperation Agency	1.2
JIT	Just-in-time	3.3.1

Term	*Meaning*	*Defined in Section*
Joint Venture or JV	FIDIC definition (Red/Yellow/Emerald/Silver Books)	FIDIC GCs
JV Undertaking	FIDIC definition (Red/Yellow/Silver/Emerald Book)	FIDIC GCs
KfW	*Kreditanstalt für Wiederaufbau*	1.2
Knowledge Area (PM)	an identified area of project management defined by its knowledge requirements and described in terms of its component processes, practices, inputs, outputs, tools, and techniques	1.1
KPIs	Key Performance Indicators	2.1
lag	the amount of time a successor activity has to be delayed with respect to a predecessor activity	8.4
Laws	FIDIC definition (Red/Yellow/Gold/Emerald/Silver/Green Books)	FIDIC GCs
lead	the amount of time a successor activity can be advanced with respect to a predecessor activity	8.4
lean principles	identify value, map the value stream, create flow, establish pull, and seek perfection	3.3.1
Lean Six Sigma	a methodology that combines the focus on waste elimination of lean with the Six Sigma methods to solve quality issues	10.5
Letter of Acceptance	FIDIC definition (Red/Yellow/Gold/Emerald Books)	FIDIC GCs
Letter of Tender	FIDIC definition (Red/Yellow/Gold/Emerald Books)	FIDIC GCs
Lining	FIDIC definition (Emerald Book)	FIDIC GCs
Local Currency	FIDIC definition (Red/Yellow/Silver/Gold/Emerald Books)	FIDIC GCs
LPS®	Last Planner System®	3.3.1
lump sum contracts	the Contractor takes the risk for the price	5.7.5
Materials	FIDIC definition (Red/Yellow/Silver/Gold/Emerald/Green Books)	FIDIC GCs
MCC	Millennium Challenge Corporation	1.2
MDBs	Multilateral Development Banks	1.2
Measurement Performance Domain	addresses activities and functions associated with assessing project performance and taking appropriate actions to maintain acceptable performance	2.1
MEAT	most economically advantageous tender	5.8
MIDP	master information delivery plan	5.12.4
Milestone	FIDIC definition (Emerald Book)	FIDIC GCs
Milestone Certificate	FIDIC definition (Emerald Book)	FIDIC GCs
MoM	Minutes of Meetings	12.4.4
month	FIDIC definition (Red/Yellow/Silver/Emerald Books)	FIDIC GCs
MW	megawatt	9.2.2.1
NCR	non-conformity report	10.4.2
net resource	circular use of resources	3.1.4
net zero	carbon neutrality	1.3

Term	*Meaning*	*Defined in Section*
NGO	non-governmental organisation	3.2.3
No-objection	FIDIC definition (Red/Yellow/Silver/Emerald/ Green Books)	FIDIC GCs
NOD	Notice of Dissatisfaction. FIDIC definition (Red/ Yellow/Silver/Emerald Books)	FIDIC GCs
Notice	FIDIC definition (White/Red/Yellow/Gold/Em erald/Silver/Green Books)	FIDIC GCs
NPP	nuclear power plant	4.3.5
O&M	operation and maintenance	6.4.4
OPEC	Organization of the Petroleum Exporting Countries	9.1.2.4
Orange Book	Conditions of Contract for Design-Build and Turnkey	4.1.5
Owner	May be a government, a public company, a corporation, a private body, or an individual developer. The Owner is usually called: the 'Client' employing a 'Consultant'; the 'Employer' employing a 'Contractor'; the 'Buyer' buying goods from the 'Seller'	5.1
Part	FIDIC definition (Red/Yellow/Emerald Books)	FIDIC GCs
Particular Conditions	FIDIC definition (Red/Yellow/Emerald/Silver/ Green Books)	FIDIC GCs
Party	FIDIC definition (White/Red/Yellow/Gold/Em erald/Silver/Green Books)	FIDIC GCs
Payment Certificate	FIDIC definition (Red/Yellow/Emerald Books)	FIDIC GCs
PBA	project bank account	9.2.4
PDCA	plan-do-check-act approach to project quality	3.3.2
Performance Certificate	FIDIC definition (Red/Yellow/Emerald/Silver/ Green Books)	FIDIC GCs
Performance Damages	FIDIC definition (Yellow/Emerald/Silver Books)	FIDIC GCs
performance domain	defined in the seventh edition of the PMBOK® Guide as a group of related activities that are critical for the effective delivery of project outcomes	2.1
Performance Security	FIDIC definition (Red/Yellow/Emerald/Silver/ Gold Books)	FIDIC GCs
Permanent Works	FIDIC definition (Red/Yellow/Emerald/Silver/ Gold/Green Books)	FIDIC GCs
PFI	Private Finance Initiative	9.2.5
PGC	Prolongation Costs	13.9.4
PI	professional indemnity	13.7
PIM	project information model	5.12.4
Pink Book	FIDIC MDB Harmonised Edition of the 1999 Red Book	4.1.7
Planning Performance Domain	addresses activities and functions associated with the initial, ongoing, and evolving organisation and coordination necessary for delivering project deliverables and outcomes	2.1

Term	*Meaning*	*Defined in Section*
Plant	FIDIC definition (Red/Yellow/Emerald/Silver/Gold/Green Books)	FIDIC GCs
PM	project management: the application of knowledge, skills, tools, and techniques to project activities to meet the project requirements	1.1
PM plan	the document that describes how the project will be executed, monitored, controlled, and closed	6.3.1
PMBOK	project management body of knowledge	1.1
PMBOK® Guide	A Guide to the Project Management Body of Knowledge published by PMI	1.1
PMI	Project Management Institute	1.1
PMIS	project management information system	6.3.6
PPP	public–private partnership	4.1.15
predictive approach	a development approach in which the project scope, time, and cost are determined in the early phases of the life cycle	3.2.1
Programme	FIDIC definition (White/Red/Yellow/Silver/Emerald Books)	FIDIC GCs
Project Claims Management	PMBOK Knowledge Area in construction: covers the processes for claims management, dispute prevention, and dispute resolution	2.2
Project Communications Management	PMBOK Knowledge Area: includes the processes required to ensure timely and appropriate planning, collection, creation, distribution, storage, retrieval, management, control, monitoring, and ultimate disposition of project information	2.2
Project Cost Management	PMBOK Knowledge Area: includes the processes involved in planning, estimating, budgeting, managing, and controlling costs so the project can be completed within the approved budget	2.2
Project Financial Management	PMBOK Knowledge Area in construction: covers the processes to acquire and manage the financial resources for the project	2.2
Project HSSE Management	PMBOK Knowledge Area in construction addressing the project health, safety, security, and environmental (HSSE) management of the project and the product of the project	2.2
Project Integration Management	PMBOK Knowledge Area: includes the processes and activities to identify, define, combine, unify, and coordinate the various processes and PM activities	2.2
Project Procurement Management	PMBOK Knowledge Area: includes the processes necessary to purchase or acquire products, services, or results needed from outside the project team	2.2

Term	*Meaning*	*Defined in Section*
Project Quality Management	PMBOK Knowledge Area: includes the processes for incorporating the organisation's quality policy regarding planning, managing, and controlling project and product quality requirements in order to meet stakeholders' expectations	2.2
Project Resource Management	PMBOK Knowledge Area: includes the processes to identify, acquire, and manage the resources needed for the successful completion of the project	2.2
Project Risk Management	PMBOK Knowledge Area: includes the processes of conducting risk management planning, identification, analysis, response planning, response implementation, and monitoring risk on a project	2.2
Project Schedule	PMBOK definition	PMBOK Glossary
Project Schedule Management	PMBOK Knowledge Area: includes the processes required to manage the timely completion of the project	2.2
Project Scope Management	PMBOK Knowledge Area: includes the processes required to ensure the project includes all the work required, and only the work required, to complete the project successfully	2.2
Project Stakeholder Management	PMBOK Knowledge Area: includes the processes required to identify the people, groups, or organisations that could impact or be impacted by the project, to analyse stakeholder expectations and their impact on the project, and to develop appropriate management strategies for effectively engaging stakeholders in project decisions and execution	2.2
Project Work Performance Domain	addresses activities and functions associated with establishing project processes, managing physical resources, and fostering a learning environment	2.1
prolongation	extended duration of the works during which time-related costs are incurred as a result of a delay	9.2.8
Provisional Sum	FIDIC definition (Red/Yellow/Emerald/Silver/Gold Books)	FIDIC GCs
Purple Book	FIDIC Model Representative Agreement	4.2
PVP	photovoltaic park	9.2.1.1
QA	quality assurance	5.12.4
QBS	quality-based selection	5.4
QC	quality control	10.2
QM	quality management	6.4.4
RAP	Resettlement Action Plan	15.2.2.1
Red Book	FIDIC Conditions of Contract for Construction	1.2

Term	*Meaning*	*Defined in Section*
remeasurement contracts	the Contractor takes the risk for the unit prices, while the Employer takes the risk for the quantities	5.7.5
RES	renewable energy sources	15.4.3
Retention Money	FIDIC definition (Red/Yellow/Emerald/Silver/Gold Books)	FIDIC GCs
Review	FIDIC definition (Red/Yellow/Emerald/Silver Books)	FIDIC GCs
RFI	request for inspection	10.4.2
RIBA	Royal Institute of British Architects	3.2.2
Schedule of Baselines	FIDIC definition (Emerald Book)	FIDIC GCs
Schedule of Contractor's Key Equipment	FIDIC definition (Emerald Book)	FIDIC GCs
Schedule of Payments	FIDIC definition (Red/Yellow/Emerald/Silver/Gold Books)	FIDIC GCs
Schedule of Performance Guarantees	FIDIC definition (Yellow/Emerald/Silver Books)	FIDIC GCs
Schedule of Rates and Prices	FIDIC definition (Red/Yellow/Emerald/Silver Books)	FIDIC GCs
Schedules	FIDIC definition (Red/Yellow/Emerald/Silver/Gold/Green Books)	FIDIC GCs
S-curve	a chart displaying cumulative costs during the Time for Completion	9.1.3
SDGs	UN Sustainable Development Goals	1.3
SEA	sexual exploitation and abuse	15.2.2.1
Section	FIDIC definition (Red/Yellow/Emerald/Silver/Gold/Green Books)	FIDIC GCs
Services	FIDIC definition (White Book)	FIDIC GCs
Silver Book	FIDIC Conditions of Contract for EPC/Turnkey Projects	1.2
Site	FIDIC definition (Red/Yellow/Emerald/Silver/Gold/Green Books)	FIDIC GCs
Six Sigma	a set of methodologies and tools used to improve business processes by reducing defects and errors, minimising variation, and increasing quality and efficiency	3.3.1
SLAs	service level agreements	5.12.2
smart contracts (iContracts)	intelligent contracts: a computer program that incorporates the terms and conditions of a contract into computer code working on a network	5.12.5
SMEs	small and medium-sized enterprises	5.12.6
Special Provisions	FIDIC definition (Red/Yellow/Emerald/Silver/Green Books)	FIDIC GCs
Specification	FIDIC definition (Red/Green Books)	FIDIC GCs
SPV	Special Purpose Vehicle	9.2.1
SRO	senior responsible owner	4.3.1

Term	*Meaning*	*Defined in Section*
Stakeholder Performance Domain	addresses activities and functions associated with stakeholders	2.1
Statement	FIDIC definition (Red/Yellow/Emerald/Silver/Gold Books)	FIDIC GCs
Subcontractor	FIDIC definition (Red/Yellow/Emerald/Silver/Gold Books)	FIDIC GCs
Team Performance Domain	addresses activities and functions associated with the people who are responsible for producing project deliverables that realise business outcomes	2.1
TED	Tenders Electronic Daily	5.12.6
Temporary Works	FIDIC definition (Red/Yellow/Emerald/Silver/Gold/Green Books)	FIDIC GCs
Tender	FIDIC definition (Red/Yellow/Emerald/Silver/Gold Books)	FIDIC GCs
Tests after Completion	FIDIC definition (Red/Yellow/Emerald/Silver Books)	FIDIC GCs
Tests on Completion	FIDIC definition (Red/Yellow/Emerald/Silver Books)	FIDIC GCs
TG	FIDIC Task Group	4.1.6
TIDP	Task Information Delivery Plan	5.12.4.4
Time for Completion	FIDIC definition (White/Red/Yellow/Gold/Emerald/Silver/Green Books)	FIDIC GCs
TOC	Taking-Over Certificate: FIDIC definition (Red/Yellow/Emerald/Silver/Gold/Green Books)	FIDIC GCs
TQM	Total QM	3.3.1
UK	United Kingdom	5.8.5
UN	United Nations	1.3
Uncertainty Performance Domain	addresses activities and functions associated with risk and uncertainty	2.1
Underground Works	FIDIC definition (Emerald Book)	FIDIC GCs
Unforeseeable	FIDIC definition (Red/Yellow/Gold/Emerald/Silver/Green Books)	FIDIC GCs
USA	United States of America	9.1.2.3
Variation	FIDIC definition (White/Red/Yellow/Gold/Emerald/Silver/Green Books)	FIDIC GCs
VAT	value added tax	9.2.15.3
VBM	value based management	3.3.1
VE	value engineering	1.3
VR	virtual reality	3.4.23
VSM	value stream mapping	3.3.1
waterfall development	a breakdown of project activities into linear sequential phases, meaning that they are passed down onto each other, where each phase depends on the deliverables of the previous one and corresponds to a specialisation of tasks	3.2.1

Term	*Meaning*	*Defined in Section*
WBIF	Western Balkans Investment Framework	9.2.2.1
WBS	work breakdown structure	6.3.3
White Book	FIDIC Client/Consultant Model Services Agreement	1.2
WHO	World Health Organization	11.4.1.1
Works	FIDIC definition (Red/Yellow/Gold/Emerald/Silver/Green Books)	FIDIC GCs
Yellow Book	FIDIC Conditions of Contract for Plant and Design-Build	1.2

INDEX

Italics page references indicate figures. **Bold** page references indicate Tables